C++ Toolkit for Engineers and Scientists

Second Edition

Springer
New York
Berlin
Heidelberg
Barcelona
Hong Kong
London
Milan
Paris
Singapore
Tokyo

James T. Smith

C++ Toolkit for Engineers and Scientists

Second Edition

With 12 Illustrations

 Springer

James T. Smith
Department of Mathematics
San Francisco State University
1600 Holloway Avenue
San Francisco, CA 94132-1722
USA
smith@math.sfsu.edu

Windows 95, 98, NT are registered trademarks of the Microsoft Corporation.
Intel486 is a registered trademark of Intel.

Library of Congress Cataloging-in-Publication Data
Smith, James T., 1939–
    C++ toolkit for engineers and scientists/James T. Smith.
      p. cm.
    Includes bibliographical references.
    ISBN 0-387-98797-5 (soft cover: alk. paper)
    1. C++ (Computer program language)   I. Title.
QA76.73.C153S654   1999
005.13'3—dc21   99-18595

Printed on acid-free paper.

First Edition © 1997 International Thomson Computer Press.

© 1999 Springer-Verlag New York, Inc.
All rights reserved. This work may not be translated or copied in whole or in part without the written permission of the publisher (Springer-Verlag New York, Inc., 175 Fifth Avenue, New York, NY 10010, USA), except for brief excerpts in connection with reviews or scholarly analysis. Use in connection with any form of information storage and retrieval, electronic adaptation, computer software, or by similar or dissimilar methodology now known or hereafter developed is forbidden.
The use of general descriptive names, trade names, trademarks, etc., in this publication, even if the former are not especially identified, is not to be taken as a sign that such names, as understood by the Trade Marks and Merchandise Marks Act, may accordingly be used freely by anyone.

Production managed by Frank McGuckin; manufacturing supervised by Joe Quatela.
Camera-ready copy provided by the author.
Printed and bound by Hamilton Printing Co., Rensselaer, NY.
Printed in the United States of America.

9 8 7 6 5 4 3 2 1

ISBN 0-387-98797-5 Springer-Verlag New York Berlin Heidelberg   SPIN 10713930

Contents

Preface xiii
About the Author xv

1 Introduction 1

1.1 WHO NEEDS THIS BOOK? 2
　　Prerequisites 4
1.2 WORKING WITH SOFTWARE TOOLS 5
1.3 THIS BOOK'S STRUCTURE 6
　　What Would Come Next? 8
　　The Accompanying and Optional Diskettes 8
　　MSP Revisions and Extensions 9
1.4 C, C++, AND BORLAND C++ 9
　　Why Use C++? 9
　　Is C++ a High-Level Language? 10
　　Software Tools 12
　　Minor Considerations 12
　　Borland C++ 13
1.5 THIS BOOK'S MSP SOFTWARE TOOLKIT 14
1.6 REFERENCES AND FURTHER TOPICS 16

2 Software Organization 21

- 2.1 SOFTWARE LAYERS 21
  - Basic Input/Output System (BIOS) 22
  - DOS/Windows Operating System 24
  - Borland C++ Library and Run-Time Support 26
  - Mathematical Software Package (MSP) 27
  - Client Programs 28
- 2.2 C++ MINOR FEATURES 28
  - // Comments 29
  - Local Variable Declarations 29
  - new and delete Operators 29
  - Mandatory Function Prototypes 30
  - Initialized Parameters 30
  - Reference Parameters 30
  - C++ Version 2 Stream Input/Output 31
- 2.3 OVERLOADING 33
  - Function Templates 34
  - Using Templates for Library Code 35
- 2.4 ABSTRACT DATA TYPES AND CLASSES 36
  - Abstract Data Types (ADTs) 37
  - Parallel with Higher Mathematics 38
  - C++ Classes and Objects 38
  - Class Templates 39
  - Member Functions 40
- 2.5 MEMBER AND FRIEND FUNCTIONS 40
  - Principal Argument Syntax 41
  - Selectors 41
  - Operator Functions 42
  - Membership Criteria and Friend Functions 42
- 2.6 CONSTRUCTORS AND DESTRUCTORS 43
  - Default Constructor 43
  - Copy Constructor 44
  - Converters 45
  - Assignment Operator 45
  - Destructors 46
- 2.7 DERIVED CLASSES 46

    2.8 EXCEPTIONS 48
            The Stack and Function Invocation 49
            The Stack and Exception Handling 50

3   Computing with Scalars 55

    3.1 MSP General MODULE 57
            #ifndef and #endif 58
            Definition of This 58
            Identifier Capitalization 59
            The Boolean Type 60
            Empty Strings and Blank Characters 61
            Pausing to Inspect Output 62
            The Digits Function 63
            max and min 63
            Templates for Derived Functions and Operators 64
            Exceptions 64
    3.2 INTEGER ARITHMETIC 65
            C++ Integer Types 65
            Addition, Subtraction, and Multiplication 66
            Order and Absolute Value 67
            Integer Division 68
            Additional Integer Operations 69
            Formatting Output 70
            Integers of Arbitrary Length 70
    3.3 FLOATING-POINT ARITHMETIC 70
            Real and Floating-Point Numbers 71
            IEEE Binary Floating-Point Standard 72
            Arithmetic and Rounding 74
            Input/Output 76
    3.4 Math.H LIBRARY FUNCTIONS 76
            Real Arithmetic Functions 77
            Real Powers, Logarithms, and Hyperbolic Functions 78
            Real Trigonometric Functions 79
    3.5 Complex.H LIBRARY FUNCTIONS 81
            The complex Class 81
            complex Arithmetic 82
            Complex.H Elementary Functions 83

Exponential, Hyperbolic and Trigonometric Functions 84
Square Roots, Logarithms, and Powers 86
Inverse Trigonometric Functions 88
Need for `complex(z)` Casts 88
3.6 TESTING LIBRARY FUNCTIONS 89
Internal Validation Experiments 90
3.7 HANDLING `Math.H` EXCEPTIONS 93
`Math.H` Error Handling 93
MSP Function `_matherr` 95
`Complex.H` Errors 98
3.8 HANDLING FLOATING-POINT EXCEPTIONS 99
Interrupts 99
Hardware Exceptions 101
Coprocessor Exception Masking 102
Borland C++ Initialization and Exit Routine 104
Borland C++ Hardware Exception Handling 104
Signaling 105
MSP `SIGFPE` Service Function `FPError` 107
3.9 MSP `Scalar` MODULE 110
Codas for Generating Function Template Instances 112
Type Identification Functions 113
Displaying Scalars 114
Constant `ii = complex(0,1)` 116
`abs`, `real`, `imag`, and `conj` for Noncomplex Scalars 116
Random Scalars 117
3.10 MSP `ElemFunc` MODULE 117
Factorials and Binomial Coefficients 118
Functions `Root` and `logbase` 118
Hyperbolic Functions 119
Trigonometric Functions 120

4 Solving Scalar Equations 123

4.1 PRELIMINARIES 124
`Equate1` Module and Header File 124
Trace Output 124
Bracketing Intervals 125
Checking Solutions 125

        4.2 BISECTION METHOD   126
        4.3 FIXPOINT ITERATION   133
        4.4 NEWTON-RAPHSON ITERATION   139

5   **Vector Classes   151**

        5.1 Vector MODULE INFRASTRUCTURE   152
               The Vector Data Structure   152
               Selectors   154
               Constructors   155
               Destructor   158
               Copiers   159
               Assignment   160
               Equality   163
               Keyboard Input and Display Output   163
               Stream Input/Output   166
               Coda   169
        5.2 VECTOR ALGEBRA FUNCTIONS   170
               Constructing Special Vectors   171
               Vector Norms   173
               Plus, Minus, and Conjugate   174
               Addition and Subtraction   175
               Scalar Multiplication   177
               Scalar and Inner Products   177
               Cross and Triple Products   179
               Replacement Operators   180
        5.3 Vector MODULE DEMONSTRATION   180
               Contours and Surfaces   180
               Stokes' Theorem   186

6   **Polynomial Classes   189**

        6.1 Polynom MODULE INFRASTRUCTURE   190
               The Degree Selector   192
               Constructors   193
               Copiers, Assignment, and Type Conversion   195

6.2 POLYNOMIAL ALGEBRA FUNCTIONS 196
    Negation, Addition, and Subtraction 197
    Multiplication and Powers 199
    Division 202
    Differentiation and Integration 206
    Horner's Algorithm and Evaluation 207
    Testing with Legendre polynomials 210
6.3 POLYNOMIAL ROOTS 212
    Implementing Newton-Raphson Iteration 213
    Cauchy's Bound 214
    Multiplicity of a Root 215
    Finding Real Roots 218
    Finding *All* Real Roots 219
    Testing with Legendre Polynomials 222
    Finding Complex Roots 223

7 Matrix Classes 229

7.1 `Matrix` MODULE INFRASTRUCTURE 230
    The `Matrix` Data Structure 230
    Selectors 232
    Constructors 234
    Destructor 236
    Converters 237
    Copiers 238
    Assignment 240
    Equality 241
    Keyboard Input and Display Output 241
    Stream Input/Output 243
7.2 MATRIX ALGEBRA FUNCTIONS 245
    Constructing Special Matrices 248
    Matrix Norms 248
    Plus and Minus 250
    Conjugate and Transpose 250
    Addition and Subtraction, Trace 252
    Scalar Multiplication 253
    Matrix Multiplication 253
    Replacement Operators 255

8 Matrix Computations 257

- 8.1 SYSTEMS OF LINEAR EQUATIONS 257
- 8.2 GAUSS ELIMINATION 260
  - Downward Pass 261
  - Upward Pass for Nonsingular Square Systems 264
- 8.3 DETERMINANTS 266
- 8.4 GAUSS ELIMINATION SOFTWARE 268
  - Basic Gauss Elimination Routine 268
  - Pivoting 271
  - Using a Row Finder 273
  - Downward Pass 274
  - Upward Pass 275
  - LU Factorization 278
- 8.5 GENERAL RECTANGULAR SYSTEMS 280
  - Nonsquare or Singular Systems 280
  - Function GaussJordan 283
  - Homogeneous Linear Systems 284
- 8.6 MATRIX INVERSES 286
- 8.7 TESTING WITH HILBERT MATRICES 289
- 8.8 EIGENVALUES 292
  - Example Application: Compartmental Analysis 293
  - Characteristic Polynomial 294
  - Newton's Formula 296
  - Computing Characteristic Polynomials 297
  - Computing Eigenvalues 298
  - Eigenvectors and Eigenspaces 299

9 Iterative Solution of Systems of Equations 303

- 9.1 FIXPOINT ITERATION 305
  - Implementing the Multidimensional Fixpoint Algorithm 307
  - Examples: The Need for Error Analysis 309
  - Jacobians 314
  - A Multidimensional Mean Value Theorem 316
  - Convergence Condition, Speed of Convergence 317

9.2 ITERATIVE SOLUTION OF LINEAR SYSTEMS 320
    Jacobi Iteration 321
    Gauss-Seidel Iteration 323
    Example: A Random System 325
    Example: PDE Boundary Value Problem 326
    Example: Second-Order Linear ODEBVP 329

9.3 MULTIDIMENSIONAL NEWTON-RAPHSON ITERATION 334
    Quadratic Convergence 334
    Newton-Raphson Formula 336
    Complex Newton-Raphson Iteration 338
    Equate2 Template NR 340
    Examples 343

Appendix: Selected Source Code 347

A.1 VECTOR AND MATRIX DISPLAY OUTPUT FUNCTIONS 347
A.2 MSP HEADER FILES 350
A.3 EXAMPLE MATHEMATICAL FUNCTION WITH INTERMEDIATE OUTPUT 370
A.4 OPTIONAL AND ACCOMPANYING DISKETTES 371

Bibliography 375
Index 379

Preface

This book describes the design, construction, and use of a numerical analysis software toolkit. It's written in C++, Version 2.0, and makes essential use of that language's Object-Oriented Programming (OOP) features. Its development environment is the Borland International, Inc., Borland C++ compiler, Version 5.02, for IBM-compatible personal computers. However, the book describes few features specific to that product.

The toolkit and its description and background discussions cover the most fundamental aspects of numerical analysis. At the core of most scientific or engineering application programs are some of the concepts and techniques presented here. The most basic include details of computation with floating-point real and complex numbers; mathematical functions in the C++ Library; and a general OOP framework for vector, polynomial, and matrix algebra. On this foundation routines are constructed for solving nonlinear equations, linear and nonlinear systems of equations, and eigenvalue problems.

The book is heavily weighted toward software development. What's new here is the emphasis on software tools and on OOP techniques for handling vectors, polynomials, and matrices. Rather than describing programs implementing specific numerical techniques to solve specific application problems, the book constructs reusable tools with which you can implement many techniques for solving broad classes of problems. Examples are included to demonstrate their use. The tools are organized into layers. The deepest is formed by the C++ library functions for computing with real and complex numbers. The entire package is constructed so that you can change between real and complex scalars at will. Resting on these scalar routines is vector algebra software. Vectors are implemented as OOP objects, and a program can handle vectors of any dimensions simultaneously. This feature is critical for manipulating polynomial objects, which inherit properties of vectors but boast specific properties of their own. For example, you can add polynomials of different degrees, handling them as vectors of different dimensions; and you can multiply polynomials, but that operation isn't useful for vectors in general. Many higher-level numerical analysis tools can be built on the vector and polynomial algebra routines. For others, however, another software layer for defining and handling matrix objects is built on top of the vector class.

While this book emphasizes OOP features in implementing vector, polynomial, and matrix algebra, it actually takes a middle course in using OOP techniques. The organization of higher mathematics into the theories of vector spaces, polynomial and matrix algebras, and so forth, is the result of many decades of work ultimately aimed at scientific and engineering application problems. It's an efficient, elegant, and highly successful methodology. Through similar organizational devices, OOP is beginning to let developers attain comparable generality and grace in software toolkits. This book employs OOP techniques only where they obviously further that goal. Here, OOP is used to organize numerical application programs and make them look like the mathematics they represent.

It is possible to write C++ numerical software *entirely* in an OOP framework, where *every* computational entity is an object of some class. Borland C++ is shipped with a class library to encourage that process. But those techniques are under current development. It may take years to determine how *best* to represent various kinds of numbers, functions, and so forth, as OOP objects. *C++ Toolkit for Engineers and Scientists* keeps to the middle course, using OOP techniques just for vector, polynomial, and matrix algebra, because it's not now clear what further steps would take it closer to the goal: making programs easier to construct, understand, and maintain.

The book explains and makes thorough use of two C++ features recently implemented by Borland C++: *templates* and *exception handling*. Where classes differ only in the types of their data components or member function arguments, they're defined by *class* templates. Where functions differ only in the types of their arguments, they're defined by *function* templates. Thus, a single template defines the classes of real, complex, and integer vectors. Another template defines the functions that invert real and complex matrices. A program written with this book's software tools *throws an exception* if it asks the underlying hardware/software system to do something impossible or illegal, such as adding numbers whose sum is too large to represent in your computer, or inputting numerical data past the end of an input file. Each function then executing can *catch* and *rethrow* these exceptions: Each has an opportunity to identify itself and take emergency action. Thus, you can ascertain at run time exactly where and how the exception occurred—for example, in adding scalars during computation of a vector norm while solving a nonlinear system by Newton-Raphson iteration. You can easily adapt this technique to take whatever emergency action is appropriate in *your* application.

All MSP header files and compiled object code (large memory model) is included on the diskette accompanying this book, as well as source and executable code for several test and demonstration programs. Source code for all MSP toolkit routines and several more test programs is available from the author at reasonable cost.

About the Author

James T. Smith, Professor of Mathematics at San Francisco State University (SFSU), holds degrees in that subject from Harvard, SFSU, Stanford University, and the University of Saskatchewan. His mathematical research lies mostly in the foundations of geometry, with emphasis on multidimensional transformational geometry. His computer experience began with the UNIVAC I. During the early 1960s, Smith helped the Navy develop mainframe and hybrid digital-analog simulators for radiological phenomena. He's taught geometry and computing techniques since 1969 and was instrumental in setting up the SFSU computer science program. During the early 1980s, he worked in the IBM PC software industry. Smith has written seven research papers on geometry and five books on software development.

1

Introduction

This book is an introduction to C++ techniques for scientific and engineering application software development. It will show you how to use C++ object-oriented programming (OOP) methods to define and work with classes of mathematical structures, or objects, commonly involved in these applications—vectors, matrices, polynomials, and complex numbers. C++ ultimately gives you more freedom than most other languages in handling objects such as these but at considerable cost: You must provide the underlying software that constructs, maintains, manipulates, and dismantles these structures. The book concentrates on those matters and implements the basic algorithms of vector, matrix, and polynomial algebra and analysis. It shows how to base powerful application software on that foundation.

The compiler used for all the work in the book is Borland C++, Version 5.02, *EasyWin* environment, by Borland International, Inc. [4], for IBM-compatible personal computers. Most of the language and C++ library features used here are included in C++, Version 2.0, developed at AT&T Bell Laboratories [11, 52]. C++ includes a version of the C language, differing only slightly from ANSI C as described in American National Standard X3.159-1989.

This book is really a report on a research project: How can you use C++ OOP techniques to facilitate scientific and engineering applications software development? How can these techniques enhance the quality of your software? The project required restricting the scope of the software considered. The book describes in detail the development of software for these areas of numerical analysis:

- Real and complex arithmetic
- Elementary functions
- Vector and matrix algebra
- Polynomial algebra
- Solution of polynomial and transcendental equations

- Solution of linear systems of equations
- Eigenvalue problems
- Solution of nonlinear systems of equations

Most of this software works equally well with real and complex scalars. In addition, the book implements standard C++ and specific Borland C++ programming methods for reacting to arithmetic and library function error situations (exceptions).

A comprehensive introduction to scientific and engineering applications software would cover several other areas, too. But this book stresses software development problems more than a comprehensive introduction would. You'll meet the major OOP software development problems while implementing algorithms in the areas just listed. Once you can manage those, C++ OOP-based implementation of general algorithms to solve specific problems in other areas—for example, integration and initial value problems—should be straightforward. In fact, most really complicated programming problems in numerical analysis occur in the areas covered in this book. Constraints of time and size limit it to these central themes.

The book is built around a Mathematical Software Package called MSP, which I am developing to study C++ OOP implementation of numerical analysis algorithms. Some typical application problems are described to show why the algorithms are needed in scientific and engineering practice. The text discusses the underlying mathematics as required. It's emphasized mainly where an understanding of the algorithms requires mathematics beyond the usual introductory courses. The book describes MSP routines in detail and displays much of the source code with profuse documentation. Example test programs and output show how MSP is used in practice.

You'll find all MSP header files and compiled object code (large memory model) on the diskette accompanying this book, as well as source and executable code for several test and demonstration programs. You may use the coupon at the back of the book to obtain—at reasonable cost—source code for all MSP toolkit routines and several more test programs.

This introductory chapter describes the book, its purpose, and its audience in greater detail. It considers general software toolkit methodology and pros and cons of C++ and Borland C++ for scientific and engineering application software development. Finally, it lists a number of references and resources for further work in this subject.

1.1 WHO NEEDS THIS BOOK?

> **Concepts**
> *Why use C++?*
> *C++ is a research tool grafted onto a commercial base*
> *MSP is a research tool*

> *OOP and higher mathematics*
> *Future trends*
> *C++ is now used in production software*
> *Prerequisites for this book*

Who needs this book? You do, if you're developing scientific or engineering application software and must use C++ to remain compatible with some other software. Or perhaps you want to experiment with C++ object-oriented programming (OOP) techniques to facilitate software development or to see how they can enhance software quality in this area.

AT&T Bell Laboratories extended the popular commercial system programming language C to form C++, a language designed to foster the use of OOP techniques in diverse applications. Thus, C++ is a compromise. OOP techniques restrict programming practices to enhance efficiency of program development and ensure program robustness. On the other hand, *freedom* from such restrictions had been a major aim in designing C.

OOP techniques for defining classes of data objects parallel the organization of higher mathematics. When you study mathematical structures, you consider operations on data *together* with the data—for example, in studying a vector space, you consider the vector addition and scalar multiplication operations along with the vectors. Moreover, you regard some structures as particular instances of others, perhaps enhanced by additional features—for example, a polynomial algebra is a vector space with some additional operations, including polynomial multiplication and evaluation. This organization was refined over decades to handle some deep mathematical and logical problems that originated in applications. By applying it in programming practice, OOP will exert a strong influence on mathematical software development. In particular, it will facilitate construction of very general routines that are easily adaptable to specific situations, and the connections between software and mathematics will be transparent.

Many, but not all, OOP techniques are implemented in C++. This, plus the widespread dissemination of the Borland C++ compiler and the extreme flexibility of the underlying C language, makes it an ideal vehicle for OOP experimentation in mathematical software development. This book's MSP software package represents the current status of such an experiment.

This book is an argument for incorporating OOP techniques in scientific and engineering application software development, but it doesn't push for adoption of C++ as a standard language for developing production software. Twenty years from now, the next generation of programmers will view OOP as ordinary, common-sense application programming methodology. But C++ may be remembered mainly as a curious language, which grafted OOP features onto a host of another species to produce a vehicle for experimental development in a commercial environment. This book is written to help you carry on those experiments.

C++ requires you to provide software to create, maintain, manipulate, and dismantle the mathematical data structures needed for your application. Application programmers using other languages don't encounter so many low-level tasks. The book's MSP modules for vector, matrix, and polynomial algebra allow you to perform these tasks easily, usually with efficient grace. Its higher-level modules for solving equations and systems of equations show you how to build powerful application software based on the low-level OOP routines.

You see here just a snapshot of the ongoing MSP project. This package, and your own MSP modifications and enhancements, will continue to evolve. The original MSP version [49, 50] was hampered because Borland C++ didn't yet implement two essential C++ features—class templates and exception handling. In *The Annotated C++ Reference Manual* [11], C++ designer Bjarne Stroustrup described those features as experimental; they hadn't yet become standardized. They are now, however, and *the current MSP thoroughly utilizes templates and exception handling*. Evolution of C++ continues, but more slowly, and will no doubt trigger some future changes in MSP. Finally, OOP itself is an area of active computer science research. During the next few years, new ideas will result in new and modified languages and ultimately produce major trends in commercial software development. Further MSP development will likely depend on my projected study of OOP techniques for handling functions as data, which seems necessary for graceful and general implementation of integration and differential equations techniques.

Thus, you should approach C++ application software development with enthusiasm but with caution, too. The language, its OOP techniques, and this book will allow you to write programs more effectively and gracefully and to easily implement algorithms that would present major difficulties with other languages. On the other hand, your programs might not remain current for long. Your new techniques will be supplanted with newer ones, you'll develop even greater power as a programmer, and you'll soon want to redo your work to achieve the ultimate in effectiveness and grace. But that won't be the end, either.

The evolution of C++ as a research tool notwithstanding, this language *will* be used for developing production application software. The widespread availability of Borland C++ and comparable compilers, and the industry's continued support of these powerful and convenient tools, will ensure that. Therefore, you may well need to develop scientific or engineering application software in C++ to achieve compatibility with other C++ software. If that's the case, this book will clarify the issues and help you get started with an exciting programming project.

Prerequisites

What must you know *before* reading this book? It assumes you're a skilled programmer in some higher-level language, and you're familiar with C. Most C++ features that go beyond the C language are described in this book, but not in complete detail. You'll need a general C++ manual or text. With hardware concepts you need only minimal acquaintance: There's occasional reference to memory organization, the stack, and registers. To become adept with Borland C++, you should be familiar with the various hardware components of IBM-compatible PCs and with DOS/Windows user services. Finally, to understand the numerical analysis algorithms covered in this book, you should have mastered university-level mathematics through multivariable calculus and elementary linear algebra.

1.2 WORKING WITH SOFTWARE TOOLS

> **Concepts**
> *Software toolkits*
> *Borland C++ Library*
> *Commercial library supplements*
> *Writing software tools*

C++ requires you to provide software to create, maintain, manipulate, and dismantle the mathematical data structures needed for your application—for example, vectors, matrices, and polynomials. Application programmers using other languages don't encounter so many low-level tasks. As a C++ programmer, you'll need a toolkit of routines for performing them. That can be expanded to include routines for real and complex arithmetic and elementary functions, vector and matrix algebra, and polynomial manipulation. The resulting low-level toolkit can be expanded further with higher-level routines for more complicated tasks such as computing determinants, solving equations or systems of equations, and computing integrals. The higher-level routines can be organized in the same manner as the theory and algorithms presented in numerical analysis texts. Your toolkit contains known solutions to common problems, so you can concentrate on *new* problems as they occur in your work.

The Borland C++ Library is organized as a toolkit. You'll need its functions occasionally for the services it offers to programmers in all application areas, particularly for input/output, string handling, mathematical functions, and so forth. You should become familiar with its organization, or else you may waste time trying to solve programming problems for which the Library already provides convenient solutions. The Library consists of several thousand service functions covering the following programming areas:

data structure manipulation	directory management
error handling	executing concurrent programs
executing separate programs	file input/output
interactive input/output	keyboard input
low-level graphic output	manipulating memory areas

mathematical functions
text-mode screen handling
time calculations
memory management
text string processing

Many of these services are specified by ANSI standards, so they are available for other C or C++ compilers. Many are described in great detail in other books [48, 41]. This book covers the Library's mathematical functions and its error-handling features.

Although the Borland C++ Library is comprehensive and professionally implemented, it falls short of your needs in some areas. One is, of course, numerical analysis. That's the main reason for this book. Others that may be critical for your work are asynchronous communication and high-level graphic output. Commercial products can provide further support for your work in these areas; some are mentioned in Section 1.6.

What principles should govern toolkit development? First, *isolate mathematical and programming problems*, and attack them independently while they occupy your entire attention. That's a much more likely way to produce useful tools than ad hoc solutions by a programmer really interested only in getting some particular application to work. Second, *strive for generality*. Each application will differ a little from previous ones. A tool that's too particular will need modification again and again and may lose a little robustness each time. Next, *test the code!* Have others use it in as many different situations as possible. Finally, *write readable, understandable, trustable code and documentation*. Programmers who can't understand how a tool works may not want to use it in products on which their jobs depend.

The tools in the Borland C++ Library and in its commercial supplements were generally written with those guidelines in mind. The tools in this book's MSP package were, too, but there's a difference. Commercial products are intended to assist in building production software and are tested and modified accordingly over a long period of time. MSP is an experimental toolkit, with no such history.

1.3 THIS BOOK'S STRUCTURE

> **Concepts**
> *The book's organization*
> *Why stop here?*
> *What would come next?*
> *The accompanying and optional diskettes*
> *MSP revisions and extensions*

The chapters in this book are organized into three groups:

Introduction

1 Introduction
2 Software Organization

Borland C++ Library and MSP Foundation

3 Computing with Scalars
5 Vector Classes
6 Polynomial Classes
7 Matrix Classes

High-level MSP Routines

4 Solving Scalar Equations
8 Matrix Computations
9 Iterative Solution of Systems of Equations

Chapter 2 considers programming fundamentals. It discusses first the layers of software that are simultaneously active when an application program is executing—from the program itself down through layers of tools like this book's MSP software, through Borland C++ and DOS/Windows layers, to the bottom layer: your machine's Basic Input/Output System (BIOS). The remainder of Chapter 2 is devoted to OOP and C++ fundamentals on which are based the data structures—complex numbers, vectors, matrices, and polynomials—used in this book.

The second group of chapters considers the mathematical and error-handling features of the Borland C++ Library and the low-level MSP routines. Low-level MSP routines construct, maintain, manipulate, and dismantle the fundamental structures and carry out algebraic computations with real and complex numbers, vectors, matrices, and polynomials.

The chapters in the third group are more typical of books on scientific and engineering application programming. They describe various algorithms, based on calculus and linear and polynomial algebra, for solving common mathematical problems occurring in applications. These methods are then implemented in C++, based on the low-level routines discussed in the second group of chapters. The algorithms for the third group were selected for two reasons:

- They are fundamental ideas of numerical analysis.
- They provide a test bed for the low-level routines—major design and implementation problems should surface here, if at all.

The high-level material was limited to the third group of chapters solely for reasons of space and time. Chapter 4 was placed early in the book because it uses only part of MSP's low-level routines to solve scalar equations. It can give you an idea of the process before you dive into the more involved programming techniques for handling vectors, polynomials, and matrices.

Most scientific and engineering application programming books concentrate on the high-level algorithms and their implementations, de-emphasizing or ignoring low-level aspects. But the main reason for using C++ must be to construct the data structure classes and their low-level maintenance routines to allow the use of OOP techniques in applications. Therefore, the book must emphasize those aspects. Once you have mastered the techniques for implementing the high-level algorithms in Chapters 4, 8, and 9, applying them to specific problems in other areas of numerical analysis should be straightforward.

What Would Come Next?

In this book you see a snapshot of the MSP software package at a particular state of development. Were the book to continue with additional material on high-level algorithms, it would include chapters on:

- Interpolation and approximation
- Numerical integration and differentiation
- Ordinary differential equation initial value problems (ODEIVP)

You can see that there's no obvious place to draw the line. An earlier MSP version [49] incorporated techniques for definite integrals, including line and surface integrals and complex contour integrals. Space considerations prevented their inclusion here, except as a limited example in Section 5.3. Moreover, in my judgment, really graceful and general implementation of integration and ODEIVP algorithms requires a system for handling functions as data, analogous to the function space concept in higher mathematics. How to do that in C++ is an intriguing problem—a target for future study.

The Accompanying and Optional Diskettes

For your convenience, all MSP header files and compiled object code (large memory model) are included on the diskette that accompanies this book. The headers are listed in Appendix A.2. The diskette also contains source code for all the demonstration programs in this book, and source and executable code for programs that test the MSP `ElemFunc`, `Polynom`, `Gauss`, and `Equate2` modules.

You may use the coupon at the back of the book to obtain—at reasonable cost—an optional diskette containing the source code for *all* MSP toolkit routines and several more test programs.

MSP Revisions and Extensions

MSP revisions and extensions will be announced on Internet home page

   `http://math.sfsu.edu/smith`

If you have no Internet access, use the address on the coupon at the back of the book or send e-mail to `smith@math.sfsu.edu`.

1.4 C, C++, AND BORLAND C++

> **Concepts**
> *Why use C++?*
> *Is C++ a high-level language?*
> *Software tools*
> *Portability*
> *Program size and speed*
> *Accessing and controlling hardware details*
> *Borland C++*
> *Borland C++ EasyWin feature*
> *C++ memory models used*

This section discusses some of the benefits and costs of programming in C and C++ and the benefits of using the Borland C++ compiler.

Why Use C++?

Since C++ is an extension of C, reasons for and against using those languages for software development are closely related. In the programming literature, several standard arguments are often recited in favor of C:

- C is a high-level language, well suited to expressing algorithms of all sorts.
- C fosters writing elegant, readable programs.
- C programs are portable to many computer systems.
- Excellent compilers are available.

- Excellent software tools are available in many areas.
- C compilers produce small, fast executable code.
- C lets you access and control hardware features.
- C programming makes you consider low-level details.

How do these statements apply to C++? They're considered now one by one, both as stated, and as extended to apply to C++. I do not agree fully with all of them. The first and the last statements are even contradictory.

Is C++ a High-Level Language?

The original C language is really *not* a high-level language. With C you're constantly forced to consider low-level address details. Unless you become a C specialist, you'll be an inefficient C programmer, because those details are often very difficult to get right. Even if you manage to avoid most addressing problems in a simple program, during debugging you'll usually need to confront them to figure out what really happened when your program didn't do what you expected. C is not particularly well suited to expressing most algorithms, except for those explicitly involved with addressing. Low-level concerns often predominate, and the resulting C programs don't look like the algorithms. Thus, the contradiction between the first reason for choosing C and the last is resolved in favor of the last.

C does foster writing elegant, readable programs. But it's really *neutral*: It also fosters writing inelegant, unreadable programs. If anything, the scale tips negative on this criterion.

Most of the enhancements that transform C to C++ support high-level language features. Here are the chief examples:

- *The class concept:* As in higher mathematics, you can encapsulate the definition of a data structure with the definitions of its operations.

- *Regarding a particular class K as an instance of a general class G:* You can write routines to manipulate all objects in G and apply them to particular instances in K, which may have more elaborate data structures and additional or more specialized operations.

- *Class templates:* You can define several classes at once, if they differ only in the types of their data components and member function arguments.

- *Overloading function names and operators:* You can use similar, intuitive notation for like operations on objects of different classes.

- *Function templates:* You can use type variables to define several functions at once, if they differ only in their argument types.
- *Reference parameters:* You can pass function parameters by reference instead of value, without explicitly mentioning their addresses.
- *Parameter initialization:* You can assign default values to some function parameters and omit them from calling sequences unless you want to give them special values.
- *Stream input/output:* The C++ Library includes input/output operations that are somewhat less awkward than the standard input/output functions in the C Library.

If you use these features systematically, you can produce elegant, readable programs much more readily in C++ than in C. In fact, a major objective of this book's MSP software toolkit is to enable you to write application programs that look like the mathematical descriptions of the algorithms they implement. The cost of employing these C++ features is the requirement for elaborate underlying software, such as the low-level MSP routines, to construct, maintain, manipulate, and dismantle the data structures you want to use.

One aspect of the OOP data encapsulation technique is defeated in C++. A class definition can declare *private* some components of its data structures, so that only designated *member* and *friend* functions can access them. That should make debugging easier, since only a limited number of functions can change private data. However, the libertarian design of C allows easy circumvention of this restriction through low-level maneuvers.

Two other C++ features are awkward. First, one aspect of the class concept is a mandatory syntax for the principal arguments of member functions, which conflicts with standard mathematical notation and discourages their use. (For example, the real and imaginary components of the `complex` data structure are private; if the function `real` that returns the real component of a `complex` argument z were a member function, you'd have to invoke it with the expression `z.real()`. To permit the standard notation `real(z)`, it's declared a `friend`. But definitions of `friend` functions aren't encapsulated with the data structure definition.) Second, the parameter initialization feature invites class design errors that make it collide with overloading and produce bugs that are difficult to identify. Moreover, those design problems are hard to resolve.

Some high-level aspects of C++—for example, *name spaces*—are still under development. Using an evolving language gives you quick access to improved programming techniques, but you may lose stability.

In summary, C++ consists of some very powerful, still evolving, high-level features grafted onto a low-level base language. There's little reason to use C++ if you're not going to rely on those features. If you prepare the low-level foundation of your programs properly, you can write elegant, readable high-level applications programs that clearly reflect the underlying mathematics. However, the debugging and program maintenance process will continually force you to face lower-level details to find out what a misbehaving program is actually doing.

Software Tools

Excellent software tools are available for C programmers. The Borland C++ Library and some commercial supplements were mentioned briefly in Section 1.2. My earlier book, *Advanced Turbo C* [48], is an extremely detailed description of most of the Borland C Library and one (now obsolete) commercial supplement. Plauger's books [41, 42] are more current and authoritative accounts of the standard C and C++ Libraries.

Of course, C tools all work with C++. Software toolkits specifically for C++ are beginning to appear. This book's MSP is one. You should approach such products with caution. They're generally libraries of class definitions and code for member functions and friends and related functions. Since OOP is an active research area and C++ is still evolving, any such tools—particularly MSP—are experimental. The *best* ways to use OOP to solve common programming problems haven't necessarily been discovered yet.

Minor Considerations

You can write C programs that are portable to many environments, and excellent C compilers are available for many machines. C++ programs are somewhat less portable, and fewer C++ compilers are available, simply because it's a younger language. As remarked earlier, C++ was designed as a vehicle for research in programming language design. That may restrict the number of compilers available. (On the other hand, you could once make the same statement about C.) Borland C++ is an excellent compiler for both C and C++ programs. Its specific qualities are considered later. Almost all of its nonportable features are found in its Library.

C compilers generally produce small, fast executable code. Too few C++ compilers have been studied in detail to make such a statement about C++. Assessment will be difficult, because C++ is designed to enhance the efficiency of program development and maintenance, and that may be a conflicting goal. This book emphasizes the development process completely. It reports the results of an experiment in software design, for which program speed and size were irrelevant.

The last reason given earlier for choosing C as a development language is that it lets you access and control hardware features. That's certainly true for C++ as well (and for some other high-level languages). It's irrelevant for most scientific and engineering applications programming, unless your project involves reading data from sensors or communication lines or controlling some piece of equipment.

Borland C++

Why should you use the Borland C++ compiler? Here, the arguments are less controversial:

- Borland C++ has a *wonderful* user interface.
- You can configure most features to fit your habits.
- Syntax errors are immediately flagged in the editor.
- From the editor you can control a powerful source code debugger.
- Its Library is comprehensive, well documented, and reliable.
- Borland C++ includes standard features that facilitate development of very large software systems.
- It runs under Microsoft Windows and includes a toolkit for developing Windows applications.
- It will compile DOS applications.

In summary, Borland C++ is generally a delight to use.

All MSP test programs were compiled using the Borland C++ *EasyWin* feature. This automatically provides a window for the program but activates only the minimum repertoire of Windows features. This technique has several pros:

- You don't need to know *anything* about programming specially for the Windows environment. (In fact, I don't!)
- All Borland C++ user interface features are active: You don't have to leave it, as you would to run a DOS application.
- *EasyWin* programs run *almost* exactly like DOS programs.

It has one con:

- *EasyWin* programs don't run *exactly* like DOS programs.

In particular, *EasyWin* programs use the Windows standard character set, which is less appropriate than that of DOS. Some commonly used mathematical characters aren't available: ≈, ϕ, ≠, ≤, ≥, Σ, √, and the two halves of the integral sign. Moreover, you can't redirect standard input/output.

All low-level MSP functions were compiled first with the C++ *small* memory model. That was helpful, because subscripting and other pointer errors are common when working at that level. Those often cause your program to commit NULL pointer assignment errors, which Borland C++ will identify after a test run terminates, if it's compiled with the small model. Test programs for higher-level routines required both data and code space larger than 64 KB, so at that point I switched to compiling with the *large* model.

Don't confuse *Borland C++* with its big sister *Borland C++ Builder*. In some packages they're shipped together. C++ Builder does have a "console application" environment comparable to *EasyWin*. But C++ Builder is really designed to facilitate large-scale Windows interactive input/output programming. Moreover, the C++ Library that accompanies C++ Builder is different from the Borland C++ Version 5.02 Library.

These differences make porting MSP Software to the C++ Builder environment a substantial project:

- The C++ Builder Library is implemented with name spaces, and expects auxiliary libraries such as MSP to use name spaces too. You'll have to learn the technique, then make minor changes to every MSP source code file.
- C++ Builder implements complex number classes with templates, to permit real and imaginary components of types `float`, `double`, and `long double`. This requires very many minor modifications, scattered throughout MSP.
- C++ Builder handles floating-point exceptions differently: it uses structured C exceptions instead of signals. That simplifies the code in the MSP Scalar module, but adds minor complexity to the exception-handling features of the higher-level modules.

Other compilers will handle these aspects of C++ differently, too—they're not really standard yet. This book was limited to one implementation: the Borland C++ Version 5.02 *EasyWin* environment.

1.5 THIS BOOK'S MSP SOFTWARE TOOLKIT

> **Concepts**
> *Chief goal*
> *MSP modules*
> *What would come next?*

This book reports on a research project: How can you use C++ object-oriented programming (OOP) techniques to facilitate scientific and engineering applications software development? It's centered around a software toolkit called MSP (Mathematical Software Package), which was constructed to test these ideas. Its chief design goal was to establish the required software foundation so that client programmers using MSP can write elegant, readable programs that mirror the mathematical descriptions of the algorithms they implement.

OOP techniques permit encapsulation of definitions of classes of data structures with definitions of the operations on them, just as higher mathematics is organized. Moreover, you can regard some classes as particular instances of more general ones, just as polynomial algebras are particular types of vector spaces. Finally, you can overload operators and function names so that similar operations on different structures are denoted similarly, just as the same mathematical notation indicates multiplication of polynomials and multiplication of vectors by scalars.

Implementing this scheme in C++ requires more attention to low-level details than most application programmers enjoy. You must provide software to construct, maintain, manipulate, and dismantle the data structures you'll use, such as complex numbers, vectors, matrices, and polynomials. MSP's first task is to define those classes of data structures and provide that low-level support. Next, it includes routines to perform basic algebraic operations on these objects. On this foundation are built the high-level MSP routines, which implement fundamental numerical analysis algorithms for solving application problems.

MSP consists of ten *modules*. The first three offer general and mathematical support for all the others:

General	General support for all other modules.
Scalar	General features for handling real, complex, and integer scalars, including exception handling.
ElemFunc	Elementary functions supplementing those in the Borland C++ Math.H Library.

The next three modules complete the low-level part of MSP:

Vector	These contain the routines that support the corresponding data
Matrix	structures and the fundamental algebraic operations.
Polynom	

For example, the Vector module contains routines that construct, copy, and output vectors, and Polynom contains routines that multiply and divide polynomials. The last four modules are higher level:

Equate1	Solve equations in a single real or complex unknown.
GaussEl	Solve systems of linear equations by Gauss elimination. Compute determinants and invert matrices.
Eigenval	Construct the characteristic polynomial of a matrix and find its eigenvalues.
Equate2	Solve systems of linear and nonlinear equations by iterative methods.

Those modules implement fundamental algorithms of numerical analysis.

A comprehensive numerical analysis software toolkit would contain further modules for interpolation and approximation, numerical integration and differentiation, initial value problems, and so forth. However, this book emphasizes software development, and the major programming questions are encountered in constructing the high-level modules included here. Once you've understood these, implementing further numerical analysis algorithms to solve specific applications problems should be straightforward.

The `Vector` and `Matrix` modules are so large that they're split into *logical* and *mathematical* submodules for data handling and algebraic features. For each MSP module or submodule there are two corresponding files on the optional diskette: a `.H` header file and a `.CPP` source code file—for example, the logical part of the `Vector` module comprises files `Vector.H` and `VectorL.CPP`. The two `Vector` submodules share the same header file, as do the two `Matrix` submodules. The text describes all these files in detail.

1.6 REFERENCES AND FURTHER TOPICS

Concepts

Hardware and operating system manuals and related books
C and C++ texts and reference manuals
Borland C++ documentation and Library source code
Standard C Library references
Commercial C++ system software toolkits
Mathematical references
Other books on C++ scientific and engineering application programming
Commercial C++ mathematical software toolkits
Mathematical desktop software

To do serious scientific and engineering application software development in Borland C++, you need to find answers to hardware, software, and mathematical questions that occur constantly. This book covers only a small fraction of that material. This section, however, indicates a number of references that will help you. Some are immediately relevant—even indispensable—to this type of programming. Others provide background information, and some are texts or commercial software products whose approach is comparable to that of MSP.

This book assumes you're basically familiar with C and its use in application programming. It has to use *some* particular compiler; if you want to duplicate its test routines precisely, you'll need to gain access to and familiarity with the Borland C++ compiler and its user interface. But you don't need previous experience with C++. You'll need to be familiar with IBM-compatible hardware and the DOS/Windows operating system.

Various trade books discuss IBM-compatible hardware thoroughly. Norton [37] is at once classical and up to date. You may need to consult your system's technical reference manual to ascertain how its components conform with or depart from the standards that these books describe. Good luck: The appropriate manual may be hard to find, perhaps because the supplier is ashamed of its inadequacy!

If you need detailed information on your CPU's machine language or on your numeric coprocessor, consult the Intel 80286-80287 manual [22]. Similar volumes are available for 80386, 80486, and Pentium CPUs. They may refer to the 80287 manual for information about their coprocessors.

For information on DOS and Windows, their manuals [33] and [35] are of course important. Those delivered with the software are essential but inadequate for advanced programming use. For that, I found the Microsoft *MS-DOS Encyclopedia* [36] indispensable. That covers DOS Version 3.3; you'll have to dig deeper to find information on special features of later versions. Reference [32] updates it partially. The analogous Windows reference is Microsoft's *Windows 95 Resource Kit* [34]. Trade books by Norton [38], Duncan [9], and Petzold [40] cover some of this DOS and Windows material in a less formal style. (Duncan edited the *MS-DOS Encyclopedia*.)

The original C reference, by Kernighan and Ritchie [26], is not a good tool for learning C, but it is referred to so frequently and so religiously that you should have it. The reference work by Harbison and Steele [17] is quite useful. The original C++ references, Stroustrup [52] and Ellis and Stroustrup [11] are my mainstays. Stroustrup's history [53] provides insight into why C++ is designed the way it is. A number of text and trade books are available that introduce C, C++, and object-oriented programming techniques. I find most of them tedious and nearly useless for an experienced programmer. They often avoid discussing delicate questions that might give you trouble in programming. The elementary computer science text by Adams et al. [2] seems acceptable; its examples use C++ and the Borland compiler. More advanced is Lippmann's C++ text [29].

Borland C++ comes with a manual [4]. Its most comprehensive and expensive package includes many powerful facilities for developing large-scale commercial software, but only the Borland C++ Version 5.02 component is directly relevant to this book. Various trade books are devoted to practical details of Borland C++; I found Perry et al. [39] helpful.

My earlier book, *Advanced Turbo C* [48], contains a thorough description of most of the Borland C Library. Plauger's book [41] is *the* authoritative reference, though. He has just published a comparable book [42] on the proposed standard C++ Library. The most comprehensive Borland C++ package includes much of its Library source code. If you can't gain understanding of a Library routine through its documentation, you must conduct experiments and/or analyze its source code.

Excellent system software tools compatible with Borland C++ are available in several areas. Scientific and engineering application programmers occasionally use these to construct interactive user interfaces; to handle serial communication with data sensors, computers, or other equipment; or to display polished graphic output. For information on such products, consult advertisements and survey articles in the trade journals. Rogue Wave Software produces several such toolkits.

Scientific and engineering application programmers encounter mathematical questions as often as questions about hardware or software. I use Courant's [8] and Hille's [18] classic volumes for reference on calculus and related topics. For advanced calculus and algebraic matters, Wylie [59] and Uspensky [54] are rich sources. I can't recommend any linear algebra text: The readable ones too often ignore questions that occur in computation. Knuth's books [28] are a surprisingly good source of information on all kinds of mathematics related to computation. Of the many numerical analysis texts, two are specially good references: Burden and Faires [5], and Maron and Lopez [30]. The former is elegant; the latter has an incredible number of examples. As a reference of last resort, to find complete statements and proofs of results that most books only mention, consult Isaacson and Keller [24]. *Numerical Recipes in C* [43] by Press et al. is not a textbook, but it is a useful compendium of numerical methods, discussed from a computational viewpoint.

Several trade books, intended for the same readers as this book, marry mathematics and C++ OOP programming techniques. Flowers' approach [13] is nearest; but he doesn't cover exception handling. Next comes Shammas [47], then Klatte et al. [27], and then Reverchon and Duchamp [44].

Of the commercial software packages that implement numerical analysis techniques in C++, these stand out: Rogue Wave Software's *Math.h++* [45] and *LAPACK.h++* [46], and Dyad Software's *M++* [10]. They adapt to the C++ environment the much older, very comprehensive *LINPACK* software originally written in FORTRAN. (These companies market C++ software tools in other areas, too; Rogue Wave ventures far beyond scientific and engineering applications.)

Finally, you should become familiar with a new development in scientific and engineering application computing: *mathematical desktop software*. Several commercial products are now available to make computations on demand. You enter the problem via the keyboard, in notation approximating standard mathematics, and read the answer immediately from the display. Some products display two- and three-dimensional graphs and use graphic techniques to display formulas in true mathematical notation. Mathematical desktop products can handle problems that formerly required complicated programs in high-level languages. Most of these products can be programmed. Programming in a specialized mathematical desktop language is simpler than struggling with a general-purpose language such as C++. The most noted desktop software now is the *Mathematica* system [57, 58], which excels in symbolic algebra computations. Not needing the algebra features, I used Meredith's inexpensive but powerful *X(Plore)* program [31] for all incidental computations involved with developing and testing this book's MSP software and for drawing the graphs shown in this book. Even a modest product such as *X(Plore)* can execute—albeit slowly—all the algorithms discussed in this book.

2
Software Organization

This chapter first presents an overview of the layers of software present and running in your machine when you're executing a C++ application program. Even a high-level application program will interact with several layers, so you need to become familiar with them.

Sections 2.2 through 2.8 then discuss many new C++ features that play a role in this book's Mathematical Software Package (MSP). This chapter assumes you're familiar with C and uses basic C concepts with little comment. It is not a general introduction to C++—for that, please consult a text devoted to the language, as well as the *Borland C++* manual [4]. Rather than discuss new C++ features haphazardly in the order that they appear in MSP, this chapter considers some general C++ principles systematically. You may use it as a guide through one of the C++ texts [2, 29, 39] or the Borland manual. What's mentioned here, you need in order to understand the low-level MSP routines. What's not, you can safely ignore. Later, as you read the detailed description of MSP's lower levels in Chapters 3 and 5, you'll want to refer to this chapter to review the fundamental concepts.

2.1 SOFTWARE LAYERS

> **Concepts**
> *Basic Input/Output System (BIOS)*
> *BIOS services*
> *IBM compatibility*
> *DOS/Windows operating system*
> *User interface*
> *DOS/Windows services*

> *Borland C++ Library and run-time support*
> *This book's Mathematical Software Package (MSP)*
> `Scalar`, `ElemFunc`, `Vector`, `Matrix`, *and* `Polynom` *modules*
> *Client programs*

If you're a scientific application programmer interested in C++ programming, you're probably considering a project that involves software beyond that directly concerned with numerical analysis. Actually, any project does to some extent, particularly when implemented on a personal computer. Arguments for selecting C++ often cite the appropriateness of the C language and its offspring for detailed interaction with underlying hardware and system software and the potentiality of C++ object-oriented programming techniques for organizing complex applications.

Thus, it's important to gain an overview of the full ensemble of software present and running in your machine when your own numeric computations are executing. That software is organized into layers, with detailed hardware-conscious system software at the bottom, and an application program at the top (possibly modeling a real-world problem and written with no consideration of hardware). Here are the most apparent layers, when you're running an application program that uses this book's MSP numerical analysis software:

Top	Client application program
	Mathematical Software Package (MSP)
	Borland C++ Library and run-time support
	DOS/Windows operating system
Bottom	Basic Input/Output System (BIOS)

Your application program is called an MSP *client* because MSP provides needed mathematical *services*. Each of the upper tiers, in fact, is a client of the next lower one. This section will describe the layers, starting with the BIOS at the bottom, and explain their roles in making your application run.

Basic Input/Output System (BIOS)

Built into your PC hardware, this program starts the bootstrap process and provides low-level input/output and memory control. It's stored in Read-Only Memory (ROM) chips on your motherboard and on some optional controller boards. The hardware starts the BIOS executing when you turn on the power. Its first task is to check all your memory. This causes a pronounced delay for slow machines with large memories. The BIOS checks a few more motherboard components and initializes some tables in the low-address part of memory that describe the hardware configuration. Optional BIOS components perform similar checking and initialization on some controller boards that may be plugged into the motherboard.

Once this Power-On Self Test (POST) checking and initialization is complete, the BIOS transfers control to a tiny program that must occupy a specific place on your fixed disk. This *bootstrap* program loads the DOS/Windows operating system software—usually enhanced by including Microsoft's Windows software—into memory and transfers control to it. DOS/Windows performs a few more initializations, then transfers control to its user interface—usually the Command.Com program or the Windows Desktop—which displays a familiar screen and awaits your command. The difficulty of designing this start-up process seemed comparable to that of lifting yourself by your own bootstraps and gave rise to the "booting" terminology.

Once your computer is booted, the BIOS assumes a background role, providing low-level input/output and memory control services. Here are three typical BIOS services:

- *Disk read:* Read n sectors, starting at sector s on disk d and continuing on the same track, into memory starting at address a. Parameters n, s, d, and a are specified by the (assembly language) commands that request the service.

- *Keyboard status:* Ascertain whether the <CapsLock> key is toggled on and whether the <LeftShift> or <RightShift> key is down. The BIOS places a coded answer in a CPU register, to which the client program has access. The code also includes answers to several similar questions about other keys.

- *Display character:* Display character c at the current cursor location. Parameter c is specified by the command that requests the service.

Other typical services include reading and removing a character from the queue in memory that records recent keystrokes, setting the background color of a location on the display, and clearing the display entirely.

Most BIOS services are tied closely to standard PC hardware components. IBM designed the BIOS for the original PC and published the program listing in the PC manual [23]. Reading it is an adventure and a wonderful way to learn assembly language programming for the Intel 8086 family! Now, alternative BIOS chips are marketed by several companies for installation in IBM-compatible PCs. They must respond to the same commands in compatible ways. Generally, this cloning is successful, but compatibility problems do arise—particularly with the BIOS components for optional controllers, for which no strong standards emerged early in market history. A mathematical application programmer is unlikely to encounter these problems except perhaps in attempting to use very high resolution displays.

Successfully running DOS/Windows is the criterion for IBM compatibility of a personal computer. Most compatible PCs now run DOS/Windows as marketed by Microsoft. Some systems with nonstandard BIOS require specially adapted DOS components licensed to their manufacturers by Microsoft.

DOS/Windows Operating System

Once loaded and initialized during the bootstrap process, the operating system has charge of your machine. Users request it to execute operating system commands or programs. After a program terminates, control returns to the operating system. Besides providing a user interface, a PC operating system has three major roles:

- Organizing and maintaining the file system and providing file input/output services for clients
- Supervising the program execution process
- Providing other input/output and memory control services, much like the BIOS, but in a more general form, more convenient for client programs

Microsoft Windows now provides the most common user interface for computers running DOS. It allows you to use a pointing device for some input; arranges display output in various windows, which you can manipulate like scraps of paper on your desk; and manages fonts and other visual details for your display and printer. Also, Windows services give you closer and more consistent hardware control than DOS alone. Users unsatisfied with some Windows services can buy add-on packages that replace various Windows components with alternatives that are even more powerful or easier to use.

Microsoft periodically upgrades DOS and Windows. This book's programs were developed using Microsoft DOS 6.20 with Windows 3.10 and the add-on Norton Desktop 3.0. Microsoft's more recent Windows 95 version includes a built-in version of DOS.

Occupying one of the lower tiers of the software hierarchy in your machine, DOS/Windows is itself divided into three levels:

Top	User interface (usually the Command.Com program or the Windows Desktop)
Middle	Operating system core
Bottom	BIOS interface

The BIOS interface may also include optional *device driver* programs. These are commonly supplied by other sources, to manage your equipment in specific ways. You can adjust the bootstrap process to load them into memory with DOS/Windows and initialize them.

The user interface is a program with the same status in the software hierarchy as a client program. After initialization, the DOS/Windows core starts the interface, which displays a familiar screen and awaits your call. You can request various services, such as directory listings, or command the interface to execute another program *P*. When you do that, the interface suspends its own operation temporarily and requests the core to find, load, and execute *P*. When *P* terminates, control passes back to the core, which performs some housekeeping tasks and then reactivates the interface.

Most DOS/Windows work is done by the core. It accepts service requests from the user interface and other programs. In turn, the core may request attention from the BIOS interface. That lower-level component reformulates requests as required for a particular BIOS, obtains the service, and passes results back to the core.

Here are four typical DOS/Windows core services:

- *Open file:* Locate a file named F, set the file pointer at its start, and provide an integer h for later use as its *handle*. The service request command must specify the memory address of the character string F. The core places h in a CPU register to which the client program has access.

- *Read file:* Read n bytes from the file with handle h, starting at the file pointer, into memory starting at address a. Parameters n, h, and a are specified by the service request command.

- *Input character:* Read one character from the standard input device, waiting, if necessary, for an input to appear. The core places the input in a CPU register to which the client program has access.

- *Output character:* Send character c to the standard output device.

The DOS/Windows file system supervisor maintains directories of all disks in use by the executing software. *Locating a file* means using the directory to find the disk sectors where the file is stored. Once that information is available, DOS/Windows can translate a *read file* request into a *disk read* request as described earlier and pass it on to the BIOS. *Standard input* and *standard output* are usually associated with keyboard and display, but you can reassign them at run time (under DOS but not necessarily under Windows) to other devices or files by using DOS's redirection symbols < and > in your program execution command. DOS translates standard input/output commands into proper form for the associated device and generates appropriate BIOS service requests—for example, "send character c to the display" means "display c at the current cursor location."

Programmers using assembly language or the Borland C++ Library can write DOS/Windows or BIOS service requests. Using assembly language, or the Library routines that closely imitate it, gives you access to a vast number of services, which require books larger than this one just to describe. That's far beyond the scope of this book: Many trade books [9, 36, 37, 38] are useful guides. But few scientific application programmers gain extensive knowledge of those techniques. Instead, they use parts of the Library, which is organized according to specific needs of an application programmer—for example, you *could* use a Library routine to issue a specific BIOS disk read command. But to do that properly, you'd have to know just how DOS/Windows, as installed on *your* system, has arranged your data. You'll probably *never* discover that. Instead, you'll use Library routines to read the next data item from an open file and let the operating system take care of the details.

If you *must* resort to programming BIOS service requests—for example, to control your fancy extra-high-resolution display—your program will probably be nonportable. Another system's BIOS probably won't respond the same way, and its installed operating system won't know how to translate your requests appropriately.

Borland C++ Library and Run-Time Support

A C++ program P consists of some functions, including one called `main`. Each function can invoke any function except `main`. Source code for the functions can be apportioned among any number of files, each of which is compiled into a separate object code file. The object code files are linked into an executable file `P.Exe`. Here's the flow of control when DOS/Windows loads and executes this file:

DOS/Windows
⇒ C++ entrance routine
⇒ `main`
   ⋮
⇒ (depends on code in `main` and the other functions)
   ⋮
⇒ C++ exit routine
⇒ DOS/Windows

The entrance and exit routines are provided by the C++ compiler. They take care of housekeeping details that application programmers don't need to know about.

The functions that make up this program don't all have to be part of the project at hand. Usually, many are utility routines, completed beforehand, whose object codes are stored in a *library*. Borland C++ is distributed with a vast Library of such functions, which provide these categories of services:

- Manipulating memory areas
- Character string manipulation
- Keyboard and display input/output
- File input/output and file system maintenance
- Memory management
- Execution of DOS/Windows commands and other programs

- Timer input and time calculations
- Calculation of mathematical functions such as square root and sine
- Error handling

The C++ entrance and exit routines and the first few categories of Library functions constitute an interface—a software layer—between the BIOS and DOS/Windows and all C++ application programs. The string and mathematical functions constitute an underlying layer, as well. Except for the mathematical and error-handling functions, these services fall outside the scope of this book. They're used occasionally in its programs, with little comment. For further information, consult the Borland C++ *Library Reference* manual [4] and the books by Smith and Plauger [41, 48].

The Library was developed originally by AT&T Bell Laboratories, and then adapted by Borland International, Inc., for IBM PC-compatible computers. Many of its functions are portable to any C++ system; most work on any DOS/Windows system. A few input/output functions, however, require compatibility with the original IBM BIOS. By using those more specific functions, you gain control and speed but lose some portability. All Library functions that require DOS or the BIOS will be identified when used in this book.

Mathematical Software Package (MSP)

The numerical analysis software described in this book—the Mathematical Software Package (MSP)—is also organized in layers. Its lowest layer, the Scalar and ElemFunc modules, extends the mathematical services provided by the Borland C++ Library to provide all the required operations on *scalars*: real or complex numbers. The top layer contains what most professionals regard as numerical analysis tools: routines for solving linear and nonlinear equations and systems of equations, computing eigenvalues, and so forth. In between are layers that provide the necessary logical and algebraic manipulation of vectors, matrices, and polynomials.

Figure 2.1 provides an informal overview of the MSP structure. The middle part comprises three modules: Vector, Matrix, and Polynom. The Vector module is itself divided into two parts: logical and algebraic. The *logical* part handles vector manipulation: creation, input/output, copying, equality, and destruction. The *algebraic* part is concerned with vector norms, addition, scalar multiplication, inner product, and so forth. The Matrix module treats matrices as arrays of vectors and is split along similar lines. Polynomials are handled by module Polynom. Since they're really vectors of coefficients, their algebra is essentially an adaptation of vector algebra. However, polynomial evaluation, multiplication, differentiation, and integration are also implemented.

High-level numerical analysis routines may use the Borland C++ Library directly, as well as any functions in the Scalar, ElemFunc, Vector, Matrix, and Polynom modules.

28   C++ Toolkit for Engineers and Scientists

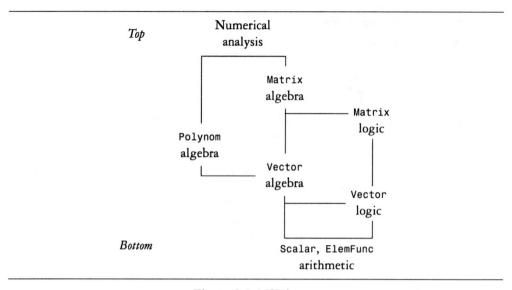

Figure 2.1. MSP layers

Client Programs

MSP client programs translate real-world problems and data into the MSP mathematics and then use MSP functions for solution. In some cases, client software might best be organized in layers, too—for example, this book doesn't directly handle approximation of data y_j at points x_j for $j = 0, \ldots, m$ by a least squares polynomial p of degree n. The coefficients of p are the solutions of a system of linear equations constructed from the x_j and y_j values [59, Section 4.6]. If your project requires this technique, you should write a general function that constructs the polynomial coefficients, given the x_j and y_j values and the degree m. This would invoke MSP routines to handle matrices and compute determinants. You should test it first on simple problems and then on more realistic ones. Finally, you can build a client program that uses this underlying routine to solve your real problem.

2.2   C++ MINOR FEATURES

Concepts

// comments
Local variable declarations

> new *and* delete *operators*
> *Mandatory function prototypes*
> *Initialized parameters*
> *Reference parameters*
> *C++ Version 2 stream input/output*
> *File input/output*

C++ is a major extension of C—a remarkable feat of grafting onto an older design a number of powerful new features based on a programming philosophy very different from the original C paradigm. The extension also includes a number of minor changes—some useful in their own right, others needed just to clear away obstacles to the major goal. Several of these, frequently used in MSP, are discussed in this section.

// Comments

One change that's inconsequential in principle but effective in practice is the introduction of // comments: whatever follows that symbol on a line is a comment. On a crowded publication page, this allows two more characters per comment than the traditional /*...*/ C comments. (However simple this change is, it's not trivial: The code

```
a = b //*check this*/c
```

has different meanings in C and C++!)

Local Variable Declarations

To improve program readability, C++ now lets you declare local variables nearer to where you use them—for example, you can usually write

```
for (int I = 0; I <= n; ++I) .
```

new and delete Operators

The C++ new and delete operators are used in MSP where Library functions malloc and free would have occurred in C code. Their syntax is simpler and less redundant.

Mandatory Function Prototypes

There are three important changes in function parameter syntax and semantics. First, prototypes are no longer optional: You *must* use one unless you include the complete function definition before all its invocations in the same compilation unit. Although major in principle, this change is minor in practice, because well written C programs always used prototypes—that's now part of the ANSI C standard.

Initialized Parameters

Second, you may give *default values* to the last one or more formal parameters in a function signature, and you may omit these parameters from invocations. You may place the initialization in a prototype; if so, it may not occur in the function definition. Here's an example:

```
void f(int I = 0);        // Prototype
void f(int I) { ... ; }   // Definition
void main() {             // All these invocations
  f(); f(0); f(1); }      // are OK.
```

Placing the initialization in a header allows a client to change it. *Caution:* An initialization in a header affects only those invocations in the same compilation unit! On the other hand, you can hide an initial value by placing it in the definition and distributing only object code for f. MSP uses initialized parameters to simplify common invocations. *Warning:* This practice tends to collide with function overloading, and invites errors that are hard to find! Refer to the discussion of overloading in Section 2.3.

Reference Parameters

The third change relating to parameters is the use of *reference* parameters. A function signature such as

```
int f(int& p)
```

specifies that formal parameter p will be passed by reference—that is, the caller passes the *address* of the actual parameter, not its *value*. This is a standard feature of other languages. (In FORTRAN, for example, it's the *only* way to pass parameters.) You can use a reference parameter p to transmit a result back to the caller. If p were a large structure instead of an integer, passing by reference would be much more efficient than passing the structure itself. That leads to a problem when p isn't intended to convey results back: f may inappropriately change the value of p (one of the most common errors in FORTRAN programming). To counter this danger, you may declare p *constant*:

    int f(const int& p).

The compiler will then catch any attempt to change p. Finally, a function may also *return* a reference:

    int& f(const int& p).

Instead of pushing a result onto the stack for the caller to pop, f pushes its address. In this case the result must be stored in static or free memory: A local variable would disappear before the address could be used to refer to it. You can achieve most of the effects of reference parameters by using pointers. But reference parameters allow the calling program to avoid pointers—a major source of error.

C++ Version 2 Stream Input/Output

Each C version seems to have new input/output conventions. They're implemented in the Library, so they aren't officially part of the language. Input/output routines constitute a *utility package*, distributed with the compiler. You may use it, or undertake the formidable task of writing your own custom input/output package. The input/output functions for previous C and C++ versions are distributed with Borland C++, but its manual recommends that you use *C++ Version 2 streams*. This facility was developed at AT&T Bell Laboratories and adapted by Borland International, Inc. It's implemented via a clever overloading (extended definition) of the C operators >> and <<. They're discussed in the Borland C++ *Library Reference* manual [4, Chapters 5 and 6] but not as thoroughly as you might like. See Lippman [29, Appendix A] and Stroustrup [52, Chapter 10] for more instructive accounts. This section will cover only those aspects of stream input/output that are actually used in MSP. Overloading in general is discussed in Section 2.3.

To use Version 2 standard stream input/output, your program must include the Borland C++ header file IOManip.H. It will automatically include another header, IOStream.H. They define types istream and ostream, which correspond to streams of data coming from or going to a memory buffer, generally associated with an input/output device. They declare some standard objects—called *streams*—of these types and associate them with standard DOS file handles. In turn, DOS associates the handles with devices:

Stream	Type	File handle	Default device
cin	istream	standard input	CON keyboard
cout	ostream	standard output	CON display
cerr	ostream	standard error	CON display

C++ stream input/output syntax is best illustrated by example. To prompt a user, then input `int` and `double` values on separate lines and display them, use statements such as

```
int I;  double x;
cout << "Enter integer, real values  i,x  on separate lines:\n";
cin  >> I >> x;
cout << "I = " << I << "  and  x = " << x;
```

Operator `>>` automatically flushes `cout` before pausing for `cin` input, so the prompt is guaranteed to be displayed.

Unparenthesized multiple `>>` and `<<` expressions are evaluated from left to right, and each of these overloaded operators returns as a value its left operand. Thus, a statement such as `cin >> I >> x` is interpreted as `(cin >> I) >> x`. First, `cin >> I` is executed, inputting I and returning the value `cin`; second, `cin >> x` is executed, inputting x.

The standard streams use default formats to input or output the various built-in types. You can change format—for example, field width, justification, scientific or fixed-point notation—by inserting a *manipulator* in the input/output stream. For example, the statements

```
cout << 17;
cout << setw(10) << 17;
```

will display the two-digit numeral 17 twice: The first exactly occupies a two-digit field (default format); the second is right justified in a ten-digit field. *Warning:* The field width changes for *one output only!* As another example, the statement `cout << endl` uses manipulator `endl` to flush the `cout` output stream and trigger a new line.

The stream output format manipulators for `double` values are not quite as general as the corresponding `StdIO` format features for C. MSP requires the omitted detail, so it uses `StdIO` function `sprintf` to convert numerical output first to an output string S. Then it executes `cout << S` to output the string.

You can also overload operators `<<` and `>>` to provide stream input/output for user-defined types. Examples are discussed in Chapter 5.

Your programs can also include the Borland C++ header file `FStream.H`, which defines types `ifstream` and `ofstream` analogous to `istream` and `ostream` but for file input/output. You can declare objects of these types and associate them with files or input/output devices. See Lippman [29, Appendix A] for examples. Some of the screen output displayed in this book's figures is produced by the following lines in my test programs:

```
#include <FStream.H>
ofstream File("e:\\Test.Out",ios::out);
//#define cout File
```

The second line declares an object File of type ofstream, associates it with file \Test.Out on disk e:, and opens that file for output. After a test program has produced acceptable cout output, the // is removed from the #define line. That sends all subsequent standard output to the open file. (This technique is less than ideal for interactive programs, because user input still echoes to the display, not to the file. The file must then be edited carefully to reproduce the normal cout display exactly. This roundabout method is necessary because the Borland C++ EasyWin output window doesn't fully support the Windows clipboard.)

The Library provides (undocumented) routines that automatically convert an input or output stream to type void* when required. A zero value indicates that the most recent operation attempted on that stream failed. This lets you use code such as

```
if (cin >> x) Process(x);
  else Terminate();
```

to process an input x or terminate if the input failed (for example, if the user entered <Ctrl-Z>). Or, if File is an ofstream object as in the previous paragraph, you could write

```
if (File << x) Continue();
  else Terminate();
```

to continue processing after outputting x or to terminate if output failed (for example, if File was not open for output).

2.3 OVERLOADING

> **Concepts**
> *Overloading*
> *Operator functions*
> *Function templates*
> *Superseding function template instances*
> *Using templates for Library code*

As mentioned earlier, a major goal of this book's MSP software is to let C++ scientific and engineering application programs look like the mathematical descriptions of the corresponding algorithms. That will make programs more understandable, and consequently less prone to error, in both the development and maintenance phases. A major OOP technique, which contributes to this goal, is *overloading*: use of the same function name or operator symbol to denote various functions or operators that differ in the types of their arguments. By overloading functions and operators, you can use familiar short names for mathematical processes. For example, the addition operators for scalars, vectors, polynomials, and matrices can all be represented by the familiar + sign and the equality tests by ==. You can still denote the value of polynomial p at scalar argument x by p(x) even though p is stored as an array of coefficients.

C++ enables overloading through two simple devices. First, in encoding a function name, the compiler considers the entire function signature, or list of argument types. Functions with similar names that differ in signature have different name codes, so the compiler can ascertain which to apply in a given situation. Second, operators are equivalent to C++ *operator functions*: int operator +, for example, corresponds to function int operator+(int i, int j). Thus, any function overloading technique applies to operators, too.

The ., .*, ::, and ?: operators cannot be overloaded. The singulary * and & operators and the binary comma operator are so tied to C++ syntax that overloading them would be inadvisable. Others, such as the binary (), [], and = operators, are delicate. MSP does overload these last three, following strict C++ guidelines.

Overloading sometimes conflicts with the provision for default function parameters. Providing a default parameter really creates an alternate form of a function with a different (shorter) signature. The compiler will flag as an error a subsequent attempt to implement the same signature through overloading.

Function Templates

You can use C++ function *templates* to create many overloaded functions at once, if they have the same source code except for argument types. Figure 2.2 is an example, simple enough to be instructive but too artificial for practical use. The keyword class in the function template suggests that you're using a C++ class, but that's not so. The (arbitrarily chosen) identifier Number stands for any C++ type, and main can invoke f for an argument of any type (including C++ classes). You don't have to templatize all arguments: Templates for functions with two arguments of the same or possibly different types could begin with lines such as these:

```
template<class Num>        template<class Num>      template<class N1,
    int g(Num x,               Num h(Num x,                   class N2>
        int y) {                   Num y) {           int j(N1 x,
                                                           N2 y) {
```

You can supersede a function template for a particular argument type by including an explicit definition for a particular instance. For example, if you included the definition

```
double f(double x) {            // This supersedes
    return x + 1; }             // the template.
```

in the program shown in Figure 2.2, then your third output line would read f(6.5) = 7.5.

```
template<class Number>           // This function template pro-
  Number f(Number x) {           // vides source code for the
    return x + 32; }             // required instances of  f .

void main() {
  cout << "\nf(65)   = " << f(65)            // Calls  f(int) .
       << "\nf('A')  = " << f('A')           // Calls  f(char) .
       << "\nf(6.5)  = " << f(6.5); }        // Calls  f(double) .
```

Output

```
f(65)   = 97
f('A')  = a
f(6.5)  = 38.5
```
The code for a lowercase character is 32 + the uppercase code.

Figure 2.2. Using a function template

Function templates are *syntactic* devices—that is, once the compiler has recognized one, it proceeds with other code until it finds an invocation of a template instance. It uses the template to *generate* the source code for that instance and finally compiles *that* into object code. It finds and reports errors within a template only during that final compilation process. Thus, a template error can go long undetected until some modification requires a new instance, which reveals an error involving a type incompatibility.

Using Templates for Library Code

Software packages like MSP are organized into header files and object code libraries. You include headers with your client program, compile it, then link your object code with code extracted from a library. Libraries are compiled without knowing which function template instances client programs such as that shown in Figure 2.2 will need. Therefore, the library source code itself must cause generation of *all* supported instances. A simple way to do that is to include in the library source code a function—not meant to be executed—that invokes (hence, causes the compiler to generate) all instances supported by the library. Library documentation must warn clients to limit their requirements to those instances. Figure 2.3 illustrates the analogous organization for the program shown in Figure 2.2 with the superseding instance described previously.

The best way to grasp overloading strategy is to see it in action. The low-level MSP modules contain many examples, which are discussed in detail in later chapters.

Header file `Template.H`

```
template<class Number>
  Number f(Number x);
```

Library source code file

```
#include "Template.H"

template<class Number>                  // This function template pro-
  Number f(Number x) {                  // vides source code for the
    return x + 32; }                    // required instances of  f .

double f(double x) {                    // This supersedes
  return x + 1.; }                      // the template.

void UseTemplate() {                    // Not to be executed:
  f(1);                                 // invoke all supported
  f('1'); }                             // template instances.
```

Client source code file

```
#include "Template.H"

void main() {
  cout << "\nf(65)  = " << f(65)        // Calls  f(int) .
       << "\nf('A') = " << f('A')       // Calls  f(char) .
       << "\nf(6.5) = " << f(6.5); }    // Calls  f(double) .
```

Figure 2.3. Using a template in a library source code file

2.4 ABSTRACT DATA TYPES AND CLASSES

Concepts
Abstract data types (ADTs)
Parallel with higher mathematics
C++ classes and objects
Class templates
Member functions

C++ was developed by grafting onto the C language some new features, which support implementation of Abstract Data Types (ADTs) and use of Object-Oriented Programming (OOP) techniques. This book doesn't attempt to illustrate all the concepts underlying ADTs and OOP. But their effect on the design of the mathematical software presented here is important. Thus, this section will describe some ADT and OOP concepts that played major roles in that design.

Abstract Data Types (ADTs)

The ADT concept developed as a way of modularizing software design and limiting the effect of programming errors. When software uses an ADT strictly, processes may not access the data unless the ADT definition *explicitly* permits. Thus, changing the design of software implementing an ADT can affect only an explicit list of processes. Moreover, if data become corrupted while you're executing this software, the definition narrows the field where you must search for the error.

The C language programming paradigm, however, conflicts with the ADT restriction. C programmers don't like to be told *no*. Therefore, ADT concepts were incorporated into C++ only loosely, so that unless an ADT explicitly allows access by a C++ function f, it's just *difficult*, but not impossible, for f to access the data. Unauthorized access should be the result of a conscious (therefore relatively safe?) effort, and should involve conspicuous programming ploys that a maintenance programmer can easily detect and understand.

The ADT concept is so general that it's hard to describe: To avoid writing something obviously false, you tend to write as little as possible. An ADT consists of a (small) set F of (related) function prototypes and a data structure definition D. D defines the set of all data handled interchangeably by the functions in F. The definition normally specifies the organization of the memory allocated to store each instance of D. (If you have to bring in further conditions—for example, conditions on what bit patterns are allowable—you probably haven't selected the set F appropriately.) The only aspect of the ADT that is shown to client programs is the set F of prototypes. The data structure definition and the implementation of the functions in F are hidden. A client program has no access to the data except by invoking functions in F. The client never knows how the data are stored in memory or how the functions in F manipulate them, except as revealed by the prototypes.

Consider, for example, a vector algebra ADT. F will contain prototypes of functions for vector addition, scalar multiplication, and so forth, and D will define structures called vectors. Clients know—through the addition prototype—that the sum of two vectors is a vector. But they needn't know *how* vectors are stored or *how* addition is performed. Conventional vector algebra software might store vectors as arrays of scalars, but software for handling very long, sparse vectors (mostly zero entries) might store them as linked lists. The corresponding addition algorithms are quite different.

Parallel with Higher Mathematics

This example shows how closely the ADT concept parallels modern higher mathematics. You study vector algebra by considering *vector spaces*. A vector space consists of a set $D = \{\alpha, \beta, ...\}$ of vectors and a (small) set F containing the addition and scalar multiplication operators $+$ and $\cdot$. A short list of axioms such as $t \cdot (\alpha + \beta) = t \cdot \alpha + t \cdot \beta$ tells what rules these vectors and operators obey. Vector space theory derives theorems from these axioms. Some of them are very deep and involved. Many different vector spaces occur in practical mathematics. Numerical analysis considers spaces whose vectors are n-tuples of real numbers, n-tuples of complex numbers, polynomials, integrable functions, power series, matrices, and so forth. A vector algebra ADT as described in the previous paragraph defines a vector space if its addition and scalar multiplication operations obey the axioms. By presenting vector algebra in the abstract context of vector space theory, we *modularize* the mathematics: We ensure that the theorems you learn about vector spaces apply equally well to *all* those concrete examples. Virtually all of modern higher mathematics uses this modularization technique.

The previous two paragraphs skirted a conflict in terminology that perplexes a mathematician. In mathematical language, a set or class D has *members*: The vectors of a vector space or a vector algebra ADT D are members of D. In standard C++ terminology, however, the functions in F are called members of the ADT. This collision hampers discussion and understanding. I choose to retain some C++ usage by calling the functions in F *member* functions: They do *apply* to members. But the instances of a data structure definition D are called *objects*. Vectors will be called *objects*, not members, of D. Mathematicians developed abstract mathematical terminology over decades, with time to reflect on its side effects. After decades, ADT and OOP terminology will mature, too. By that time, C++ will be just a historical curiosity.

As mentioned earlier, the ADT concept was built into C++ only loosely. The data structure definition isn't hidden; it lies bare for all clients to admire. (Or to scoff at, if that's warranted.) It may have several components, some of which are private, accessible only through member functions, and some public, available to all clients for use or abuse. The member function prototypes are public, of course. But their implementations can be public or compiled separately, giving clients access only to object code. Finally, there are provisions for giving specific non-member functions access to private data and for making specific member functions private. The private functions can be invoked only by other member functions. (They're for dirty work that the public must never see!)

C++ Classes and Objects

The C++ adaptation of the ADT is the *class*. The declaration of a class called Model looks like this:

```
class Model {
  public:
    :
    Declarations of public data structure components and
    prototypes or definitions of public member functions
    :
  private:
    :
    Declarations of private data structure components and
    prototypes or definitions of private member functions
    : } .
```

Public and private parts can be intermingled in any order. Once declared, `Example` is a type with the same status as `char`, `int`, `double`, and so forth. The term `Model` *object* is used for *instance of the* `Model` *data structure*.

The name of a member function `f` isn't meaningful without some clue to the class name. Usually this is given through the types of its arguments. When that's not appropriate, you may use a class name qualifier like this: `Model::f`.

For example, you could define a class `Point`, whose objects correspond to points in a coordinate plane:

```
class Point {
  public:
    double x,
    double y; }
```

Use it like a C struct: If you declare `Point P` then you can refer to its public parts `P.x` and `P.y`. But it's a C++ class, not a struct. As you'll see in this book, you can do much more with classes.

Class Templates

The `Point` objects just defined are composites, built from two components of a simpler type. It's useful in mathematics and software engineering to consider *general* processes that build composites—for example, given two sets X and Y we can build a new set S of (ordered) pairs $<x,y>$ of objects x and y in X and Y. We call S the *Cartesian product* of X and Y. The coordinate plane considered in the previous paragraph is an example: the Cartesian product of two copies of the set of real numbers. The *general* set theoretic Cartesian product *concept* lets us define and use the Cartesian product of *any* two sets.

Class *templates* are general C++ techniques for building composites from C++ objects of one or more given classes, to be specified later. For example, corresponding to the general Cartesian product concept, you can define

```
template<class xType,
         class yType>
  class Pair {
    public:
      xType x;
      yType y; }
```

To use this template, choose any previously defined type names for xType and yType. Write Pair<int,double> P to declare P an object of the class Pair<int,double>. That type is an *instance* of the Pair template. The entries P.x and P.y of P have types int and double. Your program can use as many template instances as you want.

Class templates are *syntactic* devices—that is, once the compiler has recognized one, it proceeds with other code until it finds a declaration of a template instance. It uses the template to *generate* the source code for that instance and finally compiles *that* into object code. It finds and reports errors within a template only during that final compilation process. Thus, a template error can go undetected until some modification requires a new instance, which reveals an error involving a type incompatibility.

Member Functions

Design of the public and private member functions of a class is complicated by two demands. First, as already mentioned, client programs must be able to carry out all object manipulations through the public member functions. Second, objects of a class often have varying size. Those of a polynomial class, for example, might include coefficient arrays for polynomials of various degrees. In such cases, the components of indeterminate size are generally represented by pointers to storage areas in free memory. You must provide public functions for creating, destroying, duplicating, comparing, and input/output of these components.

To satisfy these requirements, several kinds of public member functions—constructors, assignment, selectors, and so forth—frequently occur in class declarations. Moreover, some syntax considerations affect all member functions. Sections 2.5 through 2.7 consider these topics.

2.5 MEMBER AND FRIEND FUNCTIONS

> ### Concepts
> *Principal argument syntax: concatenation,* const *functions,* this *pointer*
> *Selectors*
> *Operator functions*

| *Membership criteria*
| *Friend functions*

Principal Argument Syntax

C++ has adopted a new syntax for declaring and invoking member functions of a class Example. The first argument of a member function—its *principal* argument—*must* have type Example. Traditional function invocation syntax is used for the other arguments, but the principal argument is omitted in the declaration of a member function—for example, this class definition fragment declares member functions f and g with principal arguments of type Example:

```
class Example {
    ⋮
    int f(int I);
    Example& g();
    ⋮ }.
```

f also has an int argument and returns an int value; g returns a reference to an Example object. If X is an Example object, then the expression X.f(17) calls f with arguments X and 17. Function g returns a reference to an Example object, which can be passed as principal argument to another member function. This kind of function composition has a new *concatenation* syntax: X.g().f(17) indicates that g is called with principal argument X, producing an Example value stored in a temporary variable T, then f is called with arguments T and 17.

A member function's principal argument is always passed by reference. You may want to indicate that the function doesn't change that argument's value. You can't place the const qualifier before the parameter as you would for a non-member function, so place it after the parameter list. For the example function f in the previous paragraph, you'd declare int f(int I) const.

Within the definition of a member function, the principal argument is anonymous. It doesn't appear in the formal parameter list. Occasionally, a name is necessary, though. To that end, C++ provides the pronoun *this. Actually, within a member function definition, this is the name of a pointer to the principal argument, so *this is the argument itself. The programs in this book use the abbreviation This for *this. It's not necessary to write (*this).f(17) to send *this to member function f as its principal argument: Just write f(17).

Selectors

Declaring public a data component of a class permits unlimited public access: A client program can use or change it at will. If you want to provide limited access—for example, to prevent changing a variable or to check array indices before granting access—you must use public member functions called *selectors*. A selector that provides read-only access is usually a member function with no non-principal arguments. It returns the desired component of the principal argument. The index checking facility is usually provided by overloading the subscripting operator [ ].

Operator Functions

It's also possible to declare class member *operators*. Briefly, singulary and binary operators such as ! and == are equivalent to certain *operator* functions. If these are declared as member functions, the corresponding single or left-hand operand becomes the principal argument and is omitted in certain syntactic constructs.

Membership Criteria and Friend Functions

While designing a class Example, you'll often ask whether a function f should be a member function. The answer must be *yes* if

1. some C++ rule explicitly says so, or
2. f *must* have access to private components of Example objects, or
3. you want to use the principal argument syntax for invoking f.

Since the principal argument syntax is mandatory for member functions, the second and third criteria in the previous paragraph sometimes collide. What if a function h must have access to private components of Example objects, but you don't want to use the principal argument syntax? In that case declare h a *friend* of the Example class and use traditional syntax:

```
class Example {
  public:
     ⋮
     friend int h(int I, Example X);
     ⋮ }.
```

You must declare h in the same compilation unit with the Example definition. Rely on friendship, for example, when h must have access to the private components of objects of two different classes.

What if the previous two paragraphs don't require your program to use a member or friend function? A reasonable strategy for applying a new idea is Ockham's razor: *Don't* use a member or friend function unless your design requires it. But that's too simple for this situation. C++ rules for inheritance, automatic type casting, and generation of template function instances are interrelated, delicate, fuzzy, and poorly documented. They differ for member, friend, and unfriendly non-member functions. In some instances, I have found it significantly more graceful to use member operators in MSP even though the criteria just listed don't require it. However, MSP employs friend functions only where absolutely required. Perhaps there *is* a good argument for broader use of friendship. The question is open—more experience is needed to settle it.

2.6 CONSTRUCTORS AND DESTRUCTORS

> **Concepts**
> *Default constructor*
> *Copy constructor*
> *Converters*
> *Assignment operator*
> *Destructors*

Default Constructor

When a class has a data component of large or indeterminate size, such as a long vector or a polynomial of varying degree, you normally store these data in free memory space allocated by the new operator; the class component is really a *pointer* to these data. When you declare an object as a global or local variable, the compiler will allocate space for the pointer in static memory or on the stack. You must provide code to allocate free memory storage for the data. This requires a *constructor*—a public member function that shares the class name and returns no value—for example,

Declaration	*Definition*
class Example {	Example::Example() {
public:	⋮
Example();	Code to allocate free memory, and so forth
⋮ }	⋮ }

There are different kinds of constructors; this example is called a *default* constructor. Its execution syntax is similar to that of a traditional declaration: The statement Example X constructs an Example object X, and Example Y[7] constructs an array of seven Example objects Y[0], ..., Y[6]. The first declaration executes your constructor once; the second executes it seven times.

A default constructor provides no flexibility: You can't specify how much memory to allocate. Fortunately, though, a constructor may have arguments to specify size, initial value, and so forth—for example, a constructor Example(int n) might allocate free memory for an n dimensional vector. To declare such a vector V, you might compute n, then declare Example V(n).

You can't directly declare an *array* of such vectors, however: The syntax `Example V(n)[7]` is illegal.

Instead, you may define a member function `Example& SetUp(int n)` to do the work of `Example(int n)`. To declare an array of seven vectors of length n, you'd declare an array `Example V[7]` using the default constructor and then execute `SetUp` on each entry of `V`:

```
for (int i = 0; i < 7; ++i) V[i].SetUp(n).
```

If you need this `SetUp` facility, you'll probably want to define

```
Example::Example(int n) { SetUp(n); }
```

to emphasize the relationship of the constructor and the `SetUp` function.

Copy Constructor

Occasionally, you need to construct a new `Example` object that's an exact copy of an existing `Example` object X. For this, you use a *copy* constructor, which you must declare as `Example(const Example& X)`. It should determine the size of X if necessary; use the `SetUp` function to allocate sufficient space and then copy X into the new object. You can create a new copy Y of object X by executing `Example X(Y)` or `Example X = Y`.

The compiler also generates code that invokes your copy constructor whenever appropriate to make *temporary* copies. This occurs if you pass an `Example` argument by value to a function or a function returns such a value. To invoke function `Example f(Example X)`, for example, the caller will use your copy constructor to copy X onto the stack as the actual parameter TempX for f. Then f is invoked and uses TempX in place of X in its computations. TempX disappears when f terminates. One TempX component is probably a pointer to the free memory storage allocated earlier by your copy constructor. Unless you make some provision to destroy it, that storage will persist—allocated but inaccessible because its pointer has disappeared. This provision is described in the last paragraph of this section.

The return value computed by f can't remain in place on the stack, because it's a local variable for f and would also disappear when f terminates. Therefore, the compiler generates code that uses your copy constructor to copy the return value into a temporary local variable for the caller. If your function call has the form Y = f(X), you'll assign this temporary value immediately to another variable, and the temporary variable isn't really needed. (Assignment is discussed in a later paragraph.) However, the compiler must also let you pass the return value immediately as an argument to another function g—to evaluate g(f(X)), for example. To accommodate that, a temporary variable is required.

When an argument or a function value is passed by reference, no temporary copies are needed, and your copy constructor is not involved.

Converters

Closely related to the copy constructor for a class Example is the function you'd need to convert to Example a variable of a related type SourceType—for example, you might want to convert a scalar to a polynomial with degree 0. You need to define a *converter*: a public Example member function Example(const SourceType& X), which returns no value. It should allocate—probably via the SetUp function described earlier in this section—sufficient space for the Example object and then enter into the Example components the appropriate data from X. The implementation is just like that of the copy constructor. Moreover, you can use ordinary function invocation syntax: Example(X) is simply the Example object obtained by converting X from type SourceType.

The reverse conversion, *from* Example *to* some type TargetType, requires defining a public Example member operator function

```
operator TargetType() const.
```

Although the special syntax doesn't so indicate, this function must return a TargetType value. The syntax for invoking this converter is exactly the same as that for converting in the opposite direction.

Assignment Operator

If you use a class Example with a pointer component, you must provide code to perform assignments X = Y between existing Example objects X and Y. This will resemble the code for your copy constructor, but it doesn't have to set up X in free memory: X is already there. You'll have to decide what to do if X and Y may point to free memory areas of different sizes. That may indicate an error, or you may want to copy part of Y into a smaller X or all of Y into part of a larger X. To provide this enhanced assignment facility, you overload the assignment operator = by first declaring it a public member operator:

```
Example& operator=(const Example& Source).
```

Its definition looks like this:

```
Example& Example::operator=(const Example& Source) {
    ⋮
    Code to adjust the memory allocated to the principal argu-
    ment This if necessary, and to copy Source to This.
    ⋮
    return This; }
```

If a program includes an assignment statement with mismatched types, and you provide the appropriate converter, the compiler will automatically generate code to invoke it before executing the assignment. Similarly, if you pass an argument of the wrong type by value when invoking a function, and you provide the converter, the compiler will generate code to call it. This makes programs more readable but also more hazardous. If the type conflict was unintentional, and the converter doesn't transform the offending object into an exactly equivalent form, your program can appear to execute correctly but actually give incorrect results. One way to detect such an error during debugging is to insert an output statement in the converter to announce its execution.

Destructors

C++ programmers have great flexibility in choosing processes to be carried out by constructors. In the examples under the previous heading, the constructors allocated free memory storage. When OOP techniques are used for input/output programming, for example, constructors might also open a file or open a screen window corresponding to a file buffer or a window image in memory. You must generally provide some code for housekeeping when the constructed objects pass away: release free memory storage, close a file, redraw the screen, and so forth. Otherwise, free memory might remain allocated but inaccessible because its pointer has disappeared, a file might remain open with buffer unflushed, or an inactive window might remain on the screen. To take care of such matters, C++ invites you to place such housekeeping tasks in a class *destructor* function. The compiler generates code to execute it whenever an object X passes away. This occurs when you explicitly destroy X by executing `delete X`, and—more frequently—whenever execution control leaves the block in which X was defined. The syntax is simple: The destructor for a class `Example` is simply a public member function named `~Example()`, with no nonprincipal argument and no return value.

2.7 DERIVED CLASSES

> **Concepts**
> *Class inheritance*
> *Parallel with higher mathematics*

The algebraic structures of higher mathematics are organized into an elaborate hierarchy. Here's a part of it that's heavily used in numerical analysis:

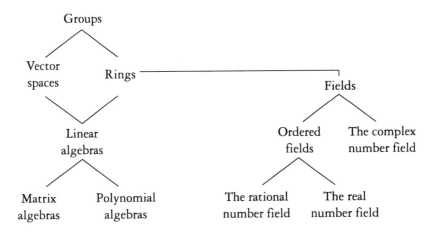

Each kind of structure inherits qualities from those above it. Vector spaces and rings are special kinds of groups; linear algebras inherit properties of both vector spaces and rings; and so forth. At the lowest level, these structures become more nearly concrete: the real number field, in which we commonly perform arithmetic, is an example of an ordered field. Computer arithmetic works with numbers that belong to the rational field. Symbolic algebra packages such as *Mathematica* [57, 58] also do arithmetic in the rational number field. At the upper levels are structures you rarely see mentioned in numerical analysis texts; they're used to describe qualities shared by widely differing concrete examples. The properties of addition shared by rational numbers and matrices are studied in group theory, for example.

Various programming specialties use similar hierarchies—for example, a display software package may define a structure called a window, of which there are several kinds, including menu. In turn, there may be several kinds of menu.

Experience with hierarchies such as these led to development of the OOP class inheritance technique. In C++ you can define a class ClassD, so that it inherits qualities from a base class ClassB. ClassD is said to be *derived* from ClassB. Its declaration looks like this:

```
class ClassD public ClassB {
   ⋮
   Declarations of additional ClassD components and of
   member functions that differ from those of ClassB
   ⋮ } .
```

The token public specifies that public ClassB components remain public when regarded as components of the derived ClassD. (Alternative access specifiers are available but not used in

MSP.) You can also specify inheritance from more than one base class. Generally, a derived class `ClassD` is similar to its base class `ClassB`, but

- `ClassD` has additional data or function components beyond those of the base class, or
- some `ClassD` member functions differ from those of the base class, or
- the values of the `ClassD` data components are more restricted than those of the base class.

The third provision is usually carried out by modifying `ClassB` member functions: You restrict those that construct or alter the data components.

Class inheritance is the basic OOP technique *least* used in MSP. Although it seems appropriate to use inheritance to model the part of the higher mathematics structure hierarchy used in numerical analysis, considerable overhead—intellectual as well as programming overhead—is incurred. Classes would need to be declared for each type of structure used, and for structures such as groups, which are not used per se but would serve as common base structures. Many of these classes would require their own constructors, destructors, and so forth, and the conversions between the large number of resulting types would have to be analyzed in detail for consistency and for conformity with mathematical usage. In my opinion, such a design could be useful if the result were a software package so easy to describe and so like ordinary mathematics that client programmers could ignore virtually all of the class structure details.

But that's impossible with C++ programming, because it retains that troubling quality of C: You just can't ignore details. A seemingly minor error can have far-flung effects. Sorting these out during debugging requires facility with the dirtiest aspects of the language. (Comparable facility with English sometimes helps, too.)

Thus, it seemed inadvisable for the MSP to rely heavily on class inheritance. The technique is used to derive a polynomial class from the corresponding vector class: Polynomials are regarded as vectors of coefficients, and new member functions are added for polynomial multiplication and evaluation. The discussion of the `Polynom` classes in Chapter 6 will give you an idea of the power of the inheritance technique.

2.8 EXCEPTIONS

Concepts

Maintaining control after an error occurs
The stack and function invocation
The stack and maintaining control
`try` *and* `catch` *blocks*
`throw` *command*

> *Exception types*
> *Interfacing with low-level error-handling features*

Over the years, the evolving C and C++ compilers supported various ad hoc methods for maintaining program control when an error situation occurred. The current *exception-handling* scheme was adopted about 1990.

Suppose you're executing a program with *no* error handling, and it crashes after asking your hardware/software system to do something impossible or illegal. Then:

- Your program and the operating system report only the sketchiest information—if any—about *what* error occurred.

- The only way you can tell *where* your program failed is to analyze which input/output occurred before the failure. Adding output to gain information makes the analysis more tedious.

- Unless you have source code for the failing component, you can't step through it with a debugger. When that *is* possible, it's tedious and confusing.

This book's MSP software package incorporates exception handling. You can see how to apply the technique to your software, including MSP client programs. In most error situations, this system tells you what failure occurred and what functions were executing then. You get a record of the nested function invocations from main down to the one that failed—even if you're using precompiled MSP library code. You know where to start stepping through source code with the debugger, if that's necessary. If you need better information about an error, you may be able to modify the MSP exception handlers to provide it: Then you won't have to analyze and discard huge amounts of data about parts of your program that *are* functioning properly. If you can't avoid an error situation, and wish to maintain control after it, you can incorporate exception handlers in your own routines so you can regroup and proceed to the next task.

The Stack and Function Invocation

To understand how C++ exception handling works, you need to know what happens when one function invokes another. That involves use of the *stack*. Dynamic memory is allocated to each function f as it's called, always stacked immediately "on top of" the dynamic memory previously allocated to the calling function. Dynamic memory is for communication between those functions and for temporary use by f as necessary. When f returns, its stack memory is freed for general use: The stack returns to its status before the invocation.

The compiler sets up the code for this process without any detailed instructions from you—for example, consider what happens when function Caller executes a statement such as z = f(x,y). The default system is described. It can be modified for special needs, particularly to interface with functions written in languages that handle function calls differently. See the *Turbo Assembler* manual [3, Chapter 18] for details.

First, `Caller` places the values of parameters `y` and `x` on the stack (or their addresses, if they're reference parameters). Then it stacks the return address. This is the address of the instruction in `Caller` that should be executed immediately after `f` completes its computation. That instruction will be identified by the letters RRR when it occurs later in this paragraph. Next, `Caller` executes a jump instruction, transferring control to `f`. `Caller` knows where `f` is, because the linker inserted into the `.exe` file the addresses of `Caller` and `f` relative to the start of your program, and the operating system provided the starting address when it loaded the `.exe` file. Now, `f` begins execution and computes a function value using the parameters, which are on the stack immediately under the return address. It places the value in a location known to `Caller`. (Various techniques are used, depending on the number of bytes in the return value.) When this computation is finished, the stack will be the same as it was when `f` began execution. Moreover, all objects created by class constructors while `f` was executing will have been destroyed by the corresponding destructors. Finally, `f` removes the return address from the top of the stack and uses it to jump back. `Caller` removes the parameters from the stack—this is instruction RRR—and copies the return value into `z`. The stack is now the same as it was before `Caller` invoked `f`.

The Stack and Exception Handling

Now, suppose `f` fails: It asks your hardware/software system to do something illegal, such as divide by zero or read past the end of a file. Just keeping control in `f`, reporting the failure, and stopping is an insufficient reaction, because `Caller` may have started something that it must finish *somehow*—for example, an input/output operation. Reporting the failure and returning to `Caller` is insufficient, too, because `Caller` expects, but may not get, an appropriate result from `f`; that could cause a worse failure of `Caller`.

When C++ exception handling is incorporated, `Caller` announces to the compiler,

- if a certain kind of error occurs during execution of the following block of code—called a try block—report that to me, restore the stack to the state it's in now, and destroy all objects created by class constructors during that execution; then I'll handle the error.

The error could be failure of any function `f` invoked in that block, failure of any function `g` called by such an `f`, failure of any function called by such a `g`, and so on. Moreover, `Caller` can agree to handle several kinds of errors or *all* kinds supported by the exception-handling system. The try block must be followed immediately by a `catch` block for each kind of error that `Caller` will handle. The last `catch` block may be a general one, which handles all remaining errors reported by the software you're using. (Borland C++ automatically provides such a `catch`-all block at the end of function `main`. If you don't handle some exception, that will—it terminates your program and displays the message Program Aborted.) A variable of any C++ type whatever can serve to identify the kind of error; it's then called an *exception*. You might not need to use the value of that variable, since its type conveys *some* information. Or your software might use that value to convey very detailed information about the error.

The requirement that all objects constructed during execution of a try block be destroyed before entering the corresponding catch block allows orderly completion of input/output operations and others that require dedicated resources. If you acquire the resource—for example, access to a file or memory—by executing a constructor, then the corresponding destructor should release it for other programs to use.

Figure 2.4 demonstrates this process. Caller invokes f, which declares an object E of class Example. The Example constructor and destructor merely announce themselves. (Their codes are so short that they're included "in line" here—not a common practice in this book.) Then f invokes g, which simulates failure and throws exception 13. The output shows that the line in g after the throw doesn't execute; nor does g return to f. But the output shows that E is destroyed, and then the int exception handler in Caller takes control.

If you want to catch all other exceptions that might be thrown during execution of a more complicated version of f, you could append a catch-all block after the catch block in Figure 2.4:

```
catch(...) {
   cout << "\nCaller\t catches an unspecified exception."; }
```

(The C++ ellipsis symbol ... is rarely used otherwise, but it is certainly handy here.)

MSP exception handling is designed to describe an exception as soon as it occurs and then report what functions were executing. To that end, each try block has a catch-all block, and each catch block terminates with a throw command. That rethrows the exception just caught. The catch-all block of the function that invoked Caller will catch the rethrow, announce itself, and pass it on. This system is described in detail with the individual MSP functions in later chapters.

One major question remains: Do the Borland C++ Library routines issue exceptions properly to interface with the MSP handlers? *Generally no*, because they have to remain compatible with C/C++ programs written before exception handling was adopted. Earlier language versions have at least three different ways of handling errors, albeit awkwardly:

- Many input/output functions return a *status code*. You must inspect it after calling the function and possibly react to an error status instead of proceeding normally.

- Some Library functions use the *signaling* technique. It's rather like throwing an exception, but the destructors aren't executed automatically, as in Figure 2.4. Moreover, there's no analogous use of types to classify error signals.

- Yet others issue *software interrupts*, and you have to provide your own interrupt service routine to handle the error.

```
class Example {
  public:
    Example() { cout << "\nExample constructed."; }
    ~Example() { cout << "\nExample destroyed. "; }};
void g() {
  cout << "\ng\t\"fails\", throws 13 .";
  throw(13);
  cout << "\nInaccesible code."; }
void f() {
  cout << "\nf\t declares Example .";
  Example E;
  cout << "\nf\t invokes g .";
  g();
  cout << "\nThis doesn't execute."; }
void main() {                          // main plays the
  try {                                // role of Caller .
    cout << "\nCaller\t invokes f .";
    f();
    cout << "\nThis doesn't execute."; }
  catch(int n) {
    cout << "\nCaller\t catches " << n << " ."; }}
```

Output

```
Caller   invokes  f .
f        declares Example .
Example  constructed.
f        invokes  g .
g        "fails", throws 13 .
Example  destroyed.
Caller   catches  13 .
```

Figure 2.4. Throwing and catching an exception

To use exceptions, each of these error-handling techniques must be interfaced to the new system. MSP does that, as follows:

- Its file input/output routines inspect the status code and throw exceptions as appropriate. These are described in Sections 5.1 and 7.1.
- It includes signal response and interrupt service routines that merely set up and throw exceptions. These are described in Sections 3.7 and 3.8.

Realistic exception-handling demonstrations are provided in those sections.

3

Computing with Scalars

The numerical analysis routines described in this book are organized into a Mathematical Software Package (MSP). Its components are C++ functions for manipulating numbers, vectors, matrices, and polynomials, and for solving equations and systems of equations frequently encountered in science and engineering. Many of these routines—for example, functions that evaluate sums and products—are individually quite simple, while others, such as some routines that solve polynomial equations and linear systems, are quite intricate. A major MSP design goal is to make the individual functions as simple as possible, while enabling client programmers to build very powerful and complicated application programs easily.

A second design guideline is flexibility. MSP uses very general structures for scalars, vectors, matrices, and polynomials, and provides a comprehensive set of general functions for manipulating them. Because of their generality and simplicity, client programmers can adapt them easily to build programs to solve problems in various application contexts.

The third design goal is to enable client programmers to write programs that resemble mathematics. As much as possible, software intricacies are handled by the low-level MSP routines discussed in this chapter and in Sections 5.1, 6.1, and 7.1, so that client programmers may concentrate on the mathematics. The more a program resembles the mathematical description of the algorithm it implements, the less likely it is to contain errors and the easier it is to maintain and adapt to related problems.

A fourth goal, minor from a scientist's or engineer's point of view, but major for writing this book, is to show how you can harness the flexibility and power of C++ to help solve mathematical problems of major scientific and engineering interest.

This chapter and Sections 5.1, 6.1, and 7.1 describe in great detail the lower levels of MSP software: its modules for manipulating scalars, vectors, polynomials, and matrices. While their mathematical aspects are generally inconsequential, their programming features are quite the opposite. In fact, MSP is designed to separate these qualities: Where the mathematics is complicated, the programming should be simple.

Using C++ emphasizes this distinction. One of its major features is its facility for overloading arithmetic operator symbols such as +, -, and * to denote operations on data such as vectors and matrices. By using C++, you can compute a product of the sum of vectors V and W by the difference of matrices A and B, for example, by writing (V + W)*(A - B). Awkward and verbose function invocation syntax disappears, and you don't have to declare temporary variables to store intermediate results such as V + W and A - B. Another C++ feature, its inheritance mechanism, enables MSP programmers to refer to polynomials in the manner common in mathematics: as the vectors of their scalar coefficients. Vector operations can be adapted for some aspects of polynomial algebra, and new features can be added to implement others.

C++ attains flexibility and power at a cost. The mathematical programming features just mentioned are *not part of* the language; C++ merely *enables* a programmer to implement them. That programmer must be familiar with many sophisticated programming concepts, including some normally regarded as belonging to system programming. This kind of programming is confined to the low-level MSP modules described in this chapter and Sections 5.1, 6.1, and 7.1, so that client programmers using the modules can concentrate on the mathematics.

Thus, some of the programming in this chapter and all of the programming in Sections 5.1, 6.1, and 7.1 is far more intricate than what you'll meet in the rest of the book. If you're interested in *developing* this kind of software, you'll want to read this chapter in detail, try alternative approaches, find out where the difficulties lie, and try to develop better modules that are still usable by application programmers. But if you're one of the latter class, mainly interested in *using* MSP to solve scientific and engineering problems, you'll want to read it more quickly to find out what's here, what it does, and how to use it. You can then proceed to the descriptions of the more mathematical functions later in the book and use the examples there to guide you in writing your own application programs. But you should anticipate reading parts of this chapter and Sections 5.1, 6.1, and 7.1 again—perhaps several times—to figure out what's going on when your programs malfunction.

This book's approach to C++ numerical analysis programming isn't meant to be definitive or even to set a precedent. Its aim is more modest: to show that the major MSP goals can be attained effectively by using C++. There's no attempt to show that C++ is the best possible language for this work. (In fact, I would probably disagree with such a claim.)

There is not even a concerted attempt to find the best possible C++ techniques. Here are several features of C++ that ensure that the MSP, in its present design, is only a brief stopping point in a much broader flow of numerical analysis software development:

- C++, like its ancestor C, was developed to enhance effectiveness in system programming, not specifically to facilitate scientific application programming.

- Some of the Object-Oriented Programming (OOP) goals of C++ are inconsistent with some of the original goals of the C language design. This results in awkwardness in some parts of C++, regardless of the type of program under development.
- Some C++ features, particularly the principal argument syntax, which are intended to enforce the OOP paradigm, actually contravene some mathematical practices developed over decades to handle gracefully some commonplace but logically complex mathematical concepts.

Finally, before you plunge into programming details, you should consider briefly the process of *testing* these low-level routines. Once the overall organization and the catalog of MSP functions was determined, and their prototypes designed, test routines were written. These exercise every function in every module in ways commonly required in scientific application programming. The functions were tested as they were written. This was the main test bed, but many additional routines were used to verify correct operation for special inputs (for example, does the MaxNorm function correctly give the norm 0 for the empty vector?). Actually, this description of the testing process is idealized. Frequent minor design changes made the development of the test bed more gradual and more chaotic. It's too long, boring, and messy to include in this book. It's certainly *incomplete*: Additional bugs will come to light. If you need to test an MSP function further or modify it or build a new one, you should replicate part of the original testing process. You'll find some neat test-bed fragments involving Legendre polynomials in Section 6.2, and many of the examples in later chapters can be used to test low-level functions. Some well organized parts of the test-bed code are included on the accompanying and optional diskettes. (See Appendix A.4.)

3.1 MSP General MODULE

Concepts

#ifndef *and* #endif

Definition of This

Identifier capitalization

The Boolean *type*

Empty strings

Pausing to inspect output

The Digits *function*

max *and* min

Templates for derived functions and operators

Exceptions

The MSP General module contains some macros and functions used by all other MSP modules and by client programs. They're somewhat unrelated, but they are rather simple. This module is the natural place to begin, if you want to understand any aspect of the lower levels of the MSP. Like the other modules, it has a header file General.H, and a source code file General.CPP, which defines functions declared in the header. Including some function definitions in the header file would have made them easier to read but would prevent Borland C++ from using a precompiled header—a time saver for large projects. Therefore, MSP functions are usually declared in .H header files and defined in .CPP source code files.

You'll find the General header and source code files with the other MSP modules on the optional diskette. They're displayed in Figures 3.1 through 3.4. General.H must be included in every MSP and client source code file. It causes several C++ Library header files to be included as well; therefore, client programs don't need to include those headers explicitly.

#ifndef and #endif

The first General.H header feature is a set of C++ preprocessor directives:

```
#ifndef General_H
#define General_H
    ⋮
(The rest of the header file)
    ⋮
#endif
```

Each time the preprocessor encounters the directive #include "General.H" in a source code file and starts reading the header file, it first asks if General_H has been entered in its list of defined tokens. If not (#ifndef), it makes the definition (#define) and reads the rest of the header file. But if the token is defined, the preprocessor has already read the file, so it skips to the end (#endif). This is a standard C technique to avoid duplicate definitions and declarations, which are often illegal. All MSP headers use it.

Definition of This

The first General.H macro defines This to stand for (*this). Described in Section 2.5, this points to the principal argument of a class member function. I found that the abbreviation helped avoid many trivial syntax errors. The parentheses prevent possible operator misassociation when This occurs in a compound expression.

```
//************************************************************
//                     General.H

#ifndef    General_H
#define    General_H

#include   <IOManip.H>
#include   <StdLib.H>

//************************************************************
// General programming aids
                                        // Shorthand for the
#define This    (*this)                 // target of a class'
                                        // self-pointer.

typedef  int       Boolean;             // Simulation of the
const    Boolean   True  = 1;           // two-valued Boolean
const    Boolean   False = 0;           // data type.

Boolean StrEmpty(const char* S);        // Is string  S  empty?
#define EmptyStr ""                     // Empty string.
const    char      Blank = ' ';         // Blank character.

void Pause(                             // Display a prompt, then
  const char* Prompt                    // await and discard a
    = "\nPress any key...\n\n");        // keyboard character.

                                        // Number of characters
int Digits(int x);                      // in the shortest numeral
                                        // for  x .

int min(int t1,                         // Return the smaller of  t1
        int t2);                        // and  t2 .  The  StdLib.H
                                        // template doesn't suffice.
```

Figure 3.1. MSP header file General.H
Part 1 of 3; continued in Figure 3.3.

Identifier Capitalization

The definition of This illustrates this book's capitalization convention: Capitalized identifiers don't occur in C++ or its Library (except for a few written entirely in uppercase). On the other hand, most MSP identifiers are capitalized. The major exception to that rule is an occasional

```
//************************************************************
//                    General.CPP

#include "General.H"
#include    <ConIO.H>

Boolean StrEmpty(const char* S) {       // Is string  S  empty?
  return *S == 0; }

int Digits(int x) {                     // Return the number of
  int p = 1;   double n = 10;           // characters in the shortest
  while (abs(x) >= n) {                 // numeral for  x .
    ++p;      n *= 10; }
  if (x < 0) ++p;
  return p; }

void Pause(const char* Prompt) {        // Display a prompt, then await
  cout << Prompt << flush;              // and discard a character
  getch(); }                            // from the keyboard.

int min(int t1,                         // Return the smaller of  t1
        int t2) {                       // and  t2 .  The  StdLib.H
  return (t1 < t2 ? t1 : t2 ); }        // template doesn't suffice.
```

Figure 3.2. MSP source code file General.CPP

generic variable of a built-in or the complex type, for which traditional mathematics notation uses lowercase.

The Boolean Type

Although most C++ compilers now have a built-in two-valued Boolean type, for historical reasons, MSP doesn't use it. if statements and various operators regard the int value 0 as representing falsehood and any other int values as truth. (It's easier to say *yes* to a C programmer than to say *no*!) Actually, most Borland C++ operators return the value 1 to represent truth. I feel this is a poor practice, so General.H includes definitions that make int the Boolean type for MSP and give the values 0 and 1 the names False and True. This rarely conflicts with Borland C++ conventions, and it helps eliminate logical errors.

Other desirable features of a Boolean type are *not* implemented. There's no protection against assignment of values other than 0 and 1 to Boolean variables. There's no provision for cin to

```
//*********************************************************************
// Templates for derived functions and operators.  To facilitate
// use with any operand types, their bodies are in this header.

template<class Operand>                 // Generic sign
  int Sign(const Operand& x) {          // function:  return
    return (x  < 0 ? -1 :               //    -1,0,1 for
            x == 0 ?  0 : 1); }         //    x <,=,> 0 .

template<class Operand>                 // Generic !=
  Boolean operator!=(const Operand& x,  // operator.
                     const Operand& y) {
    return !(x == y); }

template<class Operand>                 // Generic singulary
  Operand operator+(const Operand& x) { // + operator: it
    return x; }                         // has no effect.

template<class LOperand,                // Generic +=
         class ROperand>                // operator.
  LOperand& operator+=(LOperand& x,
                 const ROperand& y) {
    return x = x + y; }
⋮
```

(Similar code for generic -=, *=, %=, ^=, and /= operators)

Figure 3.3. MSP header file `General.H`
Part 2 of 3; see also Figures 3.1 and 3.4.

translate input strings True and False to Boolean values automatically or for cout to output them. It's possible to do these things but only at the cost of a page of complicated C++ code that uses nearly the full range of OOP features.

Empty Strings and Blank Characters

C++ strings are char arrays terminated by the null character—the one with ASCII code 0. An empty string consists of the terminator alone. Since a string variable S is a pointer to the beginning of the corresponding array, you can check whether S is empty by asking whether

```
//**************************************************************
// Exception handling

typedef enum {                               // MSP handles
    UnknownError, DivideError, EndOfFile,    // these
    FormatError,  IndexError,  IOError,      // exceptional
    MathError,    OtherError,  OutOfMemory } // situations.
  ExceptionType;

class Exception {                            // MSP exception class.
  public:
    ExceptionType Type;                      // What kind of situation?
    Exception(ExceptionType T                // Constructor.
              = UnknownError) {              // UnknownError ( 0 ) is
      Type = T; }};                          // used only as default.

#endif
```

Figure 3.4. MSP header file General.H
Part 3 of 3; continued from Figure 3.3.

*S == 0. This occurs frequently enough—in constructing output captions—that General.H defines function

```
Boolean StrEmpty(const char *S)
```

to return the answer to this question. You'll find its brief source code in file General.CPP.

The empty string has a simple name: " ". But that's hard to distinguish from the blank character ' ', so General.H defines a macro EmptyStr and a constant Blank to stand for those. MSP avoids macros when possible for two reasons. First, they can lead to serious errors that are hard to find. Second, even when correct, they complicate software development: You may *read* code containing macros, but your debugger *processes* it without them, after the preprocessor's translation. Making EmptyStr a char* constant, though, would prevent Borland C++ from using a precompiled header, so it's a macro.

Pausing to Inspect Output

Occasionally, you want to stop execution to inspect the display and prompt for any keystroke to resume computation. The General module supports this requirement with function

```
void Pause(const char* Prompt = "\nPress any key...\n\n") .
```

You may accept the Press any key... default or specify your own prompt. Pause sends the stream manipulator flush to cout after the prompt to ensure that it's displayed. Then Pause calls Borland C++ Library ConIO function getch to input a character, which it ignores. In turn, getch calls for a BIOS service, so this code might not be portable to incompatible systems. Pause didn't use cin stream or StdIO input because they affect the display unfavorably.

The Digits Function

MSP must be able to determine the smallest number of digits required to display an int value X. (It uses this to make neat subscripts.) Library functions sprintf and itoa each provide that information, but they're inconvenient, so MSP includes its own function int Digits(int X).

max and min

For consistency with earlier versions, the Borland C++ Version 4.0 Library header file StdLib.H includes two familiar but obsolete macros, which compute the maximum and minimum of two numbers:

```
#define max(a,b)  (((a) > (b)) ? (a) : (b))
#define min(a,b)  (((a) < (b)) ? (a) : (b))
```

They're dangerous, because each evaluates parameters a and b twice. If you used such a macro to evaluate max(f(x),b) for some function f that had side effects, you'd get the side effects twice instead of just once as your code would suggest. StdLib.H has replaced the macros with templates:

```
template <class T>
  inline const T& min(const T& t1, const T& t2) {
    return (t1 > t2 ? t2 : t1); }
template <class T>
  inline const T& max(const T& t1, const T& t2) {
    return (t1 > t2 ? t1 : t2); }
```

When you invoke one of these, parameters t1 and t2 must have the same type or else the compiler must find a way to convert them. Unfortunately, it doesn't always find one when you think it should, and it complains via an error message. For *one* of those situations, I couldn't find a graceful solution. Therefore, MSP declares and implements in files General.H and General.CPP one specific instance of the min function: int min(int t1, int t2). Of course, *this* specific instance *is* covered by the template. But that declaration evidently must be present for the compiler to handle variants with parameters of differing const and reference int types—which are required in later MSP modules. You'll probably encounter additional examples in your programs; you may want to fix them the same way.

Templates for Derived Functions and Operators

You *can* overload C++ operators such as != and += without regard to related operators such as == and +. That would certainly be confusing. To ensure consistent definition of such related operators, General.H includes their *templates*, shown in Figure 3.3—for example, if operator x == y is defined for operands x and y, the compiler will use a General.H template to generate a definition for x != y; if x + y is defined to return a value of type T and x = z is defined for an operand z of type T, the compiler will generate a definition of x += y. These example templates will work later for vector, matrix, and polynomial operands. Templates analogous to += define the -=, *=, %=, ^=, and /= operators. Also included is a generic Sign(x) function, which works whenever x < 0 and x == 0 are defined: it returns -1, 0, or 1 depending on whether x is negative, zero, or positive. Finally, General.H includes a template for a generic singulary + operator, so you can use an expression +x just as you'd use -x later, when x is a vector or matrix. All these template functions are defined, as well as declared, in the header, so the compiler can generate all instances as required by client programs. (If they were in the source code file, client programs could use only whatever instances were compiled ahead of time.)

Exceptions

As mentioned in Section 2.8, MSP includes a general exception-handling system. It will report the nature and location of most errors—arithmetic (for example, attempting to divide by zero), input/output (attempting to open a nonexistent file), and so on. You can generally retain control in your MSP client program. The system is easy to understand, so you can modify it if you wish to do something more elaborate. Most of the details are described in later sections and included in other MSP header files, but a small part is introduced in General.H, because you should be able to use it in *any* MSP client. It doesn't depend on the details of any of the later modules.

Figure 3.4 shows the General.H definition of the enumeration ExceptionType, whose constants name the exceptional situations that MSP handles. Their details are described in context in later sections. If you should decide to enhance MSP to handle another exceptional situation, append its name to this list. When they encounter such a situation, most MSP routines throw an object of type Exception. This class is also defined in Figure 3.4. Exception objects have a single public data component: an ExceptionType variable Type, which reports what situation occurred. The constructor Exception(ExceptionType T = Unknown) merely sets Type = T. Its source code is so short it's included with the class definition in General.H.

3.2 INTEGER ARITHMETIC

> **Concepts**
> *Integer types*
> *+, –, and * operators*
> *Modular arithmetic*
> *Order and absolute value*
> *Integer division*
> *Additional functions*
> *Formatting output*
> *Arbitrarily long integers*

Although this chapter is mostly concerned with real and complex numbers, it starts by considering the integers. There are three reasons for this. First, integers, after all, form the basis of the real number system. Second, all computations involve counting, hence integer operations, to some extent. Third, some features of integer arithmetic yield results that aren't easy to predict unless you've taken the time at least once to study them in detail. This section surveys the various integer types available to C++ programmers, stressing their similarities. It considers the standard integer operations and functions supplied with C++ and a few more that you may find useful. It concludes by mentioning some aspects of integer arithmetic that are not supported by C++ or this book's MSP software package.

C++ Integer Types

Sixteen-bit Borland C++ provides six integer types, which you can distinguish by their ranges of values:

Type	Full type name	Range	Macro
char	signed char	$-2^7 \ldots 2^7 - 1 = 127$	
int	signed int	$-2^{15} \ldots 2^{15} - 1 = 32767$	= MAXINT
long	signed long int	$-2^{31} \ldots 2^{31} - 1 = 2147483647$	= MAXLONG
	unsigned char	$0 \ldots 2^8 - 1 = 255$	
	unsigned int	$0 \ldots 2^{16} - 1 = 65535$	
	unsigned long	$0 \ldots 2^{32} - 1 = 4294967295$	

The two macro constants are provided for your convenience in Borland C++ header file Values.H. (The ranges in this table are not C++ standards but depend on the machine for which the language is implemented.) The storage required for an integer depends on the type. For sixteen-bit Borland C++ the requirements are

```
char types: 1 byte
int  types: 2 bytes
long types: 4 bytes
```

int arithmetic is generally faster than long arithmetic, but because of the architecture of the Intel 8086 family machines on which it runs, Borland C++ performs char arithmetic no faster than int. Most C++ programs, including virtually all in this book, use int arithmetic, resorting to long variables only when extremely large sets must be counted—for example, the number of bytes in the text files containing this chapter.

Addition, Subtraction, and Multiplication

The standard integer arithmetic operations are provided for each C++ integer type. Addition, subtraction, and multiplication are really *modular* arithmetic, with these moduli:

```
char types: mod 2⁸
int  types: mod 2¹⁶
long types: mod 2³²
```

char types: mod 2^8
int types: mod 2^{16}
long types: mod 2^{32}

To perform an arithmetic operation mod M, you carry out the operation as usual, obtaining an intermediate value v; then you divide v by M, return the remainder r as the result, and discard the quotient q. The following equations hold among these quantities:

$$\frac{v}{M} = q + \frac{r}{M} \qquad v = qM + r.$$

r is adjusted to lie in the appropriate range: $-\frac{1}{2}M \leq r < \frac{1}{2}M$ for the signed types, and $0 \leq r < M$ for the unsigned ones. For example, the program fragment

```
int I = MAXINT;
I = 2*I;
cout << "I = " << I;
```

outputs I = -2. (Since MAXINT = $2^{15} - 1$, the intermediate value v is $2(2^{15} - 1) = 2^{16} - 2 = M - 2$, so the quotient is 1 and the remainder is -2, the output.) Here's an analogous example with an unsigned type: The fragment

```
unsigned long u = MAXLONG + 2;
u = 2*u;
cout << "u = " << u;
```

outputs u = 2. (Since MAXLONG = $2^{31} - 1$, the intermediate value v is $2(2^{31} + 1) = 2^{32} + 2 = M + 2$, so the quotient is 1 and the remainder is 2, the output.) You can construct similar fragments for the other four integer types. (You'll have to output char values using printf format string "%d"; otherwise, char output defaults to characters, not numerals.)

These examples should convince you that there's no such thing as overflow in integer arithmetic. But if the numbers get too large for the selected integer type, you may get unexpected results!

Order and Absolute Value

When you study arithmetic mod M, you generally regard the range of values as circular like the examples of Figure 3.5 with $M = 12$. (Modular arithmetic is taught in grade school as *clock arithmetic*!) It makes no intrinsic sense to say that one value is larger than another. But the order relation < for the infinite set of integers must be imitated in modular arithmetic for each of the C++ types. In fact, given two instances j and k of an integer type, you determine whether $j < k$ by cutting the circle at the natural place, flattening out the scale, and then checking as usual to see whether $j < k$. The natural cut depends on whether the type is unsigned or signed, as in Figure 3.5.

This seems a straightforward way to interpret the order relation $j < k$, but you should realize that because of the circular nature of modular arithmetic and the violence done by the cut, some of the familiar order properties fail to hold in integer arithmetic—for example, the statement *if $i < j$ and $j < k$ then $i < k$* is still true, but the statement *if $i < j$ then $2i < 2j$* is false when applied

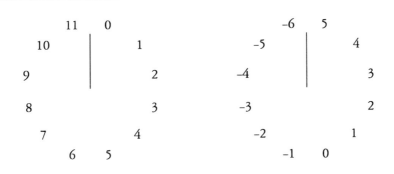

Figure 3.5. Circles for unsigned and signed modular arithmetic

```
#include <IOStream.H>
#include    <StdLib.H>

void main() {
  unsigned j = 32769;
  cout << j << ' ' << abs(j); }
```

Output

32769 32767

Figure 3.6. Improper use of abs

to type `int` arithmetic. (Try $i = 1$ and $j = 16384$: $2i = 2$ but $2j = -32768$. Following the analysis presented under the previous heading, the modulus is $M = 2^{16}$ and $j = ¼ M$, so the intermediate product of 2 by j is $v = ½ M = M - ½ M$; the quotient and remainder are $q = 1$ and $-½ M$. The latter is the final result $2j$ in type `int`.)

The absolute value function is involved with the order relation: $|j|$ is j or $-j$, whichever is ≥ 0. This operation is implemented by functions `abs` and `labs` defined in Borland C++ header file `<StdLib.H>` for `int` and `long` arguments. *Beware!* They may fail for arguments of unsigned types—for example, consider the program in Figure 3.6: Before executing `abs`, C++ automatically converts `j` to the `int` value -32767 with the same bit pattern in memory. *Beware!*

Integer Division

When you divide an integer x by a nonzero divisor d you get an integer quotient q and an integer remainder r such that

$$\frac{x}{d} = q + \frac{r}{d} \qquad x = qd + r$$

$$|r| < |d| \qquad sign(r) = sign(x).$$

Since x and r have the same sign, q is also the result of truncating the exact (real) quotient x/d toward 0. The Intel 8086-family CPUs are designed to implement this familiar process, so

Borland C++ integer division reflects these same features. For any integer type, you can obtain the quotient and remainder separately by executing

```
q = x/d;   r = x % d;
```

For the `int` type, you can get them both at once by executing the C++ Library function

```
div_t div(int x, int d)
```

The `div_t` type is defined with `div` in header file `StdLib.H` as

```
typedef struct {long quot; long rem; } div_t;
```

An analogous function `ldiv` and type `ldiv_t` are provided for dividing `long` integers.

Additional Integer Operations

As mentioned in Section 3.1, the MSP General module implements the function $\text{Sign}(x)$, whose value is $-1, 0$, or 1, depending on whether x is negative, zero, or positive. Defined by a template, it works with any argument x for which operations $x > 0$ and $x == 0$ are defined. Borland C++ Library header file `StdLib.H` includes similar templates for $\max(x,y)$ and $\min(x,y)$.

You can invoke the Borland C++ Math.H Library function

```
double pow(double b, double p)
```

with `int` arguments to compute an integer power b^p. It returns a `double` value; if that's in the `int` or `long` range, you can assign it immediately to a variable of that type, and it will be rounded correctly. More details of the function `pow` are discussed in Section 3.4. Two more functions commonly used for integer calculations are implemented in the MSP ElemFunc module:

```
long double Factorial(int    n);         // n!
double      Binomial (double x,          // Binomial coefficient
                      int    m);         // x over m .
```

The `Factorial` function returns a `long double` value because $n!$ gets so large; it overflows when $n > 1054$. ElemFunc features are discussed in detail in Section 3.10.

Formatting Output

Formatting numerical output is often unexpectedly difficult. C++ provides comprehensive facilities for this task—too many to consider here. For integer output the questions aren't really complicated. Even so, MSP needed a new function `Digits`, discussed in Section 3.1, to compute the minimum length of a numeral representing a given integer. This subject is discussed in great detail in Section 3.7 of my earlier book [48] on Borland C.

Integers of Arbitrary Length

Current developments are creating many applications for an integer data type not supported by Borland C++ or included in the MSP software of this book: *arbitrarily long integers*. These are normally stored as linked lists of integers or linked lists of digits. Devising efficient algorithms for their arithmetic is a major software engineering project. Some applications for this kind of programming are quite new, such as encryption methods for secure communication. Others originated decades or centuries ago—they stem from classical problems that require exact calculations with integer coefficients in problems involving polynomials and linear equations. Also, integers of arbitrary length form the numerators and denominators of rational numbers—a form of scalar that's coming into frequent use now to avoid round-off error. These kinds of applications are increasing in importance because the hardware necessary to perform such calculations in reasonable time is now becoming affordable. One of the first steps in developing MSP beyond the scope of this book should be to implement the arithmetic of integers of arbitrary length.

3.3 FLOATING-POINT ARITHMETIC

> **Concepts**
> *Real and floating-point numbers*
> *Significant digits and bits*
> *IEEE binary floating-point standard*
> *Intel 8087 family coprocessors*
> *8087 emulation by software*
> *Borland C++* `float, double,` *and* `long double` *types*
> *Non-normalized numbers, infinities, and NaNs*
> *Arithmetic and rounding*

The need for arithmetic testing
Input/output

Real and Floating-Point Numbers

By using decimal notation, you can represent a nonzero real number in the form $x = \pm 10^E \times d_0.d_1d_2\ldots$, where the *exponent* E is an integer; the *significand* $d_0.d_1d_2\ldots$ is an infinite sequence of digits d_k in the range $0 \le d_k < 10$ with *leading digit* $d_0 \ne 0$. The value of x is

$$x = \pm 10^E \sum_{k=0}^{\infty} 10^{-k} d_k.$$

For computation, you approximate x by truncating or rounding the significand to n *significant* digits:

$$x \approx \pm 10^E \times d_0.d_1\ldots d_{n-1} = \pm 10^E \sum_{k=0}^{n-1} 10^{-k} d_k.$$

You choose n small enough for convenient computation but large enough for the required precision. Approximations like this, and the number 0, are called *floating-point* numbers. If you want to record floating-point numbers on a physical medium, you must also limit the exponent to a specific range $E_{min} \le E \le E_{max}$. ($E_{min}$ is generally negative and E_{max} positive.)

It's instructive to contrast the sets of real and floating-point numbers available for reasoning and calculating. Consider, for example, floating-point numbers with three significant digits and exponents E in the range $-8 \le E \le 8$. The two smallest positive floating point numbers in this system are 1.00×10^{-8} and 1.01×10^{-8}; the two largest are 9.98×10^8 and 9.99×10^8. The negative floating-point numbers mirror these. In contrast, real numbers can have arbitrarily large or small magnitudes. Moreover, the real numbers form a *continuum*: there are no gaps. But the floating-point numbers form a *finite discrete* set: Nonzero floating point numbers are bounded away from zero, those nearest zero are separated by at least $0.01 \times 10^{-8} = 10^{-10}$, and those largest in magnitude can be separated by as much as $0.01 \times 10^8 = 10^6$. When you use floating-point numbers, be careful what you mean by precision, because attainable precision depends on the magnitude of the numbers you're using.

With a floating-point system like the one just considered, you may be tempted to use numbers like $u = 0.12 \times 10^{-8}$; but u isn't admissible, because its leading digit is zero. Numbers like u convey less precision (fewer significant digits) than those that really belong to the system.

Sometimes numbers like u are in fact useful; then they're called *nonnormalized* floating-point numbers.

Computers use binary notation, not decimal. Using binary notation for the significand, a nonzero real number has the form $x = \pm\, 2^E \times b_0.b_1b_2\,\ldots$, where the *exponent* E is an integer; the *significand* $b_0.b_1b_2\,\ldots$ is an infinite sequence of bits b_k in the range $0 \leq b_k \leq 1$, with *leading bit* $b_0 \neq 0$. The value of x is

$$x = \pm 2^E \sum_{k=0}^{\infty} 2^{-k} b_k.$$

For computation, you obtain a binary floating-point system by restricting the significand to n significant bits and the exponent to a range $E_{min} \leq E \leq E_{max}$.

IEEE Binary Floating-Point Standard

In the early days of computing, many different binary floating-point systems were used for numerical computation, as well as some decimal systems and some others. That was awkward, because results computed by a given algorithm would vary when transferred to a different computer or programming language.

The American National Standards Institute (ANSI) has approved a floating-point arithmetic standard. Languages and hardware that adhere to this standard produce the same results for the same arithmetic problem. The standard was prepared by a committee of the Institute of Electrical and Electronics Engineers (IEEE) and is described in its document [21].

The Intel 8087 numeric coprocessor family supports the IEEE standard. These components provide the basis for most numerical computation by IBM-compatible personal computers, the target machines for Borland C++. If no coprocessor is installed, PC software often emulates one—for example, normally compiled Borland C++ programs will use an 8087-family coprocessor for numerical computations if one is installed; if not, a rather involved software system will perform the computations as though an 8087 were in use. The only major difference is speed: The emulated computations may take as much as ten times as long to complete.

The IEEE standard includes definitions of three floating-point data types, which are implemented by the 8087 coprocessors and Borland C++: float, double, and long double. (Their IEEE names are *single*, *double*, and *extended*.) They differ in the number of bytes allocated to store a number, and consequently in the number of bits allocated to exponent and significand. These and other relevant data are shown in Figure 3.7. The following discussion explains the entries in that table. Many entries and a few other related data are given with full precision by mnemonic macro constants in Borland C++ header file Float.H. (These values may disagree occasionally with Borland C++ manuals. The Float.H constants are correct.)

	float	double	long double
Bytes for number	4	8	10
Bits for significand	23 (+1)	52 (+1)	63 (+1)
Bits for exponent	8	11	16
Exponent bias	127	1023	16383
Significant digits	7	15	19
Largest positive number ≈	3.4×10^{38}	1.8×10^{308}	1.2×10^{4932}
Smallest positive number ≈	1.2×10^{-38}	2.2×10^{-308}	3.4×10^{-4932}
$\varepsilon \approx$	1.2×10^{-7}	2.2×10^{-16}	1.1×10^{-19}

Figure 3.7. Types float, double, and long double

A double variable, for example, requires 8 bytes—64 bits—of storage. As shown in Figure 3.8, the first bit is the sign (0 for + and 1 for –). The next 11 bits store the exponent, and the final 52 store the significand. The exponent bits will accommodate up to $2^{11} = 2048$ different exponents E. They are apportioned between positive and negative E values by storing the sum e of E and the *bias* 1023. The inequality $0 \le e = E + 1023 < 2048$ implies $-1023 \le E < 1025$. But one of these bit patterns, $e = 1024$, is reserved for special use with infinities and *NaN*s (discussed later). Thus, the valid double exponents lie in the range $-1023 \le E \le 1023$.

The largest positive double value is binary $1.11 \ldots 1 \times 2^{1023} \approx 2 \times 2^{1023} = 2^{1024} \approx 1.8 \times 10^{308}$. The smallest positive value is binary $1.00 \ldots 01 \times 2^{-1023} \approx 2^{-1023} \approx 2.2 \times 10^{-308}$. (You'll see later that the binary value $1.00 \ldots 0 \times 2^{-1023}$ isn't allowed.)

A valid nonzero floating-point number is normalized: the leading bit b_0 of the significand $b_0.b_1b_2 \ldots$ is always 1. Therefore, it's not necessary to store b_0. The remaining 52 bits of a double variable are bits $b_1b_2 \ldots b_{52}$ of the significand. A double value has 53 significant bits.

Under this scheme, the value $1.00 \ldots \times 2^{-1023}$ would be stored with $s = 0$, $e = -1023 + 1023 = 0$, and $f = 00 \ldots 0 = 0$ (using the Figure 3.8. notation). All bits would be zero. But IEEE reserves that particular bit pattern, which would otherwise have represented the smallest positive value, to represent zero. The corresponding pattern with $s = 1$ represents -0. The IEEE standard has some rules and suggestions regarding the use of -0; they're beyond the scope of this book.

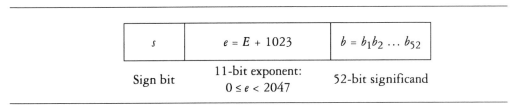

Figure 3.8. A double variable

The largest integer that can be represented as a `double` value without round-off is binary $1.11 \ldots 1 \times 2^{52} = 2^{53} - 1 \approx 9.0 \times 10^{15}$, a 16 digit number. Some, but not all 16-digit numbers are representable without round-off. This means that `double` values have 15 significant digits in general but sometimes 16.

The first floating-point number larger than 1 is $1 + \varepsilon =$ binary $1.00 \ldots 01 \times 2^0$, which gives $\varepsilon = 2^{-52} \approx 2.2 \times 10^{-16}$. The number ε, tabulated in Figure 3.7, is sometimes used in numerical analysis calculations as a threshold: Numbers δ with magnitude $< \varepsilon$ are regarded as zero, because `double` arithmetic yields $1 + \delta = 1$.

The following chart shows how all possible bit patterns of `double` variables are interpreted. Figure 3.8 showed a sign bit s, followed by an 11-bit exponent $e = E + 1023$ and then a 52-bit significand $b = b_1 b_2 \ldots b_{52}$.

e	b	Interpretation	
0	0	$(-1)^s\, 0$	(Zero values)
0	$\neq 0$	$(-1)^s\, 0.b \times 2^{-1022}$	(Nonnormalized numbers)
255	0	$(-1)^s\, \infty$	(Infinite values)
255	$\neq 0$	NaN	(Not a Number)
All others		$(-1)^s\, 1.b \times 2^E$	(Normal `double` values)

The IEEE standard requires that nonnormalized numbers be interpreted as shown but discourages their use. The infinite values and *NaN*s are used in reporting and handling error situations, such as attempts to divide by zero or take the square root of a negative number. Their use is described in somewhat more detail in Section 3.7.

All entries in the `double` column in Figure 3.7 have been explained. You can mimic this discussion for the `float` and `long double` types.

Arithmetic and Rounding

Following the IEEE standard, an 8087-family coprocessor implements the standard arithmetic operations:

$x + y$	$-x$	$x < y$	$x \times y$
$x - y$	$\lvert x \rvert$	round(x)	x / y

Further, it will extract the exponent or significand from a floating-point number and multiply by a specified power of 2 (by manipulating the exponent). Borland C++ either uses the coprocessor to perform these operations or else invokes its own routines, which emulate the coprocessor.

In order to ensure correctness of all bits of the result of a floating-point operation, the coprocessor performs every computation with some extra bits and then rounds the result. Four rounding methods are provided:

0. Round to nearest
1. Round up
2. Round down
3. Round toward zero (chop)

If a result computed with extra bits is exactly halfway between its nearest floating-point neighbors, and the coprocessor is rounding to nearest, it selects the neighbor with least significant bit zero. About half the time this rounds up, and half the time this rounds down, so some round-off errors may cancel.

The coprocessor provides a mechanism for controlling the rounding method: You store a coded *control word* in a coprocessor register. In turn, Borland C++ provides Library function

```
unsigned _control87(unsigned NewCW, unsigned Mask)
```

to set the control word. It will set the control word bits that correspond to the 1 bits in Mask. You must place the new values in the corresponding bits of NewCW. Any remaining NewCW bits are ignored. The function returns the modified control word. You'll find the _control87 prototype in Borland C++ header file Float.H, along with many mnemonic macro constants that you can use for Mask and NewCW—for example, the constant MCW_RC is the mask for the rounding control bits, and the constant RC_NEAR contains the bits to be set for rounding to nearest. To select that rounding method, execute _control87(RC_NEAR,MCW_RC). Rounding to nearest is the coprocessor's default method. To select a different method, use Float.H constant RC_DOWN, RC_UP, or RC_CHOP.

You'll probably never have to deal with rounding problems, unless you have to troubleshoot a disagreement between results of a program when run under Borland C++ and under another C system, or perhaps modify and extend features of the Borland C++ floating-point Library functions. So your lack of information about rounding will probably not be serious. But this information gap is more extensive. To what extent *does* the emulator produce the same results as the coprocessor?

Further, how do you *know* that the coprocessor really produces accurate results? The Intel literature [22] is comprehensive enough to let you control the coprocessor. (This requires low-level C++ or assembly language techniques.) But it doesn't document the underlying algorithms or any definitive tests. A software package called *Paranoia* has been developed to test floating-point arithmetic and is available in a C version [25]. The results of a comprehensive test should be published in an accessible source. Unfortunately, that's beyond the scope of this book.

Input/Output

C++ provides several methods for inputting floating-point values. They're familiar, so it's not necessary to discuss them in detail here. What's really involved in converting from a decimal floating-point input to binary notation? How do you do it by hand? Consider, as an example, the input 0.123. Write it as a ratio of two integers, first in decimal and then in binary:

$$0.123 = \frac{123}{1000} = \frac{1111011}{1111101000}.$$

Now do long division, in binary:

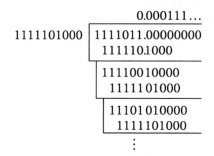

Finally, normalize: 0.123 = binary 0.000111 ... = binary 1.11 ... $\times 2^{-4}$.

Formatting floating-point *output* is often unexpectedly difficult. C++ provides comprehensive facilities for this task—too many to consider here. Unfortunately, the facilities provided by any language always seem to fall just short of what's required—for example, it was necessary to include a special formatting routine Show in the MSP Scalar module. Output of numerical data is discussed in great detail in Section 3.7 of my earlier book [48] on Borland C.

3.4 Math.H LIBRARY FUNCTIONS

> **Concepts**
>
> *Borland C++ header file* Math.H
> *Intel 8087 coprocessor instructions*
> *Arithmetic functions* frexp, ldexp, ceil, floor, modf, fmod, *and* fabs
> *Real powers, logarithms, and hyperbolic functions*
> *Real trigonometric functions*

The mathematical routines declared in Borland C++ header file Math.H include most of the so-called elementary functions covered in algebra and calculus courses. This section discusses their underlying algorithms and stresses their relationship to the Intel 8087 coprocessor instruction set. Unfortunately, in some cases neither the coprocessor algorithms nor those of the Borland C++ coprocessor emulator are documented—not even in the source code distributed with Borland C++. Section 3.6 presents simple test programs illustrating use of these functions.

The Borland C++ Library includes two versions of all these functions: for arguments and values of type double and long double. This book, however, describes only the double versions. You can generally construct the name of a long double function by appending the letter l to the name of the corresponding double function.

A few of the less common elementary functions have no Math.H implementations. You'll find those in the ElemFunc module of this book's MSP software, discussed in Section 3.10.

Real Arithmetic Functions

The first two Math.H functions are used to extract the exponent and significand E and S of a double value $x = S \times 2^E$, and to assemble x from those components:

```
double frexp(double x, int* E);    // Given  x,  return
                                   // S  and  E .
double ldexp(double S, int  E);    // Given  S  and  E,
                                   // return  x .
```

frexp merely arranges to execute the 8087 instruction FXTRACT. The next three functions are involved with rounding:

```
double ceil (double x);     // Return the smallest integer
                            // ⌈x⌉ ≥ x , in double  format.
double floor(double x);     // Return the largest integer
                            // ⌊x⌋ ≤ x , in double  format.
double modf (double  x,     // Set  c = chop(x)  and
             double* c);    // return  x − chop(x).
```

For example:

x	$\lceil x \rceil$	$\lfloor x \rfloor$	chop(x)	x − chop(x)
1.234	2	1	1	.234
−1.234	−1	−2	−1	−.234

When x is an integer, $x = \lceil x \rceil = \lfloor x \rfloor = \text{chop}(x)$. It's always true that $\text{chop}(x)$ and $x - \text{chop}(x)$ have the same sign as x. All three functions make essential use of the 8087 rounding instruction FRNDINT.

Functions fmod and fabs implement modular division and absolute value:

```
double fmod(double x,      // Divide  x / d  to get an
           double d);      // integer quotient;  return
                           // the remainder.
double fabs(double x);     // Return  |x|.
```

The value returned by fmod is determined in the same way as the remainder for integer division:

$$\frac{x}{d} = q + \frac{r}{d} \qquad x = qd + r,$$

q is an integer, $|r| < |d|$, and $\text{sign}(r) = \text{sign}(d)$. Function fmod is built around the 8087 instruction FPREM. Since the MSP template function abs described in Section 3.1 is defined for double arguments, you can use it in place of fabs to make your programs a little more readable.

Real Powers, Logarithms, and Hyperbolic Functions

Several Math.H functions implement roots, exponentials, logarithms, and powers:

```
double sqrt (double x);    // Return  √x .
double exp  (double x);    // Return  eˣ .
double log  (double x);    // Return  log x .
double log10(double x);    // Return  log₁₀x .
```

Function sqrt is merely an interface to 8087 instruction FSQRT. The algorithm that underlies the computation of $\sqrt{x}$ is undocumented. Function exp is based on the 8087 instruction F2XM1, which computes $2^x - 1$ for x in the interval $0 \leq x \leq \frac{1}{2}$. The exp code carries out some algebraic manipulations involving the constant log 2 to compute e^x after it gets $2^x - 1$ from the coprocessor. The algorithm for computing $2^x - 1$ is undocumented. Functions log and log10 are based on the 8087 instructions FYL2X and FYL2XP1, which compute $y \log_2 x$ and $y \log_2(x + 1)$. (FYL2XP1 is used to compute logarithms of numbers near 1 and restricts x to the interval $0 \leq |x| < 1 - \frac{1}{2}\sqrt{2}$.) The log and log10 codes carry out some algebraic manipulations involving the constant log 2 to compute log x and $\log_{10} x$ after they get log 2 $\log_2 x$ or $\log_{10} 2 \log_2 x$ from the coprocessor. The algorithms that underlie FYL2X and FYL2XP1 are undocumented.

Two Math.H functions compute powers:

```
double pow   (double x,     // Return  x^p .
              double p);
double pow10(int     p);    // Return  10^p .
```

For positive integral powers, pow and pow10 use the algorithm described in Section 6.2 for polynomials. For negative integers p, they compute $x^p = 1/x^{-p}$ and $10^p = 1/10^{-p}$. When $x > 0$ and p is not an integer, pow computes x^p via the formula

$$x^p = 2^{p \log_2 x}.$$

It uses the coprocessor instructions mentioned earlier to compute the power of 2 and the base 2 logarithm. You can use pow to compute most radicals via the formula

$$\sqrt[r]{b} = b^{1/r}.$$

but this won't work when $b < 0$ and r is an odd integer. MSP provides a more convenient function root, which works in all cases. It's described in Section 3.10.

Three hyperbolic functions are declared in Math.H:

$$\sinh x = \frac{e^x - e^{-x}}{2} \qquad \cosh x = \frac{e^x + e^{-x}}{2} \qquad \tanh x = \frac{\sinh x}{\cosh x}.$$

Their implementations are straightforward.

Real Trigonometric Functions

This Math.H function facilitates elementary trigonometric calculations:

```
double hypot(double x,     // Return  √(x² + y²).
             double y);
```

Its implementation uses coprocessor instruction FSQRT.

Math.H declares three trigonometric functions:

```
double tan(double A);
double cos(double A);
double sin(double A);
```

The first step in their calculation uses the coprocessor division instruction FPREM to compute $A' = A \bmod \pi/4$. Via an undocumented algorithm, the coprocessor instruction FPTAN then computes values x and y such that $\tan A' = y/x$. FPREM also returns the low-order three bits of the integer quotient of A by $\pi/4$. Function tan inspects these and negates and/or interchanges x, y so that the resulting pair represents $\tan A$ and x, y have the same signs as $\cos A$, $\sin A$. With that preparation, it's apparent how the three functions compute their values, using hypot:

$$\frac{y}{x} = \tan A \qquad \sin A = \frac{y}{\sqrt{x^2 + y^2}} \qquad \cos A = \frac{x}{\sqrt{x^2 + y^2}}.$$

Math.H also declares these inverse trigonometric functions:

```
double atan2(double y,      // Return θ with -π < θ < π
             double x);     // and  x , y = cos θ , sin θ .
double atan (double t);     // Return  arctan t .
double asin (double y);     // Return  arcsin y .
double acos (double x);     // Return  arccos x .
```

Function atan2 is based on the coprocessor instruction FPATAN, which requires that $0 \leq y < x$. When this condition is not fulfilled, atan2 must do some trigonometric manipulation. It handles negative x, y values with the equation $\arctan(-t) = -\arctan t$. If $0 \leq x < y$, it uses

$$\arctan \frac{y}{x} = \frac{\pi}{2} - \text{arccot}\frac{y}{x} = \frac{\pi}{2} - \arctan\frac{x}{y}.$$

Further trigonometric manipulations permit calculation of the remaining inverse functions, as follows. If $A = \arcsin t$, then $t = \sin A$, so $\cos^2 A = 1 - t^2$. Because A is in the range of the inverse sine, $|A| < \pi/2$, so $\cos A$ is positive and

$$\tan A = \frac{\sin A}{\cos A} = \frac{t}{\sqrt{1 - t^2}}.$$

Since A is also in the range of the inverse tangent, this equation and the definition of A yield

$$\arcsin t = \begin{cases} -\pi/2 & \text{if } -1 = t, \\ \arctan \dfrac{t}{\sqrt{1 - t^2}} & \text{if } -1 < t < 1, \\ \pi/2 & \text{if } \quad t = 1. \end{cases}$$

The inverse cosine is computed via the formula $\arccos t = \pi/2 - \arcsin t$.

3.5 Complex.H LIBRARY FUNCTIONS

> **Concepts**
> *The* complex *class*
> *Borland C++ header file* Complex.H
> complex *input/output*
> complex *arithmetic*
> *Polar decomposition*
> complex *exponential, hyperbolic, and trigonometric functions*
> complex *square roots, logarithms, and powers*
> complex *inverse trigonometric functions*
> *Need for* complex(z) *casts*

Borland C++ header file Complex.H defines a class of complex objects that correspond to complex numbers used in advanced algebra applications. It implements their arithmetic and extends to the complex domain the Math.H functions described in the previous section. This section discusses their implementation. Section 3.6 presents simple test programs illustrating their use. A few of the less common complex elementary functions have no Complex.H implementations. You'll find those in the ElemFunc module of this book's MSP software, discussed in Section 3.10.

The complex Class

Complex arithmetic is implemented by Borland C++ Library functions declared in header file Complex.H. Complex numbers $x + yi$ are represented by objects of the class complex. Each object has two data components, and the class has two member functions that perform logical services:

```
class complex {
  private:
    double re;              // Real, imaginary parts
    double im;              // of a complex number.
  public:
    complex();              // Default constructor.
    complex(double x,       // Construct the complex
            double y = 0);  // number  x + yi .
    :
```

The default constructor builds the complex zero, $0 + 0i$. Most of the other functions related to this class are declared its *friends*—they're too numerous to list all at once. Friendship allows them to manipulate the real and imaginary parts of a complex oject directly, without invoking selector functions, but doesn't force on them the alien principal argument syntax. Many of the Complex.H functions are completely defined in the header file.

Several non-member Complex.H functions perform logical services. There are two selectors:

```
double real(complex& z);     // Return real part of  z .
double imag(complex& z);     // Return imaginary part.
```

Next are two overloaded logical operators:

```
int operator==(complex&, complex&);
int operator!=(complex&, complex&);
```

The first returns True just when both the real and the imaginary parts of the two operands agree; != is its negation. Finally, there are two overloaded stream input/output operators:

```
istream& operator<<(istream&,complex&);     // Stream
ostream& operator<<(ostream&,complex&);     // i/o.
```

With the standard output stream cout, operator << uses a default format something like "(%.6g, %.6g)"—for example, if you construct complex z(sqrt(2),sqrt(3)) and execute cout << z, you'll see the display (1.41421, 1.73205). These input/output operators don't enjoy complex friendship.

Since there are no pointer types among the complex data components, there's no need for a special destructor or for special copy and assignment operators. The built-in facilities work with complex as with any other struct types.

complex **Arithmetic**

This section gives a brief summary of the complex arithmetic features defined as overloaded operator functions in Borland C++ header file Complex.H. The more involved mathematical functions corresponding to those declared in header file Math.H are discussed under later headings. Four singulary operators or functions are implemented for arguments $z = x + yi$:

$$+z = z \qquad -z \qquad \text{conj}(z) = \bar{z} = x - yi \qquad \text{abs}(z) = |z| = \sqrt{x^2 + y^2}.$$

Three versions of the binary operators +, –, *, and / are defined—for example, there are separate + operators for

```
complex + complex     complex + double     double + complex.
```

Finally, two versions of the replacement operators +=, -=, *=, and /= are defined—for example, there are separate += operators for

```
complex + complex    complex + double.
```

Complex.H provides two functions for handling complex numbers in polar form:

```
complex polar(double  r,          // Construct
              double  t = 0);     // z = r cis t .
double  arg  (complex& z);        // Principal argument of  z.
```

Execute polar(r,t) to compute r cis $t = r(\cos t + i \sin t)$. Because of the initialization t = 0, you can also use polar to convert double to complex values. Notice that

$$|\text{cis}(t)| = \sqrt{\cos^2 t + \sin^2 t} = 1.$$

Thus, if $z = r$ cis t, then $|z| = |r|$.

The *polar decomposition* of a complex number $z \neq 0$ is the unique pair r,t such that $r > 0$, $-\pi < t \leq \pi$, and $z = r$ cis t. The polar decomposition of 0 is the pair $r = 0$, $t = 0$. Clearly, $r = |z|$; t is called the *principal argument* arg z. (This old terminology is inconsistent with other uses of the mathematical term *argument*.) Complex.H functions abs(z) and arg(z) return these values r and t. The latter computes the value t using the inverse tangent function atan (or its more basic coprocessor counterpart) according to Figure 3.9.

Complex.H Elementary Functions

Through overloading, the Borland C++ Library extends to the complex domain all the Math.H functions described in Section 3.4:

```
exp      pow      sin      asin     sinh
log      sqrt     cos      acos     cosh
log10    tan      atan     tanh
```

These extended functions are prototyped in the Library header file Complex.H, so that both their arguments and values may be complex. They aren't as familiar as their real counterparts, so this section describes them and derives formulas for their real and imaginary components. The Complex.H Library functions are probably straightforward implementations of these formulas. (The source code is not readily available to check that.) For further information on the complex elementary functions, consult [59, Chapter 15].

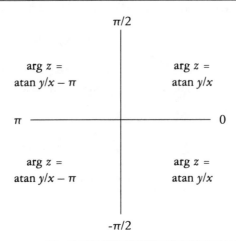

Figure 3.9. The principal argument of $z = x + yi$

Exponential, Hyperbolic and Trigonometric Functions

By using standard calculus techniques extended to complex numbers, you can show that the power series

$$\exp z = \sum_{k=0}^{\infty} \frac{z^k}{k!}$$

converges for all complex z. The function exp, which it defines, is called the *complex exponential function*, because for real z it agrees with the Maclaurin series for the real exponential function. It's easy to derive the *addition-multiplication rule* for the exponential function from the binomial expansion of powers of $z_1 + z_2$:

$$\begin{aligned}
\exp(z_1 + z_2) &= \sum_{j=0}^{\infty} \frac{(z_1 + z_2)^j}{j!} = \sum_{j=0}^{\infty} \frac{1}{j!} \sum_{j=k+l} \frac{j!}{k!l!} z_1^k z_2^l \\
&= \sum_{j=0}^{\infty} \sum_{j=k+l} \frac{z_1^k z_2^l}{k!\, l!} = \left(\sum_{k=0}^{\infty} \frac{z_1^k}{k!} \right) \left(\sum_{l=0}^{\infty} \frac{z_2^l}{l!} \right) \\
&= \exp z_1 \exp z_2.
\end{aligned}$$

Similarly, you can show that exp z is the sum of the two power series consisting of the even and the odd powers. These also converge for all complex z. The functions they define are called the *complex hyperbolic cosine* and *sine*, because for real z they agree with the Maclaurin series for the real hyperbolic cosine and sine functions:

$$\exp z = \cosh z + \sinh z$$

$$\cosh z = \sum_{k=0}^{\infty} \frac{z^{2k}}{(2k)!} \qquad \sinh z = \sum_{k=0}^{\infty} \frac{z^{2k+1}}{(2k+1)!}.$$

Clearly, $\cosh(-z) = \cosh z$ and $\sinh(-z) = -\sinh z$ for all complex z. By adding and subtracting series, you can see that

$$\cosh z = \frac{\exp z + \exp(-z)}{2} \qquad \sinh z = \frac{\exp z - \exp(-z)}{2}.$$

If you substitute $z = z_1 + z_2$ in these formulas, apply the addition-multiplication rule for the exponential function, and do the resulting algebra correctly, you'll get the *addition rules* for the hyperbolic cosine and sine functions:

$$\cosh(z_1 + z_2) = \cosh z_1 \cosh z_2 + \sinh z_1 \sinh z_2$$
$$\sinh(z_1 + z_2) = \sinh z_1 \cosh z_2 + \cosh z_1 \sinh z_2.$$

For real z, the series expansions of $\cosh zi$ and $-i \sinh zi$ agree with the Maclaurin series for the trigonometric cosine and sine functions:

$$\cosh zi = \sum_{k=0}^{\infty} \frac{(zi)^{2k}}{(2k)!} = \sum_{k=0}^{\infty} (-1)^k \frac{z^{2k}}{(2k)!}$$

$$-i \sinh zi = -i \sum_{k=0}^{\infty} \frac{(zi)^{2k+1}}{(2k+1)!} = -i^2 \sum_{k=0}^{\infty} (-1)^k \frac{z^{2k+1}}{(2k+1)!} = \sum_{k=0}^{\infty} (-1)^k \frac{z^{2k+1}}{(2k+1)!}.$$

Thus, you can use these formulas to define the *complex cosine* and *sine* functions:

$$\cos z = \cosh zi \qquad \sin z = -i \sinh zi.$$

Clearly, $\cos(-z) = \cos z$ and $\sin(-z) = -\sin z$ for all complex z. If you substitute $z = z_1 + z_2$ in these formulas, apply the addition rules for the hyperbolic functions, and do the resulting algebra correctly, you'll get the *addition rules* for the cosine and sine functions:

$$\cos(z_1 + z_2) = \cos z_1 \cos z_2 + \sin z_1 \sin z_2$$
$$\sin(z_1 + z_2) = \sin z_1 \cos z_2 + \cos z_1 \sin z_2.$$

Now it's possible to find the real and imaginary parts of the complex exponentials, sines, and cosines. First, if x and y are real, and $z = x + yi$, then

$$\exp yi = \cosh yi + \sinh yi = \cos y - (1/i)\sin y$$
$$= \cos y + i \sin y = \operatorname{cis} y$$
$$\exp z = \exp(x + yi) = \exp x \exp yi = e^x \operatorname{cis} y = e^x \cos y + i e^x \sin y.$$

Next,

$$\cosh z = \cosh(x + yi) = \cosh x \cosh yi + \sinh x \sinh yi$$
$$= \cosh x \cos y + \sinh x \left[-(1/i)\sin y\right]$$
$$= \cosh x \cos y + i \sinh x \sin y.$$

You can perform similar manipulations, occasionally using the fact that the cosines and sines are even and odd functions, respectively, to derive the remaining formulas:

$$\sinh z = \sinh x \cos y + i \cosh x \sin y$$
$$\cos z = \cosh y \cos x - i \sinh y \sin x$$
$$\sin z = \cosh y \sin x + i \sinh y \cos x.$$

If necessary, you can work with the remaining Complex.H hyperbolic and trigonometric functions directly through their definitions:

$$\tanh z = \frac{\sinh z}{\cosh z} \qquad \tan z = \frac{\sin z}{\cos z}.$$

Square Roots, Logarithms, and Powers

The Complex.H functions sqrt and log implement the complex *principal* square root and *principal* logarithm. Similar considerations lead to the definitions of these two functions. Square roots are considered first.

The *principal square root* of a complex number $z \neq 0$ is defined as the unique $w = r \operatorname{cis} t$ such that $r > 0$, $-\pi/2 < t \leq \pi/2$, and $w^2 = z$. In fact,

$$r = \sqrt{|z|} \qquad t = \tfrac{1}{2} \arg z.$$

To validate this definition, you have to verify first that $w^2 = z$ with this r and t, and, second, that only this r and t satisfy the requirements. The verification is simple:

$$w^2 = |z|\operatorname{cis}^2 t = |z|(\exp ti)^2 = |z|\exp 2ti = |z|\operatorname{cis} 2t = |z|\operatorname{cis} \arg z = z.$$

To check uniqueness, suppose $r' > 0$, $-\pi/2 < t \leq \pi/2$, and $(r' \operatorname{cis} t')^2 = z$, so that $r^2 \operatorname{cis}^2 t = r'^2 \operatorname{cis}^2 t'$. Taking absolute values, you get $r^2 = r'^2$, hence $r = r'$. That implies $\operatorname{cis} 2t = \operatorname{cis}^2 t = \operatorname{cis}^2 t' = \operatorname{cis} 2t'$, hence $2t$ and $2t'$ differ by a multiple of 2π. But that means t and t' differ by a multiple of π, hence $t = t'$: The pair r, t is indeed unique. As a special case, the principal square root of 0 is defined to be 0. Clearly, the principal square root of a real number $r \geq 0$ is $\sqrt{r}$, so it's appropriate to use the symbol $\sqrt{z}$ to denote the principal square root of any complex number z.

The *principal logarithm* of a complex number $z \neq 0$ is defined as the unique $w = u + vi$ such that u and v are real, $-\pi < v \leq \pi$, and $z = \exp w$. In fact,

$$u = \log|z| \qquad v = \arg z.$$

You can validate this definition following exactly the same steps as those in the previous paragraph. No principal logarithm is defined for 0. The principal logarithm of a positive real number u is clearly $\log u$, so it's appropriate to use the symbol $\log z$ to denote the principal logarithm of any complex number z. *Caution:* The familiar algebraic rules such as $\log ab = \log a + \log b$ don't necessarily hold for principal logarithms; you may have to adjust them by adding multiples of $2\pi i$ to one side. See [59, Section 15.7] for details.

You can show by a recursive proof that for all complex z and all integers p, $z^p = \exp(p \log z)$ — for example, $\exp(2 \log z) = (\exp \log z)^2 = z^2$. Moreover, $w = \exp(\tfrac{1}{2} \log z)$ is the same as $\sqrt{z}$: You can check that $w^2 = z$ and $-\pi/2 < \arg w \leq \pi/2$. Thus, it's appropriate to define complex powers in general through the formula $z^p = \exp(p \log z)$. This is implemented by the Complex.H function pow. *Caution:* Some of the usual algebraic rules for computation with powers don't hold for this definition, because of the difficulty with the logarithm discussed in the previous paragraph.

The complex *principal common logarithm* is defined as

$$\log_{10} z = \frac{\log z}{\log 10}.$$

Clearly, if $w = \log_{10} z$, then $10^w = z$. Conversely, if $10^w = z$, then w differs from $\log_{10} z$ by a multiple of $2\pi i/\log 10$. This function is rarely used, but it's implemented by Complex.H function log10. Take caution with ordinary algebra rules, as mentioned in the previous two paragraphs.

Inverse Trigonometric Functions

When complex numbers are permitted, you can derive algebraic formulas for the inverse trigonometric functions—for example, if

$$z = \cos w = \frac{\exp wi + \exp(-wi)}{2} = \frac{(\exp wi)^2 + 1}{2 \exp wi}$$

then $(\exp wi)^2 - 2z \exp wi + 1 = 0$. By the Quadratic Formula,

$$\exp wi = \frac{2z + \sqrt{4z^2 - 4}}{2} = z + \sqrt{z^2 - 1}$$

$$w = -i \log\left(z + \sqrt{z^2 - 1}\right) = \arccos z.$$

In this formula, the radical sign means *the principal square root plus a multiple of πi*, and the symbol log means *the principal logarithm plus a multiple of $2\pi i$*. You can define the *principal inverse cosine* arccos z by using the *principal* square root and logarithm functions. You can check that this function agrees with the usual inverse cosine when z is real. It's implemented by Complex.H function acos.

Through analogous manipulations, you can derive formulas for the complex principal inverse sine and tangent functions:

$$\arcsin z = -i \log\left(z + \sqrt{1 - z^2}\right) \qquad \arctan z = \frac{i}{2} \log \frac{i + z}{i - z}.$$

They're implemented by Complex.H functions asin and atan.

Need for complex(z) Casts

Borland's software engineers neglected to declare const the arguments for any Complex.H Library functions, even though they're generally passed by reference and don't change value. That collides with policy followed in other libraries and even in other parts of the Borland Library. To work around this difficulty, it was sometimes necessary to insert explicit `complex(z)` type casts in MSP functions with a `const complex& z` parameter. You'll probably experience the same problem with your programs that use `Complex.H` functions.

3.6 TESTING LIBRARY FUNCTIONS

> **Concepts**
> *External and internal validation*
> *Testing* `sin` *and* `cos`
> *Testing* `sqrt` *and other inverse functions*

How can you test the accuracy of the `Math.H` and `Complex.H` functions? Given the number of vital activities that depend on accurate calculations, there should be a standard way to certify basic functions. But literature on this subject is rare. The Intel coprocessor manuals and the Borland C++ manuals don't even mention it. I feel that books on numerical analysis software *should* cover this topic. Because of the lack of easily accessible material, it would be interesting to undertake and report on a certification project. Space and time permit only a preliminary study in this book.

The easiest way to proceed, of course, would be to compare Borland C++ results against some system that's already certified. Information about such systems, if they exist, is inaccessible, so this *external* validation method isn't available.

Another method would be to ascertain what algorithms are implemented by the coprocessors and the emulator software and then certify the algorithms mathematically. We would then have to check for bugs in their implementation. Again, the information is inaccessible. Obtaining it through reverse engineering with a debugger would be feasible but very arduous. Sections 3.4 and 3.5 make a start on this method, describing the `Math.H` functions in considerable detail. But critical information is lacking:

- Description of the algorithms that underlie the coprocessor instructions FSQRT, F2XM1, FYL2X, FYL2XP1, FPTAN, and FPATAN

- Assurance that the emulator software uses the same algorithms

Even with this information, considerable investigation would be necessary to check the implementation.

A third strategy would be to develop a new set of mathematical algorithms, whose accuracy is known, and through some means certify that the implementation is bug free. Then you could compare results with those from the coprocessor and emulator. This would be a major development project, comparable to designing and validating a new emulator, except that its efficiency would be unimportant.

Centuries before computers came into use, mathematicians used yet another technique—*internal* validation—to check tables. With this method, you select algebraic or trigonometric identities that relate tabulated function values and check that they're satisfied within acceptable round-off error. This method is used next to check some of the Math.H and Complex.H functions.

Internal Validation Experiments

You can select many identities to check Math.H and Complex.H functions, emphasizing various function properties. The program in Figure 3.10 uses the sine double angle formula to test sin and cos. It inputs a value x and then computes the relative error E incurred by using $2 \sin x \cos x$ in place of $\sin 2x$. If the values are accurate to 16 significant digits, then E should be approximately 10^{-16}—or larger if $\sin 2x$ is very small. The displayed output shows that E is generally smaller than this prediction.

Success for a very large sample of x values would constitute strong evidence for the validity of the sin and cos functions. Two possibilities could weaken the impact, however. First, if the double angle formula had been heavily involved in the design of the algorithms, it might not betray systematic errors. Second, software bugs often occur at case boundaries—transitions between methods that apply to arguments x in different ranges. Without source code to identify those, a sampler would probably miss the most vulnerable regions.

If an x value failed this test, you wouldn't know whether to suspect sin or cos. You'd need to try another identity that involves only one of these functions, and you'd probably want to use as little arithmetic as possible, too.

It might be more appropriate for this program to operate entirely in binary. That would prevent confusing sine and cosine phenomena with computation of 10^{-n} and conversion to decimal output.

Compiling the expression y == 0 in the complex instance of the function E template in Figure 3.10 triggers a Temporary used ... warning, because the double constant 0 must be converted to complex. You can avoid the warning by using the expression y == Scalar(0). This book's MSP software package makes that modification systematically, to cut down the number of warnings. (The ideal number is *none*!) However, that makes the code harder to read, so this ploy is avoided in some demonstration programs.

Another type of internal validation applies to inverse functions—for example, the program in Figure 3.11 involves the square root and square functions. Given inputs x and n, it computes $y_1 = \sqrt{x}$ rounded to n decimals and $y_0, y_2 = y \pm 10^{-n}$. Then it outputs $y_k^2 - x$ for $k = 0, 1, 2$. The difference for $k = 1$ should have the smallest magnitude. Sample outputs are given, showing what happens when the limit of precision is reached. In this example, function sqrt does achieve the maximum possible precision, because the difference for $k = 1$ is smallest until the y_k values become

```
//*****************************************************************
// SinTest.CPP    Testing the accuracy of sines and cosines

#include "General.H"
#include <Complex.H>

template<class Scalar>
  double E(Scalar t) {
    Scalar y = sin(2*t);
    if (y == 0) return 0;
    return abs((y - 2*sin(t)*cos(t))/y); }

void main() {
  cout << "Testing the accuracy of  sin(t), cos(t)\n"
          "Relative error of the  sin 2t  formula:\n"
          "E = |(sin 2t - 2 sin t cos t)/sin 2t| .  ";
  for (;;) {
    cout << "\n\nEnter  t :  ";
    complex t;  cin >> t;                // If  t  is real use
    cout << "E = "                       // double  E ,  else
         << (imag(t) == 0                // use  complex  E .
             ? E(real(t)) : E(t)); }}
```

Output

```
Testing the accuracy of  sin(t), cos(t)
Relative error of the  sin 2t  formula:
E = |(sin 2t - 2 sin t cos t)/sin 2t| .

Enter  t :  0.123
E = 1.13528e-17

Enter  t :  (-0.456,0.789)
E = 1.28144e-16
```

Figure 3.10. Testing sin and cos

indistinguishable. A version of the Figure 3.11 program that tested all possible inputs would constitute a *complete* validation of the square root routine (assuming validity of the squaring operation).

You could use analogous programs to validate members of other inverse pairs—for example, exp, log and tan, atan.

```cpp
#include "General.H"
#include  <Math.H>

void main() {
  cout << "            Testing accuracy of  sqrt(x)           \n"
          "Compute  y1 = sqrt(x)  rounded to  n  decimals\n"
          "and set  y0, y2 = y1 ± 10^(-n) .  Then compare\n"
          "y0², y1², and y2² with  x .                    ";
  for (;;) {
    cout << "\n\nEnter  x  and  n :  ";
    double x;  int n;  cin >> x >> n;
    double p = pow10(n);                    // p = 10n .
    double y[3];
    y[1] = sqrt(x);
    y[1] = floor(p*y[1] + .5);              // Round  y0 ... y2
    y[2] = (y[1] + 1)/p;                    // to  n  decimals.
    y[0] = (y[1] - 1)/p;                    // d = number of
    y[1] /= p;                              // digits left of
    int d = 1 + log10(y[1]);                // decimal point.
    for (int k = 0; k <= 2; ++k) {
      cout << "\ny" << k << " = ";          // n + d =
      cout.setf(ios::left,ios::adjustfield); // number of
      cout.width(n+6);                      // signifi-
      cout << setprecision(n+d) << y[k]     // cant
           << " y" << k << "² - x = "       // digits
           << setprecision(2) << (y[k]*y[k] - x); }}}
```

Output

```
        Testing accuracy of  sqrt(x)
Compute  y1 = sqrt(x)  rounded to  n  decimals
and set  y0, y2 = y1 ± 10^(-n) .  Then compare
y0², y1², and y2² with  x .

Enter  x  and  n :  7 15

y0 = 2.64575131106459      y0² - x = -4e-15
y1 = 2.645751311064591     y1² - x = 3e-15
y2 = 2.645751311064592     y2² - x = 7.7e-15

Enter  x  and  n :  7 16

y0 = 2.6457513110645907    y0² - x = 6.6e-16
y1 = 2.6457513110645907    y1² - x = 6.6e-16
y2 = 2.6457513110645907    y2² - x = 6.6e-16
```

Figure 3.11. Testing sqrt

3.7 HANDLING Math.H EXCEPTIONS

> **Concepts**
>
> Math.H *error handling function* _matherr
> Math.H *error types*
> *Installing a custom* _matherr *function*
> *MSP function* _matherr
> *MSP error handling demonstration*
> Complex.H *error handling*

Section 2.8 described the C++ *exception handling* technique, which lets your program maintain control in an error situation—reporting what happened and taking appropriate action. Basically, a function that encounters an error situation makes a report and throws an exception object E. The throw command restores the stack to its status when the most recent try block was entered, after invoking the destructors for all objects built by class constructors since that instant. The succeeding catch block expects this restored situation and processes any information encoded in E. It may attempt to recover. Or, it may decide that the error situation persists, make an additional report, and throw the same or another object for another catch block to process.

Do the Borland C++ Math.H Library functions interface properly with the exception-handling system? They can't, because exception handling was adopted as a C++ standard only recently, and C++ must remain compatible with older programs, which treat error situations differently. But there *is* a provision to *make* them throw exceptions that you can process as appropriate. This book's MSP software uses that technique. To understand it, you must first know what a Math.H function does in an error situation if MSP is *not* in use.

Math.H Error Handling

Invoked with parameters that would lead to an error situation, a Math.H function calls the Math.H error handling function

```
int _matherr(struct exception *E).
```

*E describes the situation that led to the error. Its type is defined in Math.H:

```
struct exception {
  int    type;    // Error type.
  char*  name;    // Caller's name.
```

```
    double arg1;        // Its first argument.
    double arg2;        // Second argument (if any).
    double retval; }    // Its return value.
```

You can regard `retval` as the caller's *proposed* return value. Math.H enumerates some constants for the possible values of the `type` field:

```
typdef enum {
    DOMAIN = 1,         // Argument not in domain.
    SING,               // Argument is a singularity.
    OVERFLOW,
    UNDERFLOW,
    TLOSS,              // Value has no accurate digits.
    PLOSS,              // Not used.
    STACKFAULT }        // Coprocessor stack error.
    _mexcep .
```

As examples of DOMAIN and SING errors, Math.H cites attempts to execute log(−1) and pow(0, −2)—evidently error type SING takes precedence over DOMAIN. A computation overflows or underflows when the absolute value of an intermediate or final result is too large or too small for type double. An attempt to compute sin(1e70) would cause a TLOSS error: The first step would be to compute 10^{70} mod $\pi/4$. For any digits to be accurate in that result, you'd have to use $\pi/4$ expanded to about 70 significant digits, which is impossible with type double. STACKFAULT identifies a programming error—misuse of the coprocessor or emulator stack—that has nothing to do with the mathematical properties of the Math.H functions.

The _matherr source code is shipped with Borland C++. You can see that it inspects E−>type. For UNDERFLOW errors, it replaces E−>retval by 0. For UNDERFLOW and TLOSS errors, it returns 1 to the Math.H function that encountered the error; for all other errors, it returns 0. If _matherr returns 0, the Math.H function writes an error message identifying itself and the error type. In any case it returns E−>retval to its caller. For example, if you attempt to compute double z = sqrt(−1), where sqrt is the double square root routine (not the complex one), you'll get an error message, and execution proceeds with an invalid z value. If you try to *do* anything with that, you may cause an error situation later. Figure 3.12 shows an example program and its output.

If you remove two // comment specifiers from the Figure 3.12 program, you'll include and link with the displayed two-line custom version of _matherr instead of the original. (The linker finds this _matherr code, so doesn't search the Borland C++ Library.) It returns 1 instead of 0 in this error situation, so sqrt doesn't output the DOMAIN error box. If you make it return 0 instead of 1 and run again, the box will reappear.

The previous paragraph introduces the Borland C++ provision for making Math.H functions throw exceptions in error situations: Include a custom version of _matherr. It should process the

```
#include <IOStream.H>
#include    <Math.H>
                                      // Include this to sup-
//int _matherr(struct exception *E) {  // press  sqrt   error
//   return 1; }                       // message. Returning
                                      // 0 , however, doesn't.
void main() {
  double z = sqrt(-1);
  cout << "sqrt(-1) = " << z;
  cout << "This line executes!"; }
```

Output

```
sqrt: DOMAIN error
```

```
sqrt(-1) = +NAN
This line executes!
```

Figure 3.12. sqrt error

information in *E, construct an exception object in a format appropriate to your task, and throw it.

MSP Function _matherr

The custom MSP version of int _matherr(struct exception *E) is shown in Figure 3.13—a portion of MSP source code file Scalar.CPP. It uses the data in *E to display a more informative message than the Math.H function would. The first several lines in this figure define constants used to build that message. Since they're not mentioned in the corresponding MSP header file Scalar.H, they're not accessible to client programs. The *E data aren't needed once the message has been displayed, so the MSP _matherr function constructs and throws an object of the Exception class defined in MSP header file General.H (see Figure 3.4). That object has only the ExceptionType data component Type, which _matherr sets equal to MathError, an enumerated constant defined in General.H.

The only MSP function that outputs a message specifying details of a Math.H error situation is _matherr. But each MSP function that catches *any* object outputs a message such as

```
while evaluating ... with argument ...
```

```
char* MathExceptionName[] = {
  ""         ,"DOMAIN"   ,"SING",             // Named after
  "OVERFLOW","UNDERFLOW","TLOSS",             // Math.H  _mexcept
  "PLOSS"   ,"STACKFAULT" };                  // constants.

#define NumberOfBivariates 5                  // Classes of  Math.H
char* BivariateName[] =                       // functions with two
  {"atan2","hypot","ldexp","poly","pow"};     // arguments.

int _matherr(struct exception *E) {           // Error handler for
  cerr << "\nException "                      // Math.H  functions.
       <<    MathExceptionName[E->type]
       << "\nwhile evaluating  Math.H  function  "
       <<    E->name
       << "\n  with argument";
  Boolean Bivariate = False;                       // Has the
  for (int j = 0; j < NumberOfBivariates; j++) {   // function
    char* c = BivariateName[j];                    // two
    if (strstr(c,E->name) == c) {                  // argu-
      Bivariate = True;  break; }}                 // ments?
  if (Bivariate) cerr << 's';
  cerr << " " << E->arg1;                          // Report, then
  if (Bivariate) cerr << " and " << E->arg2;       // throw an MSP
  throw(Exception(MathError)); }                   // exception.
```

Figure 3.13. MSP custom _matherr function

and rethrows it. The catches occur in order opposite that of the function invocations. Figure 3.14 demonstrates this nesting technique. Function main merely inputs a value a and then outputs $(\log \sin a)^a$. This slightly complicated function was selected to emphasize that it's often difficult to decide a priori which values belong to the domain of a function. Programs become simpler if you let the Library functions do that job. If bad input causes an error in any of the three nested functions, the MSP _matherr code outputs a distinctive error message and throws an exception. Function f catches it and throws it again to the main catch block. The output displays two unsuccessful attempts to input an appropriate a value and then a success.

Your MSP client program may need a more elaborate method for handling Math.H errors. If so, you should be able to use the MSP _matherr function, the previous example, and many other examples later in this book to guide you.

```
#include "General.H"
#include     <Math.H>

void f(double a) {
  try {
    cout << "pow(log(sin(a)),a) = " << pow(log(sin(a)),a); }
  catch(...) {
    cerr << "\nwhile executing  f(a)  with  a = " << a;
    throw; }}

void main() {
  double a;
  for (;;) {
    try {
      cout << "\n\nInput  a = ";  cin >> a; f(a); }
    catch(...) {
      cerr << "\nwhile demonstrating MSP error handling."; }}}
```

Output

```
Input  a = 1.2
Exception  DOMAIN
while evaluating  Math.H  function  pow
  with arguments  −0.0703805  and  1.2
while executing  f(a)  with  a = 1.2
while demonstrating  MSP  error handling.

Input  a = 1e70
Exception  TLOSS
while evaluating  Math.H  function  sin
  with argument  1e70
while executing  f(a)  with  a = 1e70
while demonstrating  MSP  error handling.

Input  a = 1
pow(log(sin(a)),a) = −0.172604
```

Figure 3.14. MSP error handling

Complex.H Errors

Most Complex.H functions compute the real and imaginary parts of their values by calling Math.H functions. Most error situations they encounter are triggered by the Math.H functions and handled as described earlier. Unfortunately, the Complex.H functions aren't implemented with try and catch blocks, so the MSP exception handling technique won't identify which one was executing when the error occurred. For example, executing cos(complex(1e70,1)) with MSP exception handling produces the message

```
Exception  TLOSS
while evaluating  Math.H  function  sin
   with argument  1e70
```

because Complex.H function cos is implemented via the equation

$$\cos(x + yi) = \cosh y \cos x - i \sinh y \sin x$$

discussed in Section 3.5, and the computation sin(1e70) is attempted first.

That's confusing. But one MSP design strategy was to leave Library functions unaltered. If that's an inconvenience, you can replace the Complex.H functions with your own versions—for example, Figure 3.15 contains code and output for a complex cosine function enhanced by MSP exception handling.

```
complex cos(const complex& z) {          // Complex.H  function
  try{                                    // cos  enhanced by MSP
    double x = real(z);                   // exception handling.
    double y = imag(z);
    return complex(cosh(y)*cos(x),-sinh(y)*sin(x)); }
  catch(...) {
    cerr << "\nwhile evaluating  cos(z)  with  z = " << z;
    throw; }}
```

Output from cos(complex(1e70,1))

```
Exception  TLOSS
while evaluating  Math.H  function  sin
   with argument  1e70
while evaluating  cos(z)  with  z = (1e+70,1)
```

Figure 3.15. Complex.H function cos enhanced by MSP exception handling

3.8 HANDLING FLOATING-POINT EXCEPTIONS

> **Concepts**
> *Interrupts*
> *Hardware exceptions*
> *Coprocessor exception masking*
> *Resetting the coprocessor*
> *Borland C++ initialization and exit routine*
> *Borland C++ hardware exception handling*
> *Signaling*
> *MSP* SIGFPE *service function* FPError
> *MSP error handling demonstration*

Section 3.7 described how this book's MSP software interfaces the Borland C++ Math.H Library functions with the C++ exception handling system and how you can enhance the Complex.H Library functions to do so. A closely related question remains: Do the Borland C++ floating-point arithmetic operators interface properly with the exception handling system? Like the Math.H functions, they can't, because exception handling was adopted as a C++ standard only recently, and C++ must remain compatible with older programs, which treat error situations differently. Again, though, there's a provision to *make* them throw exceptions that you can process as appropriate. MSP uses that technique. To understand it, you must first know what an arithmetic operator does in an error situation if MSP is *not* in use.

As an example, consider a program that must multiply two double values. Their product could *overflow*—it could be too large to store in a double variable. In that case, if you merely attempt to multiply them, your program will crash after displaying the error message Floating point: Overflow. What causes this?

When your hardware or coprocessor emulator detects an error such as this, it executes an *interrupt*. The interrupt concept is fundamental to system programming, but it is rarely encountered by scientific or engineering application programmers. The following paragraphs briefly describe that idea.

Interrupts

An interrupt is an event much like a function call. Section 2.8 described what happens when one C++ function invokes another—for example, when function Caller executes an expression such as f(x,y). Part of that description is repeated here to contrast it with an interrupt.

First, `Caller` places parameters x and y on the stack and then its return address. Next, `Caller` executes a jump instruction, transferring control to f. `Caller` *knows where* f *is, because the linker inserted into the* `.exe` *file the addresses of* `Caller` *and* f *relative to the start of your program*, and the operating system provided the starting address when it loaded the `.exe` file. Now, f begins execution and computes a function value v, using the parameters. It places v in a location known to `Caller`. When this computation is finished, the stack will be the same as it was when f began execution. Moreover, each object created by a class constructor while f was executing will have been destroyed by the corresponding destructor. Finally, f removes the return address from the top of the stack and uses it to jump back to `Caller`. `Caller` removes the parameters from the stack and proceeds to handle the return value v, as appropriate. The stack is now the same as it was before `Caller` invoked f.

The function invocation process depends on the *italicized* sentence in the last paragraph: The functions' object codes must be *linked*.

A *system service* function f, which provides a service to other programs, is often loaded separately: in the BIOS, with the operating system, or as an operating system extension. Its code isn't available to `Caller`, so you can't link them. Moreover, system designers can't give f a fixed address in all machines, because that would interfere with system upgrades and with your choice of optional system components. How should invocation be implemented for a system service function?

An *interrupt* is a modified function call that provides for such situations. A function f that's invoked via an interrupt is called an *Interrupt Service Routine* (ISR). ISRs must be coded according to special conventions, as you'll see. You must assign an ISR an *interrupt number* in the range 0, ..., 255. (Since your system software must provide several hundred services of this type, some ISRs merely serve as dispatchers. When you need a service, you execute the ISR after assigning a service code to one of its parameters; the ISR inspects the code and then provides the appropriate service by executing an ordinary function, which is not itself an ISR.) When an ISR with number *n* is loaded, as a part of the BIOS, of the operating system, or of some other system software package, its address must be stored in memory as *interrupt vector n*. The interrupt vectors occupy the same positions in the memory of every IBM-compatible PC. To obtain the service provided by ISR number *n*, your program must execute the machine language instruction `Int` *n*, which behaves somewhat like a function call, except that it obtains the ISR address from interrupt vector *n*. System software manuals specify the parameters of the ISR.

The interrupt concept is complicated by the circumstances in which your program—function `Caller`—needs to invoke a system function. Often, these are beyond its control. For example, data might arrive on a communication line while `Caller` is performing some low-priority computation. Or a timer may signal that it's time to back up some file. At any point in function `Caller`, an unrelated event may require execution of an ISR to obtain some higher-priority system service. Such events are often detected by your hardware: in these examples, a serial port or the clock. Those and a few other hardware components are allowed to trigger a *hardware* interrupt at any time while your program is executing. Actually, they send a signal to a hardware component called the *interrupt controller*, which checks priorities and then executes the interrupt.

How can a hardware-interrupt service avoid disrupting your program? First, ISR programmers must know how to avoid conflict with other programs over use of memory and hardware components. Second, provided no such conflicts occur, execution of your program can be suspended temporarily and reactivated later, if the contents of several CPU registers are saved at the instant of interruption and restored just before reactivation. The Int n interrupt instruction actually saves the registers in a specific order on the stack before it jumps to the ISR. Just before the ISR returns to Caller, it must use that data to restore the registers to their previous values.

The coprocessor emulator is driven by a hardware interrupt. If an 8086-family CPU encounters a coprocessor instruction but no coprocessor is installed, it executes an interrupt. The corresponding ISR is the emulator. The emulator uses the stacked return address to locate the preceding instruction in your program, which caused the interrupt, and then inspects the stacked register images and imitates the operation that a coprocessor would have executed had it been installed.

This description of the IBM PC interrupt system has been tailored to the needs of this book, and perhaps oversimplified. If you need more detail, please consult my earlier books [48, 51] and the Intel manual [22].

Hardware Exceptions

Hardware exceptions are interrupts triggered when you ask the CPU, coprocessor, or its emulator to do something it can't—for example:

- Divide an integer or a floating-point number by 0
- Perform a floating-point operation that would cause overflow or underflow (a result too large or too small to represent)
- Calculate the square root of a negative number
- Calculate the logarithm of a nonpositive number
- Calculate a trigonometric function of an angle θ so large that θ mod 2π has no accurate digits

Attempting to divide an integer by zero causes the CPU to execute an exception interrupt, equivalent to the Int 0 instruction. All the other exceptions are issued by a coprocessor or emulator. You normally encounter the last three exceptions in the above list only in connection with the Borland C++ Math.H error handling mechanism discussed in Section 3.7. Their analysis is hampered by lack of access to the source code of the Math.H functions.

Coprocessor Exception Masking

You can set the coprocessor to handle some error situations automatically without triggering exceptions: It returns *NaN* (Not *a* Number) or ±∞ values. (But it can't handle some of these values as operands in later calculations; if you ask it to do that, it *will* issue an exception.) This error-handling technique is called *masking*. It's done by setting certain bits in a coprocessor register called the *Control Word*. Bits set correspond to exceptions disabled, or *masked*. Function

```
unsigned _control87(unsigned NewCW, unsigned Mask),
```

prototyped in Borland C++ header Float.H, does it for you. For each 1 bit in Mask, _control87 sets the corresponding bit of the Control Word equal to the corresponding bit in NewCW. The function returns the previous value of the Control Word. (Thus, it's easy for you to re-establish the previous coprocessor behavior.) For your convenience, Float.H defines macro constants that specify the appropriate bits:

EM_INVALID	1	EM_UNDERFLOW	10
EM_DENORMAL	2	EM_INEXACT	20
EM_ZERODIVIDE	4		
EM_OVERFLOW	8	MCW_EM	3f

The last constant MCW_EM is the sum of all the others. For example, to disable both overflow and underflow exceptions, execute

```
unsigned CW = _control87(EM_OVERFLOW|EM_UNDERFLOW, MCW_EM).
```

To restore the previous conventions, execute _control87(CW,MCW_EM). By default, Borland C++ masks underflow exceptions. To unmask them, execute _control87(0,EM_UNDERFLOW).

You can also use function _control87 to select the coprocessor rounding method and some other options. See Section 3.3 for details.

The test program with output shown in Figure 3.16 illustrates some typical results of masking. With underflow masking on, no underflow error is noted, and the zero result is correct for most purposes. With underflow masking *off*, no error is noted, but the result is *incorrect*. With overflow masking *off*, no overflow error is noted, and the +∞ is correct for *some* purposes. If you mask exceptions this way, take care that you don't unwittingly send an infinite result to the coprocessor later as an operand. The result is hard to predict. With invalid operation masking *on* (the default), the invalid operation +∞/+∞ crashes with an inappropriate Floating-point: Overflow message, even if overflow masking was *on*. With invalid operation masking *off*, the invalid operation +∞/+∞ produces an *incorrect* result. Effective use of coprocessor exception masking thus requires systematic changes in your arithmetic software. That's beyond the scope of this book. See the Intel coprocessor manual [22] for details.

```
#include    <Float.H>
#include <IOStream.H>
#include    <Math.H>

void main() {
  double a = 1e-200;
  cout << "a = " << a;
  cout << "\nUNDERFLOW masking ON  (default)     a² = " << a*a;
  _control87(0,EM_UNDERFLOW);
  cout << "\nUNDERFLOW masking OFF              a² = " << a*a;
  a = 1e200;
  cout << "\na = " << a;
  _control87(EM_OVERFLOW,EM_OVERFLOW);
  double q = a*a;
  cout << "\nOVERFLOW  masking ON               a² = " << q;
  _control87(EM_INVALID,EM_INVALID);
  cout << "\nINVALID   masking ON              a²/a² = " << q/q;
  _control87(0,EM_INVALID);
  cout << "\nINVALID   masking OFF (default)   a²/a² = ";
  cout << q/q; }
```

Output

```
a = 1e-200
UNDERFLOW masking ON  (default)     a² = 0
UNDERFLOW masking OFF               a² = 1.48589e-224
a = 1e+200
OVERFLOW  masking ON                a² = +INF
INVALID   masking ON              a²/a² = -NAN
INVALID   masking OFF (default)   a²/a² =
```

```
Floating-point: Overflow
```

Figure 3.16. Masking floating-point exceptions

The crash depicted in Figure 3.16 leaves the coprocessor in an inconsistent state, which will probably disrupt other software later. In some noncrash situations, you might want to retain control and reset the coprocessor for later use. You can do that merely by executing function void _fpreset(), prototyped in the Borland C++ header Float.H.

If you mask a coprocessor exception and then invoke a Math.H function with an argument that causes this exception, it will still trigger _matherr. As explained in Section 3.7, _matherr returns to the Math.H function, telling it whether or not to display an error message. The Math.H function then returns to its caller a value that might be hard to predict, such as the examples in the previous paragraph. You can suppress the error message by substituting your own custom _matherr function for the original Math.H version, as described in Section 3.7.

The material on coprocessor exception masking was a digression. This technique can be useful for analyzing and handling some very special situations or perhaps as a basis for a major numerical software redesign, but it's not a vehicle for handling hardware exceptions in everyday C++ programs. Before discussing Borland C++ hardware exception handling methods in detail, it's useful to describe briefly the structure of a Borland C program.

Borland C++ Initialization and Exit Routine

The object code for your Borland C++ EasyWin program consists of two main parts: the initialization and exit routines and the code corresponding to the program itself. The initialization routine, called C0.Asm in the Borland C++ Library source code, installs the emulator ISR if no coprocessor is installed. (You can use a compiler switch to prevent installation, if you know your program won't use the emulator.) The initialization routine also sets up the heap memory allocation scheme and some other constants and addresses that your program code may use. It saves the previous values of any interrupt vectors that it changes.

After initialization, C0.Asm executes any functions you specified with #pragma startup directives in your program. Then it calls function main: This executes your program.

If main returns (that is, if your program terminates normally), C0.Asm proceeds with its next instruction, calling Library function exit(int E) with E = 0, which flushes all output file buffers and closes all open files. It also executes any functions your program registered as exit routines via Library function atexit. Finally, exit invokes function _exit(int E) with the same E parameter. This restores to their previous status any interrupt vectors the initialization routine may have changed earlier and then returns to the operating system, setting its ErrorLevel variable equal to E. If your program is using a small memory model, _exit also performs a test that reports some programming errors via the message Null pointer assignment. You may also call exit or _exit directly.

Borland C++ Hardware Exception Handling

Mathematical errors that cause hardware exceptions are detected either by the CPU or the coprocessor. The CPU detects only attempted integer division by zero; in that case it executes interrupt Int 0. What does the corresponding ISR do? This story is complicated and is given here only as an indication of the care you must take in investigating such matters. When you boot your machine, the BIOS automatically installs an Int 0 ISR that does nothing but return to the interrupted program. Midway in the boot process, the BIOS installs your operating system, replacing the BIOS ISR with one that prints a message—that is, the operating system replaces Interrupt Vector 0 by the address of the new ISR. When it loads a Borland C++ EasyWin program, your operating system transfers control to the Borland C++ initialization and exit routine C0.Asm. That replaces the Int 0 service routine by yet another, which prints the message Divide-error Exception and calls the Library routine _exit(3). This sets the operating system ErrorLevel variable = 3 to report the error. The abnormal termination doesn't invoke function exit: No registered exit routines are executed, and your files may remain in disarray.

All other hardware exceptions are issued by the coprocessor or its emulator. These also cause an interrupt. The details of the corresponding ISR are more complicated than those for the integer division exception handler just described; they even vary with the PC model. In short, the Borland C++ initialization routine installs an ISR that ascertains the type of error by looking at the coprocessor registers and then prints a message and terminates by calling _exit. For example, if you attempt to divide a floating-point number by zero, it prints Floating-point: Divide by Zero. As with the integer division exception, this abnormal termination may leave your files in disarray.

The previous paragraph was oversimplified. The Borland C++ ISR actually uses the C++ *signaling* process to call a service function that writes the message and terminates your program. It's possible for you to substitute your own service function to avoid terminating. This book's MSP software does that. The signaling process and the MSP technique are described in the remainder of this section.

Signaling

The *signaling* system for customizing exception interrupt services is based on a Borland ISR that serves several interrupts, including the hardware exceptions. This *signal dispatcher* inspects several CPU registers to classify the interrupt, choosing an identifying *flag* from this list of constants defined in Borland C++ header file Signal.H:

SIGFPE	Int 0, Int0, or coprocessor/emulator exceptions
SIGILL	The CPU was asked to perform an illegal operation
SIGSEGV	The CPU detected an illegal memory address
SIGABRT	Software requested program abortion, even if that might leave files in disarray
SIGINT	Software detected a <Ctrl-C> keystroke or the equivalent
SIGTERM	Software requested normal program termination

It may record additional information as well to facilitate interrupt service. The dispatcher then uses the flag to invoke a *signal service function* to handle whatever abnormal situation triggered the interrupt. Invoking the proper service function is called *raising* the flag. A flag can be raised *implicitly*, as described previously, in response to an interrupt. Or you can do it *explicitly* by executing function int raise(int Flag), which is declared in Signal.H. The raise function won't return to its caller unless the service function assigned to handle the signal Flag does.

What signal service functions are provided? Borland C++ declares two in header Signal.H:

SIG_IGN   Take no action (return to the caller).

SIG_DFL   Take the default action for the flag raised. Each default is described in the Borland C++ *Library Reference* manual [4] under the heading signal. Generally, it's to display a distinctive message and terminate the program, setting the operating system ErrorLevel variable appropriately. It doesn't call exit, so your files may remain in disarray. The default action may depend on additional information recorded by the signal dispatcher when the flag was raised.

But you're invited to provide your own function to service any signal flag. It must be declared like this: void Service(int Flag). To install your custom Service function to respond to signal Flag, execute

```
void* PreviousService = signal(Flag,Service) .
```

The latter parameter is a pointer to the start of your Service function. signal returns a pointer to the start of the previously active service function. Thus, you may reactivate the previous service by executing signal(Flag,PreviousService). If it fails to install the specified service function, signal will return the value SIG_ERR, a constant defined in header file Signal.H.

Whenever signal Flag is raised, the Borland signal dispatcher first executes code equivalent to signal(Flag,SIG_DFL). This ensures that if Flag is raised again while a custom function is servicing the first exception, the default service function will terminate your program. Thus, a custom Flag service function *must reinstall itself* before it completes execution.

Flags SIGFPE, SIGILL, and SIGSEGV are associated with individual codes, shown in Figure 3.17, which identify the circumstances of their invocation. The constants are all defined in header file Float.H. Each of these flags has a code to indicate that it was raised explicitly. The SIGFPE codes correspond to the Int0 interrupt; the integer zero division interrupt Int 0; and to five coprocessor exceptions described earlier. (SIGFPE also triggers a service in response to the machine language instruction Int0, a special interrupt designed for implementing integer arithmetic features not supported directly by the machine language—for example, multiple precision arithmetic.)

Flag	Codes		
SIGFPE	FPE_INTOVFLOW FPE_ZERODIVIDE FPE_INEXACT	FPE_INTDIV0 FPE_OVERFLOW FPE_EXPLICITGEN	FPE_INVALID FPE_UNDERFLOW
SIGILL	ILL_EXECUTION	ILL_EXPLICITGEN	
SIGSEGV	SEGV_BOUND	SEGV_EXPLICITGEN	

Figure 3.17. Codes associated with signal flags

When SIGFPE is raised implicitly, by a Borland C++ ISR in response to Int0 or Int 0, two parameters are stacked in addition to the one specified by the prototype

void Service(int Flag).

The stack then looks as though Service should have been prototyped as

void Service(int Flag, int Code, int* Register).

The first extra parameter is one of the Code values for SIGFPE tabulated in Figure 3.17. The second points to an array containing the values of all the CPU registers when SIGFPE is raised. A later paragraph shows how to access and use the Code parameter. The information contained in *Register enables a custom SIGFPE service function to analyze an integer arithmetic exception in more detail. That requires assembly language techniques and won't be discussed further.

When SIGFPE is raised explicitly, or implicitly in response to a coprocessor/emulator exception, the extra Code parameter is stacked but not the Register pointer. When SIGILL or SIGSEGV is raised implicitly, by a Borland C++ ISR responding to a CPU hardware exception, the additional Code and Register parameters are both stacked.

MSP SIGFPE Service Function FPError

Figure 3.18 displays the custom MSP SIGFPE service function FPError, which interfaces the Borland C++ signaling and MSP exception-handling systems. It's part of MSP source code file Scalar.CPP. The FPError prototype, with just one parameter, is specified by the prototype of function signal in header Float.H. The extra parameters stacked when SIGFPE is raised implicitly introduce a major complication in this program. Borland C++ uses a mechanism called a *variable argument* list. This feature appears in FPError as instances of macros va_list, va_start, va_arg, and va_end. They're defined in header file StdArg.H. Consult the corresponding entries in the Borland C++ *Library Reference* manual [4] for details of their use. (The va_ feature isn't

```
char* FPExceptionName[] = {                        // Named
  "FP_INTOVFLOW" ,"FP_INTDIVO"  ,"FP_???",         // after
  "FP_INVALID"   ,"FP_???"      ,"FP_ZERODIVIDE",  // SIGFPE
  "FP_OVERFLOW"  ,"FP_UNDERFLOW","FP_INEXACT",     // signal
  "FP_STACKFAULT","FP_???"      ,"FP_???",         // types in
  "FP_???"       ,"FP_???"      ,"FP_EXPLICITGEN" }; // Float.H .

void FPEHandler(int Flag) {            // Processor error
  va_list ap;                          // signal handler.
  va_start(ap,Flag);                   // Get the first op-
  int Code = va_arg(ap,int);           // tional parameter.
  Exception E;                         // Set MSP exception.
  if (Code == FPE_INTDIVO ||           // For CPU exceptions
      Code == FPE_INTOVFLOW) {         // (they don't seem
    int* Dummy = va_arg(ap,int*);      // to get here) clean
    va_end(ap);                        // up the second op-
    E.Type = OtherError; }             // tional parameter.
  else {
    _fpreset();                        // Otherwise, reset
    E.Type = MathError; }              // the coprocessor.
  signal(SIGFPE,FPEHandler);           // Reinstall this
  cerr << "\nFloating-point exception  " // handler. Report
    << FPExceptionName[Code-FPE_INTOVFLOW]; // and throw the MSP
  throw(E); }                          // exception.
```

Figure 3.18. MSP `SIGFPE` service function `FPEHandler`

discussed here because it's a standard C mechanism of general use. It's also a ludicrous programming method.)

`FPError` builds an MSP `Exception` object `E`. It first inspects `Code`, the first extra parameter after `Flag`, to see if `Flag` was raised by `Int 0` or `Int0`. If so, there's another extra parameter; if not, `Flag` was raised by the coprocessor, which therefore needs resetting by `_fpreset`. The `va_arg` macros remove one or both extra parameters from the stack. During this process, `FPError` also sets the `Type` field of `E` to `OtherError` or `MathError`. Next, `FPError` reinstalls itself: The signal dispatcher installed the default service function while raising `Flag`. Finally, `FPError` displays a message with the name of the `Code` constant and throws `E`.

To install `FPError` as the `SIGFPE` service function, MSP includes in source code file `Scalar.CPP` function

```
void MSPErrorHandling() {              // Initialize MSP
  signal(SIGFPE,FPEHandler); }         // error handling.
```

To execute it before any client's `main` function, `Scalar.CPP` includes immediately afterward the directive

`#pragma startup MSPErrorHandling.`

To obtain MSP error-handling services, a client program merely has to link with `Scalar.Obj`.

```
#include "General.H"
#include <Signal.H>

void f(double a) {
  try {
    cout << "(1/a)² = " << (1/a)*(1/a); }
  catch(...) {
    cerr << "\nwhile executing  f(a)  with  a = " << a;
    throw; }}
void main() {
  double a;
  for (;;) {
    try {
      cout << "\n\nInput  a = "; cin >> a;  f(a); }
    catch(...) {
      cout << "\nwhile demonstrating MSP error handling."; }}}
```

Output

```
Input  a = 0

Floating point exception  FP_ZERODIVIDE
while executing  f(a)  with  a = 0
while demonstrating MSP error handling.

Input  a = 1e-200

Floating point exception  FP_OVERFLOW
while executing  f(a)  with  a = 1e-200
while demonstrating MSP error handling.

Input  a = 1e200
(1/a)² = 0
```

Figure 3.19. MSP hardware exception handling

The demonstration program shown in Figure 3.19 is identical to that in Figure 3.14, except for function f, which exercises arithmetic operations instead of Math.H functions. Function main merely inputs a value *a* and then outputs $(1/a)^2$. If bad input causes an error in division or multiplication, FPError outputs a distinctive error message and throws an exception. Function f catches it and throws it again to the main catch block. The output displays two unsuccessful attempts to input an appropriate *a* value, and one success. Success for $a = 10^{200}$ reveals again that Borland C++ masks underflow exceptions: $(1/a)^2$ is too small to represent, so the output is 0.

3.9 MSP Scalar MODULE

> **Concepts**
> *MSP files* Scalar.H, Scalar.CPP, *and* Scalar.Use
> *Scalar types*
> *Codas for generating function template instances*
> *Determining the* Type *and* TypeName *of a scalar*
> *Scalar output functions* ForShow *and* Show
> DefaultFormat *function*
> *Constant* ii = complex(0,1)
> abs, real, imag, *and* conj *for noncomplex scalars*
> *Scalar functions* MakeRandom

This book's MSP software implements methods for:

- Evaluating elementary functions of scalars
- Computing solutions of scalar (single-variable) equations
- Evaluating vector and matrix algebra expressions
- Computing solutions of vector (multivariable) equations

For each of these areas, MSP supports complex and double scalars. The vector and matrix algebra modules also support int scalars. The others don't, because their techniques require scalar *division* as well as addition, subtraction, and multiplication. (Matrix inversion is not regarded as a matrix algebra feature but rather as a technique for solving vector equations.) The MSP Scalar module includes features common to all scalar computations. These are the ones you must supplement if you decide to extend MSP to support an additional scalar type. Some are implemented separately for each type; some are implemented with templates for all scalar types simultaneously.

Code for the Scalar module is contained in header file Scalar.H and source code file Scalar.CPP. The header is displayed in Figures 3.20 and 3.21 and included on the accompanying diskette. It contains several function declaration templates; you'll find the corresponding function definitions in Scalar.CPP on the optional diskette.

```
//**********************************************************************
//                           Scalar.H

#ifndef    Scalar_H
#define    Scalar_H

#include "General.H"
#include <Complex.H>

//**********************************************************************
// Logical features

#define Double   0              // MSP supports these
#define Complex  1              // Scalar  types,
#define Int      2              // but  int  only partly.

template<class Scalar>          // Return  t  type.
  int Type(const Scalar& t);

template<class Scalar>          // Return  t  type name,
  char* TypeName(const Scalar& t);   // lowercase, unpadded.

template<class Scalar>          // Return the shortest
  char* DefaultFormat(          // format.
            const Scalar& t);

template<class Scalar>          // Convert  t  to a string for
  char* ForShow(                // output as if by  printf .
    const Scalar& t,            // Call DefaultFormat(t)  for
    char* Format = NULL);       // the default format.

template<class Scalar>          // Standard output;  return
  unsigned Show(                // output length.  Call
    const Scalar& t,            // DefaultFormat(t)   for the
    char* Format = NULL);       // default format.
```

Figure 3.20. Header file Scalar.H
Part 1 of 2; continued in Figure 3.21.

```
//*************************************************************
// Mathematical features

#define ii complex(0,1)         // i² = –1 .

double abs (const double& x);   // Return |x| .
double real(const double& x);   // Return the real and imagin-
double imag(const double& x);   // ary parts and conjugate
double conj(const double& x);   // ( x , 0 , and  x ) of  x .

void MakeRandom( double&  x);   // Make  x , z  random:  0 <=
void MakeRandom( complex& z);   // x, real(z), imag(z) < 1 .
void MakeRandom(    int&  k);   // Make  k   random.

#endif
```

Figure 3.21. Header file `Scalar.H`
Part 2 of 2; continued from Figure 3.20.

Codas for Generating Function Template Instances

For your convenience, the object code generated from `Scalar.CPP` will be placed in library file `MSP.Lib` on the accompanying diskette. It must contain codes for all supported template function instances. To ensure that the compiler generates them, source code file `Scalar.CPP` ends with a *coda* containing a function template and a function:

```
template<class Scalar>              void UseScalar();
    void UseScalar(Scalar t);
```

They are shown in Figure 3.22. They are *not for execution*. The first invokes with a `Scalar` argument each template function declared in `Scalar.H` but defined in `Scalar.CPP`. The second invokes `UseScalar(t)` for `double`, `complex`, and `int` arguments t: all the supported types.

If you should add a function declaration template to `Scalar.H` with corresponding definition in `Scalar.CPP`, you must add its invocation to the definition of `UseScalar(Scalar t)` in the coda. If you should add support for a new `Scalar` type, you must also add a corresponding line to the `UseScalar()` definition in the coda.

Analogous codas are included in the source code files of all MSP modules discussed later in this book.

```
template<class Scalar>              // Invoke all  Scalar.CPP
  void UseScalar(Scalar t) {        // template functions.
    Type(t);
    TypeName(t);       ForShow(t);
    DefaultFormat(t);  Show(t); }

void UseScalar() {                  // Invoke UseScalar for
  double  x = 0;   UseScalar(x);    // each supported Scalar
  complex z = 0;   UseScalar(z);    // type.
  int     k = 0;   UseScalar(k); }
```

Figure 3.22. Coda for generating Scalar.CPP template instances

Type Identification Functions

Often it's useful for a template function to determine the type of a template argument. Its action may depend on that. If the dependence is too close, using a template might not have been appropriate. But sometimes dependence is slight—for example, the only difference in action for arguments of different types might be an output format. Therefore, the Scalar module includes functions

```
int   Type     (const Scalar& t)
char* TypeName (const Scalar& t).
```

The first returns one of the values Double, Complex, or Int, defined in Scalar.H. It's implemented in Scalar.CPP for the three supported Scalar types by one-line functions like

```
int Type(const double&  t) { return Double; }
```

Function TypeName returns the string "double", "complex", or "int". Its implementation is almost equally simple:

```
char* TypeNameArray[] = {"double",         // Names of supported
                         "complex",        // Scalar types, in
                         "int" };          // order.
  template<class Scalar>                   // Return t type
    char* TypeName(const Scalar& t) {      // name.
      return TypeNameArray[Type(t)]; }
```

Displaying Scalars

Scalar.H declares three template functions for standard Scalar output: ForShow, Show, and DefaultFormat. As you can see from the source code in Figure 3.23, ForShow invokes the C Library StdIO function sprintf to convert Scalar t to a string S, which it returns, ready for output by the caller. The Format code that governs the conversion is described later. ForShow allocates 80 bytes of free memory storage for S. If allocation fails, the new operator throws an object of type xalloc; ForShow catches it, displays an error message, and throws the object Exception(OutOfMemory). (Borland C++ header file Except.H defines xalloc. The MSP Exception class is described in Section 3.1.) If allocation succeeds, the function that called ForShow must deallocate memory when S is no longer needed. If the Format code is bad, sprintf returns a constant EOF defined in header file <StdIO.H>, and ForShow displays a message and throws Exception(FormatError). Most garbled Format codes aren't bad: They just produce garbled output. But a Format code for which the output would overflow the 80 bytes of memory allocated to S *is* bad. Unfortunately, sprintf will probably corrupt the free storage heap in that situation and cause a delayed crash, no matter what you do to recover. So ForShow pauses after that error message to let you plan emergency measures.

```
template<class Scalar>                  // Convert  t  to a string
  char* ForShow(const Scalar& t,        // S  for output as if by
             char* Format) {            // printf .
    if (Format == NULL)
      Format = DefaultFormat(t);        // 80 characters
    char* S;                            // should be
    try  { S = new char[80]; }          // long enough.
    catch(xalloc) {                     // The calling
      cout << "\nException  OutOfMemory"  // function has
              "\nwhile executing  ForShow";  // the responsi-
      throw(Exception(OutOfMemory)); }  // bility to
    int n = sprintf(S,Format,t);        // delete  S .
    if (n < 1 || n > 80) {
      cerr << "\nException  FormatError"
              "\nwhile executing  ForShow(t,Format)"
              "\nwith  t = " << Scalar(t)
                  << "  and  Format = \"" << Format << "\""
              "\n--you may have corrupted the heap--";
      Pause();
      throw(Exception(FormatError)); }  // Pause insures mes-
    return S; }                         // sage visibility.
```

Figure 3.23. Scalar function ForShow

```
template<class Scalar>                  // Standard output;
  unsigned Show(const Scalar& t,        // return output
                char* Format) {         // length.
    try {
      char* S;                          // Prepare output.
      S = ForShow(t,Format);            // This allocates
      cout << S;                        // memory!
      unsigned n = strlen(S);
      delete S;
      return n; }
    catch(...) {
      cerr << "\nwhile attempting to execute Show(t,Format)"
              "\nwith  t = " << Scalar(t)
                   << " and  Format = \"" << Format << '\"';
      throw; }}
```

Figure 3.24. Scalar function Show

MSP output function Show, displayed in Figure 3.24, calls ForShow to prepare an output string S then outputs it by executing cout << S. If ForShow encounters an error, there's no output. After output, Show frees the memory allocated to S and returns the length of S. The length is calculated by StdIO Library function strlen.

If you invoke ForShow or Show to output a Scalar value t without specifying a Format code, or with code NULL, MSP will use the format returned by DefaultFormat(t). It's selected so that sprintf will build the shortest output string for t that includes enough information for most debugging. Here's its implementation:

```
char* DefaultFormatArray[] = {          // Shortest formats:
    "% -#8.2g ",                         // double  ,
    "(% -#8.2g,% -#8.2g) ",              // complex ,
    "% 6i " };                           // int .

template<class Scalar>                   // Return the
  char* DefaultFormat(const Scalar& t) { // shortest format
    return DefaultFormatArray[Type(t)]; }// for  t .
```

Like all format codes, the default double format code begins with %. The symbols #8.2g signify that S should contain eight characters, including a decimal point and two significant digits, in floating-point format if $.0001 \leq |t| < 100$ and in scientific notation outside that range. The blanks and minus in the code indicate that positive numerals should begin with a blank, and all numerals should be left justified in S and padded on the right with at least one blank, as necessary. You can inspect some output samples in Figure 3.25. The default complex format code is constructed

Scalar	MSP Output
.0000765	7.6e–05
.000765	0.00076
77.5	78.
–66.5	–66.
–675.	–6.8e+02
.0000765 – 66.5i	(7.6e–05, –66.)
–675. + .000765i	(–6.8e+02, 0.00076)
12345	12345
–123	–123

Figure 3.25. Sample MSP scalar output

similarly. So is the default `int` code, except it right justifies and provides for displaying the maximum number (six) of digits needed. These format codes ensure that output columns will align nicely. You can find all possible format codes in the Borland C++ *Library Reference* manual [4] in the discussion of function `printf`.

Although C++ Version 2 stream input/output includes fairly elaborate formatting conventions, it doesn't support all the features used in the `DefaultFormat` code; hence it doesn't let you line up columns as nicely. That's why MSP uses `sprintf`.

Constant `ii = complex(0,1)`

Somehow, the `Complex.H` library header doesn't define a constant for the imaginary unit i, for which $i^2 = -1$. MSP header `Scalar.H` does, via a macro. Since identifiers such as `i` and `eye` are often employed for other purposes, MSP uses `ii`.

`abs`, `real`, `imag`, and `conj` for Noncomplex Scalars

Borland C++ implements functions `abs`, `real`, `imag`, and `conj` with `const complex&` `t` arguments to return the `double` absolute value and real and imaginary parts and the `complex` conjugate of `t`. In order to make its treatment of the three scalar types as uniform as possible, the MSP `Scalar` module overloads these functions with `const double&` `t` arguments, to return `double` values $|t|$, `t`, `0`, and `t`. Through automatic type casting, C++ also allows you to use these versions with an `int` argument `t` (but they still return `double` values).

Random Scalars

The fundamental Borland C++ random number generator is the function rand(), declared in Library header file StdLib.H. A statement int k = rand() produces a pseudorandom integer k in the interval $0 \leq k < 2^{15}$. Usually, your application must use a different interval. To generate a pseudorandom integer k in the interval $0 \leq k < n$, where n is a specified int value, you can use the StdLib.H function int random(n). That one merely calls rand and then scales the returned value linearly to fit the specified interval.

Often, an application needs a pseudorandom scalar value x in a specified region R. Since R varies so much in practice, the Scalar module provides function

Scalar MakeRandom(Scalar& t)

for each scalar type, which returns a pseudorandom Scalar value in an appropriate region. The returned value also replaces the value of parameter t. For the double version, $0 \leq t < 1$. For the complex version, t lies in the half-open unit square: $0 \leq $ real(t), imag(t) < 1. For the int version, $0 \leq k < 2^{15}$. Client programs can manipulate t as desired to cover a different region—for example, to generate a pseudorandom double value x in an interval a $\leq$ t < b , compute x = a + (b - a) t.

3.10 MSP ElemFunc MODULE

> **Concepts**
>
> *Factorials and binomial coefficients*
> *Functions* Root *and* logbase
> *Hyperbolic functions*
> *Trigonometric functions*

Although the Borland C++ Library includes a rich variety of real and complex functions, some are missing, and should be included, either because of their general usefulness or just to provide completeness. (For example, why should you have to remember *which* trigonometric functions are available? Why not have them all?) This book's MSP software package includes a number of additional functions of similar nature in its ElemFunc module. Like the other modules, this one has a header file ElemFunc.H, and a source code file ElemFunc.CPP. The header is listed in Appendix A.2 and included on the accompanying diskette. You'll find the source code file on the optional diskette. Most of its functions are straightforward implementations of their mathematical definitions. Many consist of a single return statement, wrapped in a try block, coupled with a catch block that displays an error message and rethrows an exception, and wrapped again in a template. Therefore, source code is not discussed much in this section.

Factorials and Binomial Coefficients

Borland C++ has no built-in factorial function, so the ElemFunc module provides one: long double Factorial(int n). Its return type was chosen to permit the largest possible argument *n*. This function causes overflow only if $n > 1054$. Don't take the result very seriously for such large values of *n*, however, since *n*! has several thousand digits and this function approximates *n*! simply by multiplying the first *n* positive integers, with 19-digit precision.

The binomial coefficients are usually defined by the equation

$$\binom{x}{m} = \frac{x!}{m!(x-m)!} = \frac{x(x-1)(x-2)\cdots(x-(m-1))}{m(m-1)(m-2)\cdots(1)}$$

where *x* and *m* are integers, with $x \geq m \geq 0$. Knuth [28, Section 1.2.6] finds utility in allowing *x* to be an arbitrary real number: the right-hand side of the previous equation still makes sense. Therefore, this definition is implemented by ElemFunc function double Binomial(double x, int n).

Functions Root and logbase

Normally, you'd compute an *r*th root of a number *b* by regarding it as a power: $\sqrt[r]{b} = b^{1/r}$. In most cases, you could use the Math.H function pow(b,1.0/r). However, there's one common situation where pow doesn't work: when $x < 0$ and *n* is an odd integer. Then, you must use the equation $\sqrt[r]{b} = -\sqrt[r]{-b}$.

MSP provides function Root to handle *all* cases. Its source code is displayed in Figure 3.26. Left-to-right evaluation of the conjunction in the conditional statement is essential!

Often in mathematical work you need logarithms $\log_b$ relative to various bases *b*. The Math.H natural and common logarithm functions log and log10 provide for bases *e* and 10 but you may need others, too, particularly base 2. $l = \log_b x$ means $x = b^l = e^{l \log b}$; hence $\log x = l \log b$, and

$$\log_b x = \frac{\log x}{\log b}.$$

MSP implements this equation with function double logbase(double b,double x). (This is inefficient, since the Borland C++ log *x* function first invokes the coprocessor or its emulator to

```
double Root(double b,                 // Return b^(1/r) .
            double r) {               // This function provides
  try {                               // for odd integral (type
    double nn;                        // long ) roots of a
    if (b < 0                         // negative b . Math.H
        && modf(r,&nn) == 0           // function pow handles
        && abs(nn) <= MAXLONG         // all other cases.
        && long(r) % 2 != 0)
      return -exp(log(-b)/r);
    else return pow(b,1/r); }
  catch(Exception) {
    cerr << "\nwhile attempting Root(b,r) with b = "
         << b << "  and  r = " << r;
    throw; }}
```

Figure 3.26. Function Root

compute $\log_2 x$ and then converts that to $\log x$ using a similar equation. But Borland C++ doesn't provide direct access to its $\log_2 x$ function.)

Hyperbolic Functions

Of the hyperbolic functions, Math.H and Complex.H implement only sinh, cosh, and tanh. (See Sections 3.4 and 3.5 for their definitions.) MSP includes the hyperbolic cotangent, secant, and cosecant as well:

$$\coth t = \frac{1}{\tanh t} \qquad \operatorname{sech} t = \frac{1}{\cosh t} \qquad \operatorname{csch} t = \frac{1}{\sinh x}.$$

Straightforward templates implement double and complex versions simultaneously.

MSP also includes the inverse hyperbolic functions. Since these aren't as familiar, it's appropriate to derive one of their formulas. The real inverse hyperbolic cosine, arccosh, is typical. Suppose $t = \cosh u$. Then $t \geq 1$ and

$$t = \frac{e^u + e^{-u}}{2} = \frac{(e^u)^2 + 1}{2e^u} \qquad (e^u)^2 - 2te^u + 1 = 0.$$

By the Quadratic Formula

$$e^u = \frac{2t \pm \sqrt{4t^2 - 4}}{2}.$$

Since the radical is < $2t$, both signs give possible values for e^u. By mathematical convention, the + sign is used. This amounts to reflecting the graph of $t = \cosh u$ across the line $t = u$ so that the resulting catenary opens to the right, and then selecting its upper half as the graph of the inverse function. The desired formula is

$$\operatorname{arccosh} t = u = \log\left(t + \sqrt{t^2 - 1}\right).$$

Here are the formulas for the other real inverse hyperbolic functions:

$$\operatorname{arcsinh} t = \log\left(t + \sqrt{t^2 + 1}\right) \qquad \operatorname{arccoth} t = \operatorname{arctanh}(1/t)$$

$$\operatorname{arcsech} t = \operatorname{arccosh}(1/t)$$

$$\operatorname{arctanh} t = \tfrac{1}{2}\log\frac{1+t}{1-t} \qquad \operatorname{arccsch} t = \operatorname{arcsinh}(1/t).$$

The same algebraic derivations work with the complex hyperbolic functions, except you must take more care to specify *which* logarithm and *which* square root. Templates in `ElemFunc.H` implement `double` and `complex` versions simultaneously for all six inverse hyperbolic functions using `Math.H` and `Complex.H` functions `log` and `sqrt`. Thus, the `complex` versions use the *principal* complex logarithm and square root functions. The MSP functions are spelled

```
asinh      atanh      asech
acosh      acoth      acsch
```

Trigonometric Functions

Of the trigonometric functions, `Math.H` and `Complex.H` implement only sin, cos, and tan. MSP includes the cotangent, secant, and cosecant as well:

$$\cot t = \frac{1}{\tan t} \qquad \sec t = \frac{1}{\cos t} \qquad \csc t = \frac{1}{\sin t}.$$

Straightforward templates implement `double` and `complex` versions simultaneously.

Similarly, the Borland C++ Library implements only three of the inverse trigonometric functions: arcsin, arccos, and arctan, corresponding to Math.H and Complex.H functions asin, acos, and atan. MSP implements arccot, arcsec, and arccsc as well, with ElemFunc functions acot, asec, and acsc. Their templates define double and complex versions simultaneously. *Warning:* Authorities disagree on the definitions of these functions. Here are definitions sanctioned by the United States National Bureau of Standards [1]:

$$\operatorname{arccot} t = \arctan(1/t)$$
$$\operatorname{arcsec} t = \arccos(1/t)$$
$$\operatorname{arccsc} t = \arcsin(1/t).$$

These all disagree with corresponding definitions [19], which were used for the earlier editions of this book and of MSP. (This choice of definition also affects the formulas for derivatives and integrals of these functions.)

4

Solving Scalar Equations

Chapter 4 describes a set of high-level software tools for solving equations in a single scalar variable. It fills a practical need, for such equations are commonplace in scientific and engineering application problems. At the same time, it provides a realistic test bed for much of the low-level MSP software introduced in Chapter 3. These tools implement several general methods for finding roots of a function. Some work with either real or complex scalars, while others are limited to the real case. When an algorithm works with either type of scalars, MSP implements it with a template, to ensure and emphasize that exactly the same computation is used for each case.

Two connecting threads run through the discussion: the bisection method and fixpoint iteration. Bisection works with a very general class of real equations, but it has limited efficiency and doesn't extend to complex equations or to multidimensional systems of equations. Fixpoint iteration is more general and flexible. Introduced first as a simple but inefficient method for solving a limited class of real equations, it's enhanced to yield the more general and faster Newton-Raphson technique—first for real equations and then for the complex case as well. Further enhancements incorporate bisection features or step size limitation to prevent the wild behavior that Newton-Raphson iteration exhibits in some situations. Later, in Chapter 9, fixpoint methods are extended further, to solve systems of equations in several variables.

All algorithms in this chapter apply to an important case: polynomial equations. Some example equations, in fact, are polynomials, but no use is made here of features particular to polynomials. Chapter 6 is devoted to *special* techniques for handling polynomials and culminates in three sections devoted to real and complex polynomial equations. Newton-Raphson iteration is adapted and enhanced to provide *all* roots of a specified polynomial, real or complex, with multiplicities. In turn, this facility is applied in Section 8.8 to compute matrix eigenvalues.

4.1 PRELIMINARIES

> **Concepts**
> MSP *module* Equate1 *and header file*
> *Trace output*
> *Bracketing intervals*
> MSP *function* Bracket
> *Always check an alleged root!*

Equate1 Module and Header File

The MSP functions for solving equations in a single variable constitute its Equate1 module. Like the modules described earlier, this one has header and source code files Equate1.H and Equate1.CPP. Much of the latter is displayed in this chapter, and it's included, of course, on the optional diskette. The Equate1 header file isn't particularly interesting: it consists only of function declarations, which are reproduced in the source code file. Therefore, it's not shown here, but it is displayed in Appendix A.2 and included on the accompanying diskette.

Trace Output

This chapter contains source code listings for many functions that solve equations $f(x) = 0$. Often they're accompanied by output listings, which trace a function's execution. These were *not* produced by the functions as listed, but by the versions in the source code file. The statements required to produce well formatted intermediate output are usually elementary, but they are so long that they tend to overshadow more important parts of the routines. Therefore, those output features are purged from the code displayed in the text. Most of the Equate1 functions have a parameter int& Code, which is used to convey an option code to the function and a status code back to the caller. If you invoke such a function with a nonzero Code value, you'll get intermediate trace output like that displayed in the text. Different nonzero Code values may produce different levels of detail; they're documented in the source code on the optional diskette. The source code displayed in the text simply omits many statements like if (Code != 0) {...}, where the dots represent very detailed output operations. As a result, the code displayed in the text is about half the size of that in the source code file, and is much easier to read and understand. Appendix A.3 displays the complete source code for function Bisect, with provisions for intermediate output. You can compare that with the version purged of output code, shown later in Figure 4.4.

Bracketing Intervals

When a function $f(x)$ is continuous on an interval $x_L \leq x \leq x_R$, and $f(x_L)$ and $f(x_R)$ differ in sign, then you know f has *at least* one root x in the interval: Its graph must cut the x axis somewhere there. Such an interval $[x_L, x_R]$ is called a *bracketing interval* for f. Finding a bracketing interval for f is often the first step in solving an equation $f(x) = 0$. MSP function Bracket will search a specified interval for a bracketing subinterval by checking the signs of f at its endpoints, dividing the interval into two parts, checking again, dividing the interval into four parts, and so forth. It stops when it finds a sign change or when the next subinterval length would be smaller than a specified tolerance T. The code for Bracket is shown in Figure 4.1. Since the first positive roots of the cosine are $1/2\pi \approx 1.5$ and $3/2\pi$, executing Bracket(xL,xR,cos,0.1) with xL = 0 and xR = 6 returns True and sets xL = 0 and xR = 3; but Bracket(0,1,cos,0.1) returns False.

Checking Solutions

All equation solving algorithms fail for some inputs. Various things can happen when a corresponding MSP function is invoked with such input. Ideal robust software would report all

```
Boolean Bracket(double&  xL,        // Find a bracketing in-
                double&  xR,        // terval  [xL , xR]  of
                double   F(double), // length  <= T  for  F
                double   T) {       // within the specified
  try {                              // interval.  Return True
    int S = Sign(F(xL));             // or False  to report
    for (double dx = xR - xL;        // success or failure.
                dx >= T;             // dx = current interval
                dx /= 2) {           // length.
      for (double x = xL + dx;       // Try intervals of this
                  x < xR;            // length.  You already
                  x += 2*dx)         // know that  F  has sign
        if (S != Sign(F(x))) {       // S  at half of the
          xL = x - dx;  xR = x;      // subdivision points.
          return True; }}
    return False; }
  catch(...) {
    cerr << "\nwhile executing  Bracket ;  now  xL,xR,T = "
         << xL << ',' << xR << ',' << T;
    throw; }}
```

Figure 4.1. Equate1 function Bracket

failures back to the caller, giving the client an opportunity to recover. In some cases, failure is signaled by an exception in the floating-point software/hardware system, such as an attempted square root of a negative number. MSP handles that sort of exception routinely. In other cases, it's not clear what constitutes an error in solving an equation $f(x) = 0$. Suppose an algorithm computes x such that $f(x)$ is small but different from zero. Is it close enough? A client might find it difficult or impossible to make such a decision in advance, as completely robust software would require. MSP puts the responsibility on the client: Some MSP functions may return a result x that's not an approximate solution of $f(x) = 0$. You must check x before using it, by evaluating $f(x)$.

4.2 BISECTION METHOD

> **Concepts**
> *Example: cam function*
> *Bisection method*
> *MSP function* `Bisect`
> *Efficiency*
> *Examples of failure*
> *Always check an alleged root!*

When a function $f(x)$ is continuous on an interval $x_L \leq x \leq x_R$, and $f(x_L)$ and $f(x_R)$ differ in sign, then you know f has *at least* one root x in the interval: Its graph must cut the x axis somewhere there. If you know such a bracketing interval $[x_L, x_R]$ and you know that there's *at most* one root in the interval, then you're in an ideal situation for solving the equation $f(x) = 0$. If you don't know a bracketing interval, but suspect there's one inside a specified interval $[x_L, x_R]$, you may be able to use MSP function `Bisect`, described in this section, to find one.

As an example, consider the cam in Figure 4.2, whose outline has polar equation

$$r = r(x) = \frac{1 + e^{-\frac{x}{2\pi}} \sin x}{2}.$$

Suppose you need to find the smallest positive angle x for which $r = 0.7$. From the graph in Figure 4.3, you can see that $f(x) = r(x) - 0.7$ is continuous on the interval $[x_L, x_R] = [0,1]$, changes sign, and has one root there. This *cam function f* is used as an example to demonstrate the MSP root finding routines described in this chapter.

The most general root finding technique is known as the *bisection method*. First, you choose an error *tolerance T*—a small positive number. You'll accept any approximation within distance T of

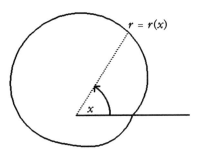

Figure 4.2. Outline of a cam

the root. Now, start with the known interval $[x_L, x_R]$, ascertain whether its left or right half contains the root, discard the other half, and repeat this process with the remaining half. After each iteration, the length of the interval containing the root decreases by half, so eventually it will become smaller than T, and you can stop, accepting either interval endpoint as an approximation to the root. Figure 4.4 shows MSP function Bisect, which implements the bisection method. As you'd expect, it has parameters corresponding to the function f, the left and right estimates x_L and x_R, and the tolerance T. In addition, Bisect uses a reference parameter Code to report the outcome of its computation: Code = 0 indicates that the iteration terminated satisfactorily with an interval shorter than T, whereas Code = –1 signifies that the specified tolerance was unattainable. The routine begins operation by testing the interval length dx. If $dx \geq T$, then Bisect computes the interval midpoint x, and compares the signs of the function

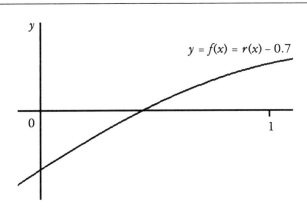

Figure 4.3. The cam function

```
double Bisect(double f(double),    // Find a root of  f .
              double xL,            // Left estimate.
              double xR,            // Right estimate.
              double T,             // Tolerance.
              int&   Code) {        // Status code.
  try {
    int   n = 0;                    // Iteration number.
    double yL = f(xL);
    double dx,x,y;
    do {                            // Iterate.  If the inter-
      dx = xR - xL;                 // val is too long to
      if (dx >= T) {                // accept, compute its
        x = (xL + xR)/2;  y = f(x); // midpoint x . If  f
        if (Sign(yL) == Sign(y)) {  // doesn't change sign on
          xL = x;  yL = y; }        // the left half, discard
        else                        // it; else discard the
          xR = x; }}                // right half.  Do at most
      while (++n < 2049 && dx >= T);// 2049  iterations.
    if (dx < T) Code = 0;           // Report success, or
      else Code = -1;               // failure to converge.
    return xL; }                    // Return left endpoint.
  catch(...) {
    cerr << "\nwhile executing Bisect ; now xL,xR,T = "
         << xL << ',' << xR << ',' << T;
    throw; }}
```

Figure 4.4. Equate1 function Bisect

values y_L and y at the left end and midpoint. If they agree, there's no sign change in left half interval, so that's discarded and the midpoint becomes the new left end: $x_L = x$. If they disagree, the right half is discarded: $x_R = x$. The process iterates by returning back to the interval length test. For reasons explained later, the number of iterations is limited to 2049.

Figure 4.5 shows execution of function Bisect to locate the root of the cam function $f(x) = r(x) - 0.7$ illustrated in Figure 4.3. The bisection method's efficiency can be estimated as shown in this chart, which corresponds to Figure 4.5:

n	Length of nth interval
0	1
1	1/2
2	1/4 = $1/2^2$
3	1/8 = $1/2^3$
4	1/16 = $1/2^4$
5	1/32 = $1/2^5$

The initial interval $[x_L, x_R]$ had length 1 and the specified tolerance was 0.05. In that case, the interval length after the nth iteration is $1/2^n$; in general, it's $(x_R - x_L)/2^n$. The number of iterations required to achieve the desired accuracy is the first integer n such that the interval length is less than T. This can be reformulated in terms of logarithms as the first n, such that

$$\frac{x_R - x_L}{2^n} < T, \quad \text{i.e.,} \quad n > \log_2 \frac{x_R - x_L}{T}.$$

If the criterion for stopping the bisection process were simply $x_R - x_L < T$, the do loop in function Bisect would repeat forever if you specified the tolerance $T \le 0$, or if you made it positive but so small that round-off error prevented computing $x_R - x_L$ accurately. To avoid that situation, Bisect enforces a limit on the number of iterations. The limit is the number required to reach an interval of the smallest possible size starting from one of maximum possible size. Since double values range in size from -2^{1023} to $+2^{1023}$, and the smallest positive double value is 2^{-1024}, the maximum number of iterations is

$$1 + \log_2 \frac{2^{1023} - (-2^{1023})}{2^{-1024}} = 2049.$$

If function Bisect returns before achieving $x_R - x_L < T$, it sets the error signal Code = -1.

If you don't cause a round-off problem by setting the tolerance too small, the bisection method will succeed whenever function f is continuous on the interval $[x_L, x_R]$, has exactly one root there, and $f(x_L)$ and $f(x_R)$ differ in sign. What if f doesn't meet all these conditions? If there's no root,

---

```
Finding a root of the cam function  f  by bisection
Left  estimate   xL :  0
Right estimate   xR :  1
Tolerance        T  :  0.05
```

Iteration	xL	xR	f(xL)	xR - xL
0	0.000000	1.000000	-0.20	1.0
1	0.000000	0.5000000	-0.20	0.50
2	0.2500000	0.5000000	-0.081	0.25
3	0.3750000	0.5000000	-0.027	0.12
4	0.4375000	0.5000000	-0.0024	0.062
5	0.4375000	0.4687500	-0.0024	0.031

```
Root   x =  0.44
Code     =  0
```

Figure 4.5. Executing function Bisect

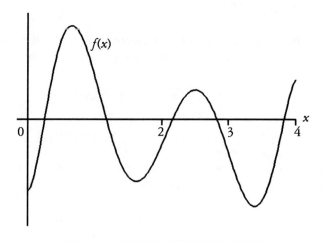

Figure 4.6. Function with five roots

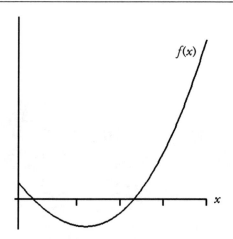

Figure 4.7. Bisection locates the left root in this example

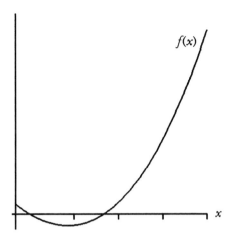

Figure 4.8. Bisection returns the right endpoint in this example

so that f doesn't change sign, bisection will repeatedly discard the left half interval and ultimately return a value between $x_R - T$ and x_R.

Figures 4.6 through 4.9 show four more situations where f fails to meet the bisection conditions. Figure 4.6 shows a function with an initial interval [0,4], which contains five roots. Bisection discards the left half interval; the remaining interval [2,4] contains three roots. A second bisection yields the interval [3,4], which contains a single root. Further bisections will locate this root as precisely as desired. This is typical: If there's an odd number of roots in the initial interval, the bisection method will locate one of them, but exactly which one depends on their position. If there's an even number of roots, bisection might or might not locate one, as shown in Figures 4.7 and 4.8.

The bisection method can also fail in various ways if the function f is discontinuous—for example, consider the function

$$f(x) = \begin{cases} \dfrac{x}{x^2 - 1} & \text{when } x \neq \pm 1 \\ \text{anything} \neq 0 & \text{when } x = \pm 1 \end{cases}$$

with the initial interval $[x_L, x_R] = [-2, +3]$, shown in Figure 4.9. The trace output in Figure 4.10 shows that function Bisect locates the pole $x = 1$ instead of the root $x = 0$!

Failures such as these emphasize a principle that should govern your use of Bisect—in fact, of any root-finding routine: After you've used it to compute an approximation x to a root of a function f, *always verify* that $f(x) \approx 0$!

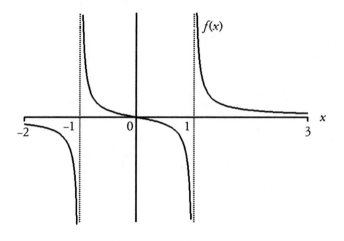

Figure 4.9. Here, bisection locates a pole, not a root

```
Finding a root of   f(x) = x/(x²-1)   by Bisection

Left  estimate   xL :  -2
Right estimate   xR :   3
Tolerance        T  :  .1

Itera-
tion         xL              xR           f(xL)    xR - xL
  0      -2.000000       3.000000         -0.67      5.0
  1       0.5000000      3.000000         -0.67      2.5
  2       0.5000000      1.750000         -0.67      1.2
  3       0.5000000      1.125000         -0.67      0.62
  4       0.8125000      1.125000         -2.4       0.31
  5       0.9687500      1.125000         -16.       0.16
  6       0.9687500      1.046875         -16.       0.078

Root    x ≈ 0.97
Code       = 0
```

Figure 4.10. Here, bisection locates a pole, not a root

4.3 FIXPOINT ITERATION

> **Concepts**
> Fixpoints and fixpoint iteration
> Good and bad examples
> Lipschitz constants
> Conditions for convergence
> Linear convergence
> MSP function Fixpoint
> Example: the cam function
> Inverse functions

The problem of computing a solution of an equation is often transformed into one of finding a *fixpoint* (fixed point) of a function g—that is, of solving a special kind of equation $x = g(x)$. Solving $x^2 = A$ for positive x and A, for example, is equivalent to solving $x = g(x)$ for any of the three functions

$$g(x) = \frac{A}{x} \qquad g(x) = \frac{x+A}{x+1} \qquad g(x) = \frac{1}{2}\left[x + \frac{A}{x}\right].$$

You can sometimes compute a fixpoint by *fixpoint iteration*: Guess an initial approximation x_0, and then calculate successive approximations

$$x_1 = g(x_0), \; x_2 = g(x_1), \; \ldots .$$

If the sequence $x_0, x_1, x_2, \ldots$ approaches a limit p and g is continuous at p, then p is a fixpoint of g because

$$g(p) = g(\lim_{n \to \infty} x_n) = \lim_{n \to \infty} g(x_n) = \lim_{n \to \infty} x_{n+1} = \lim_{n \to \infty} x_n = p.$$

For the example functions and $A = x_0 = 2$, the fixpoint is $p = \sqrt{2} \approx 1.414214$. Fixpoint iteration yields the following approximations:

$g(x) = 2/x$		$g(x) = (x+2)/(x+1)$		$g(x) = \frac{1}{2}(x + 2/x)$	
n	$x_n =$	n	$x_n \approx$	n	$x_n \approx$
0	2	0	2.000000	0	2.000000
1	1	1	1.333333	1	1.500000
2	2	2	1.428571	2	1.416667
3	1	3	1.411765	3	1.414216
Bad!		Good		Excellent!	

Apparently, proper choice of g is vital! This choice and its consequences are analyzed next.

Suppose the fixpoint p and initial approximation x_0 lie in a closed interval $I = [a,b]$. In order to apply g to the successive approximations $x_1, x_2, \ldots$, you may require that g map I into itself. Analysis of the convergence of sequence x_n involves the idea of a *Lipschitz constant* for g on I: a number L, such that

$$|g(x) - g(x')| < L|x - x'|$$

for all x and x' in I. (This concept can be defined for functions with vector, matrix, and complex arguments or values, too: Just replace the absolute value signs with appropriate norms.) Functions with Lipschitz constants are continuous, and Figure 4.11 shows that a continuous function g that maps I to itself has at least one fixpoint p there.

If $L < 1$, there can be *only* one fixpoint: if q were another, then $|p - q| = |g(p) - g(q)| < L|p - q| < |p - q|$—a contradiction!

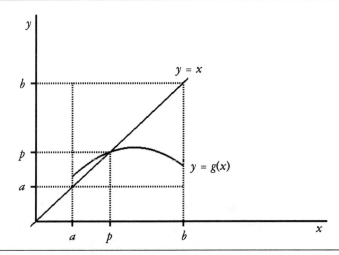

Figure 4.11. g maps interval $[a,b]$ to itself

Now, define the *error* $\varepsilon_n = x_n - p$ of the approximation x_n. Then

$$|\varepsilon_{n+1}| = |x_{n+1} - p| = |g(x_n) - g(p)| \leq L|\varepsilon_n|,$$

and applying this inequality repeatedly yields

$$|\varepsilon_n| \leq L^n |\varepsilon_0|$$

for every n. If $L < 1$, then $\varepsilon_n \to 0$, hence $x_n \to p$. The type of convergence displayed here has a special name: If a sequence, $x_0, x_1, x_2, \ldots$, approaches a limit p and successive error terms $\varepsilon_n = x_n - p$ are related by an inequality $|\varepsilon_{n+1}| \leq L|\varepsilon_n|$ with a constant $L < 1$, then x_n is said to converge to p *linearly*. (This concept can be extended to vector, matrix, and complex sequences by using norms.)

This analysis has demonstrated that you can use fixpoint iteration to approximate a fixpoint $p = g(p)$ if g maps a closed interval I containing p into itself, has a Lipschitz constant $L < 1$ there, and your initial guess lies in I. The successive approximations converge to p linearly. Moreover, suppose you know a bound $E > |\varepsilon_0|$ for the error in the initial approximation. Then you can use the inequality $|\varepsilon_n| \leq L^n |\varepsilon_0|$ to estimate the number n of iterations required to ensure that the error is less than a specified tolerance T: Stop when $L^n E < T$—that is

$$n > \frac{\log(T/E)}{\log L}.$$

This analysis partially explains the difference between the three functions g suggested at the beginning of this section to solve the equation $x^2 = A$. For the first one, $g(x) = A/x$, the required inequalities $g(a) \leq b$ and $a \leq g(b)$ would imply $A \leq ab \leq A$; hence, $A = ab$ and $g(a) - g(b) = b - a$. Thus, there's no Lipschitz constant $L < 1$ for any interval $[a,b]$ that g maps to itself. The condition for using fixpoint iteration fails, and the successive approximations don't converge. This function won't be considered further.

The difference between the good and excellent results given earlier for fixpoint iteration with the other two functions

$$g(x) = \frac{x + A}{x + 1} \qquad g(x) = \frac{1}{2}\left[x + \frac{A}{x}\right]$$

stems from the values of their derivatives at the fixpoint $p = \sqrt{A}$. If g' is continuous on the interval I under analysis and L is a number such that $|g'(\xi)| < L$ for all ξ in I, then L is a Lipschitz constant for g on I: For each x and x' in I, the mean-value theorem provides ξ between x and x', such that

$$|g(x) - g(x')| = |g'(\xi)(x - x')| \leq L|x - x'|.$$

In fact, a closer analysis [7, Section 3.3] shows that if g' is continuous at p and $|g'(p)| < L < 1$, then you can find some closed bounded interval I containing p that g maps into itself, for which L is a Lipschitz constant. Thus, you can use fixpoint iteration to approximate a fixpoint $p = g(p)$, provided $|g'(p)| < 1$ and your initial approximation x_0 is close enough to p to lie in I. (Unfortunately, this analysis doesn't tell *how* close you must be.) Further, you can use L with the inequality $|\varepsilon_n| \leq L^n |\varepsilon_0|$ to estimate the number of iterations required.

Here are the derivatives of the two functions g still under consideration:

$$g(x) = \frac{x + A}{x + 1} \qquad g(x) = \frac{1}{2}\left[x + \frac{A}{x}\right]$$

$$g'(x) = \frac{1 - A}{(1 + x)^2} \qquad g'(x) = \frac{1}{2}\left[1 - \frac{A}{x^2}\right]$$

$$g'(\sqrt{A}) = \frac{1 - \sqrt{A}}{1 + \sqrt{A}} < 1 \qquad g'(\sqrt{A}) = 0.$$

For $A = 2$, the left hand function has $g'(\sqrt{A}) \approx -0.17$; so $L = 1/5$ is a reasonable Lipschitz constant, and the error ε_n will decrease by at least a factor of 5 with each successive approximation. The right-hand function has $g'(\sqrt{A}) = 0$ for any A; hence, *any* positive L, no matter how small, can serve as a Lipschitz constant. This function is clearly the best choice for solving $x^2 = A$ by fixpoint iteration. It displays a phenomenon called *quadratic* convergence, which will be considered in Section 4.4.

Implementing fixpoint iteration requires a criterion for stopping the process. A client could find a Lipschitz constant and an initial error estimate and then determine the number of iterations required. But that's unrealistic, because those values are often difficult to estimate. The usual criterion is to stop when the distance between an approximation x and the corresponding function $g(x)$ becomes less than a specified tolerance T. Since that's not guaranteed to happen, the implementation must include an upper limit on the number of iterations.

The MSP implementation is function `Fixpoint`, shown in Figure 4.12. It has parameters corresponding to the function g, the initial approximation x, and the tolerance T. It's implemented by a template for scalars of any type, because fixpoint iteration works with complex functions as well as real. An additional reference parameter `Code` reports the outcome of the computation: `Code` = 0 indicates that the iteration terminated satisfactorily with $|x - g(x)| < T$, whereas `Code` = -1 signifies that this condition was not attained even after one thousand iterations. (This upper limit was chosen arbitrarily.)

```
template<class Scalar>
  Scalar Fixpoint(Scalar g(Scalar),    // Find a fixpoint of  g .
                  Scalar x,            // Initial estimate.
                  double T,            // Tolerance.
                  int&   Code) {       // Status code.
    try {
      int    n = 0;                    // Iteration count.
      Scalar p,dx;
      do {                             // Fixpoint
        p  = g(x);                     // iteration:
        dx = x - p;                    // do  <= 1000
        x  = p; }                      // steps. Re-
      while (++n < 1000 && abs(dx) >= T);   // port success
      if (abs(dx) < T) Code = 0;       // or failure to
      else Code = -1;                  // converge.
      return x; }
    catch(...) {
      cerr << "\nwhile executing  Fixpoint ;  now  x,T = "
           << x << ',' << T;
      throw; }}
```

Figure 4.12. Equate1 function Fixpoint

As a more realistic example to demonstrate Fixpoint, consider the cam function $f(x)$ of Figure 4.3. The equation $f(x) = 0$ is equivalent to

$$\frac{1+e^{-\frac{x}{2\pi}}\sin x}{2} = 0.7$$

$$e^{-\frac{x}{2\pi}}\sin x = 0.4$$

$$\sin x = 0.4e^{\frac{x}{2\pi}}$$

$$x = \arcsin\left(0.4e^{\frac{x}{2\pi}}\right) = g(x).$$

Thus, the root of the cam function is the fixpoint of this function g. Figure 4.13 shows execution of function Fixpoint to locate it.

In Figure 4.13, the ratios of successive values of $x - g(x)$ are approximately constant ≈ 0.075. There's a reason for that:

$$\lim_{n \to \infty} \frac{x_{n+1} - g(x_{n+1})}{x_n - g(x_n)} = \lim_{n \to \infty} \frac{g(x_n) - g(x_{n+1})}{x_n - x_{n+1}} = \lim_{n \to \infty} g'(\xi_n) = g'(p).$$

In this equation, ξ_n is a point (provided by the mean-value theorem) between the approximations x_{n+1} and x_n. Now differentiate g, compute $g'(p) \approx 0.076$ for the fixpoint p, and thus explain the values of the ratios.

Earlier, you saw that one of the conditions for successful fixpoint iteration is that $|g'(p)| < 1$, where $p = g(p)$ is the fixpoint. The following ratio limit, similar to the one just computed, shows what happens when $|g'(p)| > 1$. If g' is continuous in a neighborhood of p, then

$$\lim_{n \to \infty} \frac{\varepsilon_{n+1}}{\varepsilon_n} = \lim_{n \to \infty} \frac{g(x_{n+1}) - p}{g(x_n) - p} = \lim_{n \to \infty} \frac{g(x_{n+1}) - g(p)}{x_{n+1} - p} = \lim_{n \to \infty} g'(\xi_n) = g'(p).$$

Here, ξ_n lies between x_{n+1} and p. If $|g'(p)| > 1$, then $|\varepsilon_n|$ increases without bound: The sequence x_n diverges.

An inverse function g^{-1} has the same fixpoints as g: $p = g(p)$ if, and only if, $g^{-1}(p) = p$. If these functions are differentiable, then their derivatives at p are reciprocal. Except in the rare case when both have absolute value 1, one will have absolute value < 1 and the other, > 1. Thus, you'll be able to use fixpoint iteration with *exactly one* of them to find p—for example, the inverse of the Figure 4.13 function g is $g^{-1}(x) = 2\pi \log(2.5 \sin x)$. If you perform fixpoint iteration on g^{-1} starting with the final approximation $x_5 = 0.4436807$ in the figure, you'll just get the original values x_4, x_3, ... in reverse order. The sequence diverges.

```
Finding the fixpoint  x  of  g(x) = asin(.4 exp(x/2π))
by Fixpoint iteration

Initial estimate  x :  0
Tolerance         T :  .000001

Itera-
tion      x          g(x)        x - g(x)
 0      0.0          0.41         -0.41
 1      0.41         0.44         -0.030
 2      0.44         0.44         -0.0022
 3      0.44         0.44         -0.00017
 4      0.44         0.44         -1.3e-05
 5      0.44         0.44         -9.7e-07

Fixpoint  x =  0.4436807
Code         =  0
```

Figure 4.13. Executing function Fixpoint

4.4 NEWTON-RAPHSON ITERATION

Concepts

Quadratic convergence

Newton-Raphson iteration

MSP function `NR`

Example: the cam function

Finding the address of an instance of a template function

Example: bouncing polynomial

MSP hybrid Newton-Raphson/bisection function `NRBisect`

Secant method

MSP function `Secant`

Complex equations

In Section 4.3 you saw how the problem of solving an equation $f(x) = 0$ for a root $x = p$ of a function f can be transformed in various ways into a problem of solving an equivalent equation $x = g(x)$ for a fixpoint $x = p$ of a related function g. If g' is continuous near p, $|g'(p)| \leq L < 1$, and your initial approximation x_0 is close enough to p, then the successive approximations $x_1, x_2, \ldots$ computed by fixpoint iteration $x_{n+1} = g(x_n)$ converge linearly to p, and the approximation errors $\varepsilon_n = x_n - p$ satisfy the inequality $|\varepsilon_{n+1}| \leq L|\varepsilon_n|$. Clearly, you should choose g so that $|g'(p)|$ is as small as possible. The example

$$A > 0, \quad f(x) = x^2 - A, \quad p = \sqrt{A}, \quad g(x) = \frac{1}{2}\left[x + \frac{A}{x}\right], \quad g'(p) = 0$$

indicated that very rapid convergence might result if you choose g so that $g'(p) = 0$.

In fact, this is generally true. If g'' is continuous on a neighborhood N of p that g maps into itself, then

$$\frac{\varepsilon_{n+1}}{\varepsilon_n^2} = \frac{x_{n+1} - p}{(x_n - p)^2} = \frac{g(x_n) - g(p)}{(x_n - p)^2}$$

$$\frac{g'(p)(x_n - p) + \frac{1}{2}g''(\xi)(x_n - p)^2}{(x_n - p)^2} = \frac{1}{2}g''(\xi)$$

Here, ξ is a number between x_n and p, provided by Taylor's theorem. If $g'(p) = 0$ and $|g''(\xi)| \le M$ for *all* ξ in N, then $|\varepsilon_{n+1}| \le \tfrac{1}{2} M |\varepsilon_n|^2$ in general. If the error terms ε_n of a convergent sequence $x_0, x_1, x_2, \ldots$ satisfy an inequality such as this for any constant M, then the sequence is said to converge *quadratically* to p. (This concept can also be defined for vector, matrix, and complex sequences: Just replace the absolute values by appropriate norms.)

To achieve quadratic convergence, you need a way to transform functions f with roots p into functions g with fixpoints p such that $g'(p) = 0$. The most common method is to set

$$g(x) = x - \frac{f(x)}{f'(x)}.$$

Then you can compute

$$g'(x) = \frac{f(x) f''(x)}{f'(x)^2}$$

$$g''(x) = \frac{f'(x)^2 f''(x) + f(x) f'(x) f'''(x) - 2 f(x) f''(x)^2}{f'(x)^3};$$

hence, $g'(p) = 0$ and g'' is continuous near p provided $f'(p) \ne 0$ and f''' is continuous near p. Fixpoint iteration with this function g is called *Newton-Raphson* iteration. Under these assumptions, the successive approximations converge quadratically to p, provided your initial estimate x_0 is close enough to p. (A more precise analysis shows that only the continuity of f'' is required for this result [20, Chapter 4].

With $A > 0$ and $f(x) = x^2 - A$, the Newton-Raphson iteration function is

$$g(x) = x - \frac{f(x)}{f'(x)} = x - \frac{x^2 - A}{2x} = \frac{1}{2}\left[x + \frac{A}{x}\right].$$

This is just the example given at the beginning of this section. It's known as *Newton's method* for computing square roots.

Newton-Raphson iteration has a vivid graphical representation, shown in Figure 4.14. Consider one of the successive approximations x_n to a root p of a function f. The tangent to the graph of $y = f(x)$ at the point $x, y = x_n, f(x_n)$ has equation $y - f(x_n) = f'(x_n)(x - x_n)$. Set $y = 0$ and solve for the intercept x. You get

$$x = x_n - \frac{f(x_n)}{f'(x_n)}.$$

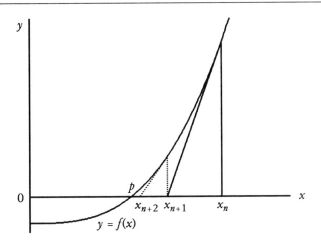

Figure 4.14. Newton-Raphson iteration

This is the next Newton-Raphson approximation x_{n+1}. Apparently, if x_n is close enough to p and the graph of $y = f(x)$ is sufficiently steep and smooth near $x = p$, the next approximation will be *much* closer to p. Figure 4.14 also shows the *next* approximation x_{n+2}; it's much closer to p than even x_{n+1}! (If x_n had been slightly left of p, the next iteration would have brought x_{n+1} to the right of p, as in this example. If the graph of f had been concave *down* near p instead of *up*, you'd have to reverse the roles of left and right in this discussion.)

The MSP implementation of Newton-Raphson iteration is function NR, shown in Figure 4.15. Since Newton-Raphson iteration is valid for both real and complex scalars, NR is a template function. It has parameters x and T, corresponding to the initial estimate x and the tolerance T, and a parameter corresponding to a C++ function FdF that computes values of both f and f'. FdF must have prototype

```
double  FdF(double  x, double&  dF)
complex FdF(complex x, complex& dF).
```

It returns the value $f(x)$ after computing dF $= f'(x)$. By packaging the code for both f and f' in a single C++ function, you can avoid duplicate computation of expressions common to f and f', as in the next example. An additional NR reference parameter Code reports the outcome of the computation. Code = 0 indicates that the iteration terminated satisfactorily when the difference of two successive approximations became less than the tolerance T, whereas Code = −1 signifies that this condition was not attained even after one thousand iterations. (This upper limit was chosen arbitrarily.)

```
template<class Scalar>                  // Find a root of  f  by
  Scalar NR(Scalar FdF(Scalar x,        // Newton-Raphson
                      Scalar& dF),      // iteration.
            Scalar x,                   // Initial estimate.
            double T,                   // Tolerance.
            int&   Code) {              // Status code.
  try {
    int   n = 0;                        // Iteration count.  FdF
    Scalar Fx,dF,dx;                    // must set  dF = f'(x)
    do {                                // and return  f(x) .
        Fx = FdF(x,dF);                     // Newton-Raphson
        if (Fx == Scalar(0)) dx = 0;        // iteration: do at
          else dx = Fx/dF;                  // most  1000  steps.
        x -= dx; }
      while (++n < 1000 && abs(dx) >= T);
    if (abs(dx) < T) Code = 0;          // Report success or
      else Code = -1;                   // failure to
    return x; }                         // converge.
  catch(...) {
    cerr << "\nwhile executing   NR ;   now x,T = "
         << x << ',' << T;
    throw; }}
```

Figure 4.15. Equate1 function NR

Figure 4.16 shows execution of function NR to locate the root of the Figure 4.3 cam function. It also contains the code for function FdFforCam used to evaluate the cam function and its derivative. You can see that expressions common to $f(x)$ and $f'(x)$ are computed only once. FdFforCam is implemented with a template, because later in this section it's used to demonstrate the Newton-Raphson method for a complex equation. The demonstration program's source code includes the template but no explicit double or complex version of the function definition. FdFforCam doesn't point to anything, so you *can't* execute double x = NR(FdFforCam, 0,1e-14, Code) to produce the output shown in Figure 4.16. Instead, the demonstration program included the code

```
double (*FdF)(double,double&) = FdFforCam
double x = NR(*FdF,0,1e-14,Code)
```

to construct a pointer FdF to the double instance of FdFforCam and invoke NR with that pointer.

The difference in accuracy between the three equation-solving methods discussed so far is remarkable:

Method	$\lvert x_5 - x_4 \rvert$	$\lvert f(x_5) \rvert$
Section 4.1 Bisection	3.1×10^{-2}	2.4×10^{-3}
Section 4.2 Fixpoint	9.7×10^{-7}	2.8×10^{-8}
Section 4.3 Newton-Raphson	4.0×10^{-17}	1.6×10^{-17}

(For the Fixpoint method, $\lvert f(x_5) \rvert$ was computed by hand.)

The rapid convergence of Newton-Raphson iteration for ideal cases, such as that shown in Figure 4.16, is attained at some cost: It's not as general, nor as predictable as the bisection method.

```
Finding a root  x  of the real cam function  f
  by Newton-Raphson iteration

Initial estimate  x :  0
Tolerance         T :  1e-14

Itera-
tion    x        f(x)      f'(x)       dx
 0    0.0       -0.20      0.50      -0.40
 1    0.40      -0.017     0.40      -0.043
 2    0.44      -0.00029   0.39      -0.00076
 3    0.44      -9.4e-08   0.39      -2.4e-07
 4    0.44      -9.6e-15   0.39      -2.5e-14
 5    0.44      -1.6e-17   0.39      -4.0e-17

Root   x  ≈  0.443680771333451
Code      =  0
```

This template function computed $f(x)$ and $f'(x)$:

```
template<class Scalar>
  Scalar FdFforCam(Scalar  x,
                   Scalar& dF) {
    Scalar T = 2*M_PI;
    Scalar E = .5*exp(-x/T);
    Scalar S = sin(x);
    Scalar C = cos(x);
    dF       = E* (C - S/T);
    return -.2 + E*S; }
```

Figure 4.16. Executing function NR with function FdFforCam

Is it ever as slow as that method? When does it fail altogether? Clearly, the iteration will halt with approximation x_n if $f'(x_n) = 0$ or fails to exist. Moreover, the condition derived earlier for quadratic convergence requires that the initial estimate be sufficiently close to the root p, f'' be continuous near p, and $f'(p) \neq 0$. When the latter conditions fail, convergence slows—for example, if $f(x) = x^2$, so that $p = f(p) = f'(p) = 0$, then

$$x_{n+1} = x_n - \frac{x_n^2}{2x_n} = \frac{x_n}{2}.$$

Thus, in this case the ratio of successive approximation errors is always ½, so the approximations converge linearly, not quadratically. That generally happens when $f'(p) = 0$ but the other conditions hold. If p is a root of order n, then the ratio is $1 - 1/n$ [30, Section 2.3B]. For high-order roots, this ratio makes convergence so slow that numerical analysis software must sometimes use very special mathematical techniques.

A more serious failure can occur when the initial estimate is too far from p. The sequence of Newton-Raphson approximations may approach a different root. Or, if there's a local extremum at a point x near p, the approximations may cycle indefinitely or at least bounce around for a while before converging. For example, execute NR with initial estimate $x = -12$ to find the root $p \approx 1.46$ (the only real root) of the polynomial $f(x) = x^5 - 8x^4 + 17x^3 + 8x^2 - 14x - 20$, shown in Figure 4.17. NR will iterate more than 50 times before settling down to converge rapidly to p. When an approximation x_n falls near the local maximum or minimum, x_{n+1} jumps far away, as

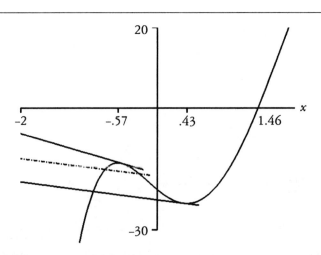

Figure 4.17. $f(x) = x^5 - 8x^4 + 17x^3 + 8x^2 - 14x - 20$

shown by the solid tangent lines. Several steps are then needed to return to the region near the root. (The dotted line is referred to later.)

Figure 4.18 shows MSP function NRBisect, a hybrid root finder, which combines features of the Newton-Raphson and bisection methods to combat this bouncing problem. You invoke it as you do the bisection method, with bracketing estimates x_L and x_R of a root p of a function f: $[x_L, x_R]$ must contain p, and $f(x_L)$ and $f(x_R)$ must differ in sign. NRBisect uses the interval

```
double NRBisect(double FdF(double,        // Find a root of  f.
                           double&),
                double xL,                // Left estimate.
                double xR,                // Right estimate.
                double T,                 // Tolerance.
                int&   Code) {            // Status code.
  try {
    Boolean Done;
    int n = 0;                            // Iteration count.
    double x,h,yL,yLp,y,yp;
    x = (xL + xR)/2;                      // The initial approxima-
    h = x - xL;                           // tion is the midpoint.
    yL = FdF(xL,yLp);
    do {                                  // Repeat until  h , the
        y = FdF(x,yp);                    // Newton-Raphson
        Done = (h < T);                   // correction or half the
        if (!Done) {                      // current interval
          if (Sign(yL) == Sign(y)) {      // length, is  < T .
              xL = x;  yL = y; }
            else                          // Adjust the bracketing
              xR = x;                     // interval.
          h = (y == 0 ? 0 : y/yp);        // Try the Newton-Raphson
          x -= h;                         // correction  h .
          h = fabs(h);
          if (x < xL || xR < x) {         // If  x  would go out of
            x = (xL + xR)/2;              // bounds, bisect instead.
            h = x - xL; }}}               // Do at most  1000  iter-
      while (++n < 1000 && !Done);        // ations. Set Code  to
    Code = -!Done;                        // report success/failure.
    return x; }
  catch(...) {
    cerr << "\nwhile executing  NRBisect ;  now  xL,xR,T = "
         << xL << ',' << xR << ',' << T;
    throw; }}
```

Figure 4.18. Equate1 function NRBisect

```
Finding a root of the bouncing function  f
  by the hybrid Newton-Raphson bisection method

Left    estimate   xL :   -26
Right   estimate   xR :     2
Tolerance          T  :    .01

Itera-
tion         xL              x              xR          f(x)        dx
  0     -26.00000       -12.00000       2.000000      -4.4e+05     14.
  1     -12.00000        -9.334385      2.000000      -1.4e+05      2.7
  2      -9.334385       -7.215922      2.000000      -4.7e+04      2.1
            :                :              :             :          :
  9      -1.311093       -0.9262393     2.000000      -20.          0.38
 10      -0.9262393      -0.4665007     2.000000      -14.          0.46
 11      -0.4665007       0.7667497     2.000000      -21.          1.2
 12       0.7667497       1.383375      2.000000       -3.3         0.62
 13       1.383375        1.466876      2.000000        0.087       0.084
 14       1.383375        1.464772      1.466876        4.9e-05     0.0021

Root   x  =  1.465
Code      =  0
```

Figure 4.19. Executing function NRBisect

midpoint as an initial estimate x, and successively improves x by performing either Newton-Raphson or bisection steps. After each step it adjusts x_L and x_R, as in the bisection method, so that $[x_L, x_R]$ becomes a smaller bracketing interval, but doesn't necessarily shrink by half. If a Newton-Raphson iteration would throw the next approximation x outside the interval, NRBisect sets x equal to the interval midpoint. The iteration terminates when a Newton-Raphson step changes x by an amount less than the specified tolerance T, or when a bisection step yields the midpoint of a bracketing interval shorter than $2T$.

Figure 4.19 shows execution of NRBisect to find the root of the Figure 4.17 polynomial that caused Newton-Raphson iteration to bounce around so much. The tabulated dx values are either the size of a Newton-Raphson step from the previous approximation or the length of the current interval. NRBisect takes Newton-Raphson steps through Iteration 10. Convergence is somewhat slow, because the graph is terribly steep and nonlinear. Then the hybrid function performs three bisections in a row, to avoid bouncing. After that, the approximation is close enough to the root for Newton-Raphson iteration to take over again and converge quickly.

Another device for avoiding bouncing is simply to limit the size of the Newton-Raphson steps, relative to the size of the current approximation. That's particularly appropriate for complex polynomial equations; it's implemented in Section 6.4 as function NRHStep. Operating on the Figure 4.19 polynomial with the same initial estimate, it required 21 iterations—a few more than NRBisect.

One disadvantage of the Newton-Raphson method and its hybrid is their requirement for a value of the derivative $f'(x_n)$ at each approximation x_n of the root p of f. That might be difficult or impossible to provide, especially if values of f are given only by a table or by the solution of some other substantial problem such as a differential equation or a multiple integral. One way to avoid this problem is to replace $f'(x_n)$ in the Newton-Raphson iteration formula

$$x_{n+1} = x_n - \frac{f(x_n)}{f'(x_n)}$$

by its approximation

$$f'(x_n) \approx \frac{f(x_n) - f(x_{n-1})}{x_n - x_{n-1}}.$$

You can check that the resulting expression

$$x_{n+1} = x_n - \frac{x_n - x_{n+1}}{f(x_n) - f(x_{n-1})} f(x_n)$$

is the formula for the x intercept of the secant line through points $x, y = x_{n-1}, f(x_{n-1})$, and $x_n, f(x_n)$. For this reason, the resulting root finding algorithm is called the *secant method*. It requires *two* initial estimates: x_0 and x_1. They need not bracket the root.

The secant method attains greater generality than the Newton-Raphson method at a cost. Two initial estimates are required, and convergence is somewhat slower. Its behavior in difficult situations is sometimes problematic, as discussed in the next paragraph. Its convergence conditions are similar to those of Newton-Raphson iteration, but the rate of convergence falls between linear and quadratic: If p is a simple root, then the approximation errors $\varepsilon_1, \varepsilon_2, \varepsilon_3, \ldots$ satisfy the equation

$$\lim_{n \to \infty} \frac{\varepsilon_{n+1}}{\varepsilon_n^r} = K, \qquad r = \frac{1 + \sqrt{5}}{2} \approx 1.6$$

for some constant K [20, Section 4.2].

While the secant method is conceptually similar to Newton-Raphson iteration, its implementation involves an extra complication: You need two initial estimates and you must keep two approximations current. You can see how this is done by inspecting the MSP implementation, function Secant, shown in Figure 4.20. Executing it to find the root of the cam function shown in Figure 4.16 produces results comparable to the Newton-Raphson method: With initial estimates 0 and 1, or 0 and 0.1, the secant method achieves the same accuracy with five or six steps. With the troublesome Figure 4.17 polynomial f, the results are mixed. With initial estimates –12 and 5 bracketing the root, Secant attains the accuracy 0.01 after only seven iterations. But with initial estimates –12 and –11 and tolerance 0.01, it falsely reports a root $x \approx -0.34$ after 16 iterations, even though $f(-0.34) \approx -15$. This happens because two successive approximations x_{14} and x_{15} straddle the local maximum, so that $f(x_{14}) \approx f(x_{15})$. The resulting secant—the dotted line in Figure 4.17—is nearly horizontal. Thus, x_{16} jumps away from the region shown and $f(x_{16})$ is so large that the secants determined by x_{15} and x_{16} and by x_{16} and x_{17} are nearly vertical. Their intercepts x_{17} and x_{18} are closer than the tolerance, so Secant terminates successfully and returns x_{18} even though that's not close to the root. With the smaller tolerance

```
template<class Scalar>                  // Find a root of  f .
  Scalar Secant(Scalar f(Scalar),
                Scalar x0,              // Initial
                Scalar x1,              // estimates.
                double T,               // Tolerance.
                int&   Code) {          // Status code.
    try {
      int    n = 0;                     // Iteration count.
      Scalar dx = x1 - x0;
      Scalar y0 = f(x0);
      Scalar y1,dy;
      do {                              // Secant method
          y1 = f(x1);                   // iteration.  Do at
          dy = y1 - y0;                 // most  1000  steps.
          dx = -y1*dx/dy;
          x0 = x1;  y0 = y1;
          x1 += dx; }
        while (++n < 1000 && abs(dx) >= T);    // Report
      if (abs(dx) < T) Code = 0;                // success
        else Code = -1;                         // or failure
      return x1; }                              // to converge.
    catch(...) {
      cerr << "\nwhile executing  Secant ;  now  x0,x1,T = "
           << x0 << ',' << x1 << ',' << T;
      throw; }}
```

Figure 4.20. Equate1 function Secant

10^{-6} this anomaly disappears, though the computation takes longer: Secant computed the root appropriately after 27 iterations.

Some variations on the secant method are easy to consider. First, you could construct its hybrid with the bisection method, analogous to function NRBisect. Second, you could remove the need for two initial estimates of the root of function f by selecting x_1 close to x_0, so that $f(x_0)$ and $f(x_1)$ aren't too close. (However, this would prevent you from using bracketing estimates, which might be safer, if you have them at hand.)

The secant and Newton-Raphson methods apply to complex equations, too, because the mathematics that underlies their convergence properties generalizes to complex functions. You can check that functions Secant and NR work with complex scalars. With initial estimate $100 + 100i$, NR will find the root $111.87 + 17.582i$ of the Figure 4.16 cam function. With estimates $\pm 5 \pm 5i$, it will find the roots $3.969 \pm 1.430i$ and $-0.701 \pm 0.524i$ of the Figure 4.17 bouncing function. You can test Secant similarly—it may be hard to find successful initial estimates. (For either method, in order to converge to a nonreal root, you *must* start with a nonreal initial estimate; otherwise, the algebra will result only in successive real approximations.)

Most complex equations that you'll encounter involve polynomials such as the bouncing function. Complex polynomial roots are so important in practice that Newton-Raphson iteration is specially adapted for that case in Section 6.3. That process, in turn, is applied to eigenvalue problems in Section 8.8.

5

Vector Classes

The MSP `Vector` class template implements an abstract data type to support vector algebra for each supported scalar type. It's also the underlying structure for matrix and polynomial computations. Thus, it has many applications, which can place widely differing requirements on the data structures. No single class definition can possibly meet all these demands. The MSP implementation is merely an example, showing how C++ can be used to tailor numerical analysis software to general or particular needs. It provides for

- Scalars of any type supported by the MSP `Scalar` module (`double`, `complex`, or `int`)
- Vectors with varying starting indices
- Vectors of varying dimension
- Bounds checking

The first provision allows you to perform real, complex, and integer vector algebra in the same program using formulas that closely resemble conventional mathematics. The second provision means, for example, that you can use vectors $X = [x_1, x_2, \ldots]$ and $A = [a_0, a_1, \ldots]$ in the same context. That's particularly convenient for polynomial algebra applications, where use of the zero subscript for the constant coefficient often collides with other notation. Other first index choices are possible, too, and you can use vectors with various index upper bounds. The `Vector` class provides complete, but flexible, bounds checking: Vector operations check their operands for consistency of lower and upper index bounds. It's often redundant and too time-consuming to verify for *each* reference to a vector entry x_k that k lies within the index bounds for X. Therefore, you can easily turn off that one bounds-checking feature.

MSP functions that manipulate vectors can be classified as *logical* or *mathematical*. Counterparts of the logical functions would be defined for most any class that has a data component of indeterminate size. The logical functions are described in detail in Section 5.1. The mathematical functions implement vector algebra; they're discussed in Section 5.2. Section 5.3 demonstrates the use of several vector algebra features to compute line and surface integrals. You'll find many more examples in the later chapters on polynomial and matrix algebra.

5.1 Vector MODULE INFRASTRUCTURE

> **Concepts**
> *Logical and mathematical functions*
> *Files* Vector.H, VectorL.CPP, *and* VectorM.CPP
> Vector *class template*
> *Selectors, subscript operator*
> *Constructors and destructor*
> *Copiers and assignment*
> *Equality and inequality*
> *Keyboard input and display output*
> *Stream input/output*
> *Coda*

As with other MSP modules, the Vector module has a single header file Vector.H. The source code is so long, however, that it's split into two files VectorL.CPP and VectorM.CPP for logical and mathematical functions. You'll find the header file on the accompanying diskette and in Appendix A.2, and the source code files on the optional diskette. The logical part of the header file is first shown in Figure 5.1. It automatically includes the MSP header file Scalar.H and other MSP and Borland C++ header files that Scalar.H loads. Parts of VectorL.CPP are displayed later in this section. The mathematical part of the header file and its corresponding source code will be discussed in Section 5.2.

The Vector Data Structure

The Vector module is designed to handle vectors of scalars of any type supported by the MSP Scalar module, so the Vector classes are implemented via a template, as shown in Figure 5.1. (The template variable Scalar doesn't really refer to that module: You could use any identifier there. Using this one makes the code easier to read.) The template defines a different class Vector<Scalar> for each Scalar type. It has private and public data and member function components. For the Vector classes, the data components happen to be private, whereas the

```
//*******************************************************
//                        Vector.H

#ifndef   Vector_H
#define   Vector_H

#include "Scalar.H"

template<class Scalar>
  class Vector {                     // The private parts of
                                     // class  Vector  are its
    private:                         // data components.
      int     Low;                   // Index bounds for the
      int     High;                  // vector entries.
      Scalar* Array;                 // Entry storage.

    public:

      // Selectors  -------------------------------------------

      int LowIndex()  const;         // Return the index lower
                                     // bound.
      int HighIndex() const;         // Return the index upper
                                     // bound.
      Scalar& operator[](int k) const; // Entry selector:  adjust
                                     // for  Low != 0.
```

Figure 5.1. Vector.H logical components
Part 1 of 3; continued in Figures 5.3 and 5.9.

member functions are all public. You can access private Vector components *only* by invoking public member or friend functions.

The Vector class design must permit vectors whose entries have varying index ranges. To facilitate ascertaining a vector's lower and upper index bounds, they form data components Low and High of Vector objects. The entries themselves are stored in a scalar Array. Since the Array size is indeterminate, it must be located in free memory. The Array component is *not* the array itself, but merely a *pointer* to its first entry in free memory. In order to enforce bounds checking, the index bounds and Array must be private components, and member functions must be provided to manipulate them as needed. In designing the class, no other private components proved necessary.

MSP uses a standard representation for *empty* vectors: Low = 0, High = –1, and Array = NULL.

According to the Chapter 2 overview, the functions necessary for logical manipulation of Vector objects include

- *Selectors*, to provide limited access to the data components
- *Constructors*, to initialize the index bounds and array pointer, and—usually—to set up the array in free memory
- A *destructor*, to deallocate the array when a vector is deleted explicitly or passes out of scope
- A *copy constructor*, for the compiler to use when it needs copies for initialization or function calls
- An *assignment* operator, overloaded to copy the array
- A general *copier*, for manipulating pieces of vectors
- Overloaded *equality* and *inequality* operators, to compare vector index bounds and arrays
- *Keyboard input* and *display output* functions
- *Stream input/output* functions.

These features are described in detail in the rest of this section. Sometimes, source code is given here in the text. For the display output function, you're referred to Appendix A.1. Three (slightly) mathematical functions, whose names begin with Make..., may be needed to test this section's other features; but they're described in Section 5.2.

Selectors

Selectors are public member functions that provide limited access to private data components. For instance, function LowIndex merely returns the Low component of its principal argument:

```
template<class Scalar>                  // Selector:  return
  int Vector<Scalar>::                  // the index lower
    LowIndex() const {                  // bound.
  return Low; }
```

The const qualifier is included because the overloaded operator doesn't change the value This of its principal argument. That provision allows the principal argument to be a Vector *constant*. The Low component can be changed only in a very controlled way by other member functions. Function HighIndex is similar.

Since vector entries are stored in a private array in free memory, you can't access them directly with the subscript operator [ ]. Instead, this must be overloaded so that if V is a Vector object and k is an int value within the V index bounds, then V[k] is a reference to the appropriate

```
template<class Scalar>                              // Select a
  Scalar& Vector<Scalar>::                          // vector entry.
      operator[](int k) const {
    if (k < Low || High < k) {                      // Remove this
        cerr << "Exception IndexError\n"            // bounds test
                "While trying to use entry     "    // for more
             << k << "  of a vector with      "    // efficient
                "High,Low = "                       // but risky
             << High << ',' << Low;                 // execution.
        throw(Exception(IndexError)); }     // Adjust for lower
    return Array[k-Low]; }                  // index  != 0 .
```

Figure 5.2. Vector entry selector [ ]

Array entry in free memory. The overloaded [] operator must return a *reference to*, not the value of, the entry, because it must be usable on the left-hand side of an assignment. (The reference is simply the address of the entry, but it's not directly accessible to your program as a pointer would be.) Since the V and Array indices range over intervals starting with Low and zero, respectively, the appropriate Array entry has index k–Low. Figure 5.2 is the selector's source code. It returns a reference Array[k-Low] after verifying that Low ≤ k ≤ High. If that test fails, the selector displays an error message and then throws an MSP Exception object with data component IndexError to describe the situation to any Exception handler that may catch it. It's often redundant and too time consuming to verify for *each* use of a vector entry that its index lies within bounds. Therefore, you can disable the verification by merely removing or commenting out the if block in this function.

Constructors

According to Section 2.6, each Vector class needs a default constructor, which the compiler will call *n* times whenever you declare an array of *n* vectors. The compiler would automatically allocate memory for the data components of each vector, but you must provide the default constructor to initialize them. It's declared in Figure 5.3, which continues the class template started in Figure 5.1. The default Vector constructor does as little as possible; it just initializes an empty vector.

When you declare a vector, however, you usually want to set up specific index bounds, and allocate its storage array in free memory. Section 2.6 noted that this process needs to be carried out in several different situations, so it should be implemented as a separate public member function SetUp(int Hi,int Lo). The Vector.H header file specifies the default parameter

```
template<class Scalar>                    // Default
  Vector<Scalar>::Vector() {              // constructor.
    Low = 0;  High = -1;                  // Set up an empty
    Array = NULL; }                       // vector.

      // Constructors and destructor     --------------------------

    Vector();                             // Default constructor:
                                          // set up an empty vector.
    Vector& SetUp(int Hi,                 // Set index bounds and
                  int Lo=1);              // allocate storage.
    Vector(int Hi,                        // Construct a vector
           int Lo = 1);                   // with specified index
                                          // bounds.
    ~Vector();                            // Destructor.

      // Copiers, assignment, and equality   ---------------------

    Vector(const Vector& Source);         // Copy constructor.

    Vector& operator=(const               //
      Vector<double>& Source);            // Assignment, possibly
                                          // with type conversion.
    Vector& operator=(const               // But only
      Vector<complex>& Source);           // double-to-complex
                                          // conversion is
    Vector& operator=(const               // implemented.
      Vector<int>& Source);               //

    Vector& Copy(int   Lo,                // Copy  Source[SourceLo]
           const Vector& Source,          // ..Source[SourceHi]  to
                  int   SourceLo,         // target starting at
                  int   SourceHigh);      // index  Lo .

    Boolean operator==(                   // Equality.
      const Vector& W) const;
```

Figure 5.3. Vector.H logical components
Part 2 of 3; see also Figure 5.1 and 5.9.

Lo = 1. Figure 5.4 contains the source code. SetUp will simply initialize an empty vector if Hi = Lo − 1. If Hi < Lo − 1, it displays an error message and throws an MSP Exception object that specifies an index error. Otherwise, SetUp asks operator new to allocate memory for

```
template<class Scalar>                      // Set index bounds
  Vector<Scalar>& Vector<Scalar>::          // and allocate
    SetUp(int Hi,                           // storage for a
          int Lo) {                         // vector with  n
    try { try {                             // entries.
      int n = Hi - Lo + 1;                  // Index error if
      if (n < 0) {                          // High < Low - 1 :
        cerr << "\nException  IndexError";  // report and throw
        throw(Exception(IndexError)); }     // an MSP exception.
      if (n == 0) {                         // If  High = Low - 1
        Low   = 0;  High = -1;              // set up an empty
        Array = NULL; }                     // vector. Other-
      else {                                // wise, allocate
        Low = Lo;    High = Hi;             // memory for the ar-
        Array = new Scalar[n]; }            // ray.  Return  This
      return This; }                        // for chaining.
    catch(xalloc) {                         // If memory is
      cerr << "\nException  OutOfMemory";   // exhausted,
      throw(Exception(OutOfMemory)); }}     // report and
    catch(...) {                            // throw an MSP
      cerr << "\nwhile allocating memory "  // exception.
        "for a Vector  with  High,Low = "   // After any ex-
           << High << ',' << Low;           // ception,set
      High = -1;  Low = 0;  Array = NULL;   // up an empty
      throw; }}                             // vector.
```

Figure 5.4. Vector function SetUp

Hi – Lo + 1 scalars. If new succeeds, it returns a pointer to the allocated space, which SetUp assigns to Array. If it fails, new throws an xalloc object. SetUp catches that; its inner catch block displays an error message and throws an Exception object, which specifies an out-of-memory error. All the SetUp code described so far is contained in an outer try block. The outer catch block handles any MSP Exception thrown in either error situation. It displays a message identifying the location of the error and then throws the same object to the next handler. If it succeeds, SetUp returns the value of its principal argument, to permit member function concatenation.

To construct an array of double vectors (of possibly varying dimensions), execute the default constructor n times by declaring Vector<double> V[n] and then call SetUp in a for loop n times to set the index bounds and allocate the memory. The vectors are then V[0], ..., V[n-1] and, for example, V[0][1] is the entry of V[0] with index 1.

Warning! The MSP exception handling technique may not maintain program control when memory is exhausted. Windows 95 resorts to my fixed disk for virtual memory after it uses all available RAM. A Borland C++ `EasyWin` test program runs out of virtual memory when trying to allocate memory for an array of about 6000 `double` vectors of length 7000, executing the code

```
const int n = 6000;
Vector<double> *V;
try {
  V = new Vector<double>[n];
  for (int i = 0; i < n; ++i) {        // Memory is exhausted
    V[i].SetUp(7000);                  // just before i = 6000 .
    cout << i << endl; }}
catch(Exception) {
  delete[] V; cout << "\nEnough said."; }}
```

That would require about $6000 \times 7000 \times 8$ bytes ≈ 320 MB. After Windows 95 churns the disk, the MSP exception handling system displays the `SetUp` error messages and `Enough said`, and the test program halts normally. At least, that's the situation as this book is written. But during development, the same test sometimes caused another program to crash, and the cause was never determined. Perhaps it was a consequence of disk I/O abuse, or perhaps even this limited `cout` output was impossible after exhausting memory.

Client programs use the constructor `Vector(int Hi, int Lo = 1)` most often. It merely calls `SetUp(Hi,Lo)` to set up the index bounds and allocate the memory. For example, you may invoke it with its last parameter defaulted by declaring `Vector<double> V(3)` to set up a vector with three entries `V[1]` to `V[3]`, as is customary in analytic geometry. Or you can include the last parameter explicitly—for example, declare `Vector<int> V(3,-1)` to build a vector with entries `V[-1], V[0], ..., V[3]`.

Destructor

When you construct a local `Vector` object `V` by declaring `Vector<int> V`, for example, the three components `V.Low`, `V.High`, and `V.Array` are stored on the stack. The last is the address of an array in free memory. When control passes out of the block where `V` was declared, these three components are destroyed. You must provide a way to release the array storage in free memory. Otherwise, it will constitute unusable storage, because it's allocated but no pointer points to it. The mechanism is the `Vector` class *destructor*. If you provide this function, the compiler will arrange to execute it at the proper time. Its source code is simple:

```
template <class Scalar>                    // Destructor.
  Vector<Scalar>::~Vector() {
    if (Array != NULL)
      delete[] Array; }
```

```
template<class Scalar>                        // Copy constructor.
  Vector<Scalar>::
    Vector(const Vector<Scalar>& Source) {
    try {
      int Lo = Source.Low;                    // Copy the index
      int Hi = Source.High;                   // bounds.
      SetUp(Hi,Lo);                           // Allocate storage
      int n = Hi - Lo + 1;                    // for  n  vector
      memcpy(Array,Source.Array,              // entries, and copy
        n * sizeof(Scalar)); }                // them.
    catch(...) {
      cerr << "\nwhile copying a   Vector ";
      throw; }}
```

Figure 5.5. Vector copy constructor

(It seems reasonable that applying the delete[] operator to a NULL pointer should have no effect at all. But that's not necessarily the case with Borland C++, so the if test is required. Without it, you might crash after applying the destructor to an empty vector.) The compiler will also execute this function to destroy temporary Vector objects it constructed for passing function parameters and return values and to complete the destruction when you execute delete V for a Vector object V.

Copiers

According to Section 2.6, each Vector class needs a copy constructor: a member function Vector(const Vector& Source) that constructs a new vector identical to Source. The compiler will invoke it automatically to make temporary copies for passing function parameters and return values, and you can use it explicitly in declarations such as Vector V(Source) or Vector V = Source. You'll find its source code in Figure 5.5. After calling SetUp to set the index bounds and allocate memory for the copy, the constructor invokes C++ Library function memcpy to copy the array of scalars. You'll find the memcpy prototype in Borland C++ header file Mem.H. It's implemented with a single machine-language instruction, so is very fast.

You'll need another copy function if you ever want to manipulate parts of vectors—for example, in assembling a matrix for Gauss elimination or disassembling one after LU factorization. Figure 5.6 explains the details for function

```
Vector& Copy(int  Lo,
    const Vector& Source,
         int  SourceLo,
         int  SourceHi).
```

```
template<class Scalar>                        // Copy
  Vector<Scalar>& Vector<Scalar>::            // Source[SourceLo]..
     Copy(int Lo,                             //    Source[SourceHi]
         const Vector<Scalar>& Source,        // to the target
         int SourceLo,                        // starting at index
         int SourceHi) {                      // Lo . First ensure
   Lo       = max(Lo,Low);                    //   reasonable
   SourceLo = max(SourceLo,Source.Low);       //   indices.
   SourceHi = min(SourceHi,int(Source.High));
   int n    = min(High-Lo ,SourceHi-SourceLo) + 1;
   memmove(&This[Lo],&Source[SourceLo],
     n * sizeof(Scalar));                     // n = how many
   return This; }                             // entries move.
```

Figure 5.6. Function to copy part of a Source vector

This routine, a member function of each Vector class, copies the Source entries with indices between SourceLo and SourceHi to the entries of target vector This, starting at index Lo. If Lo < This.Low, it starts at This.Low instead. The copy operation is performed by Mem.H function memmove, which is a little more elaborate (hence slower) than the function memcpy invoked by the copy constructor. Function memmove adjusts its operation if the source and target arrays overlap, which is possible in this case; memcpy takes no such precaution and could yield an incorrect result. This process can cause no exception, so no try/catch blocks are needed. The function returns This to permit member function concatenation.

Assignment

At first, you're tempted to declare the Vector assignment operator as a single member function within the Vector class template:

```
Vector& operator=(const Vector& Source).
```

That would declare

```
Vector<Scalar>& operator=(const Vector<Scalar>& Source)
```

for each supported instance of the Scalar type. But it wouldn't permit any type conversion; you couldn't do this:

```
Vector<complex> C;
Vector<double>  D;
   ⋮
C = D;
```

But you need that capability for almost any programming with complex scalars. MSP provides it by declaring *three* operator functions within the Vector class template:

```
Vector& operator=(const Vector<double>& Source);
Vector& operator=(const Vector<complex>& Source);
Vector& operator=(const Vector<int>&    Source);
```

This, of course, *declares* all possible combinations of scalar types for the source and target vectors. But that's perhaps misleading. Only four combinations are actually implemented: those where the source and target types agree and the one that assigns Vector<double> to Vector<complex>. (You probably wouldn't want to assign complex source vectors to double targets. I feel that integer types are normally used for counting and the others for measuring, so MSP shouldn't support assignment between int vectors and the other types. MSP clients can do that if they wish, at their own risk.)

The three Vector assignment operators without type conversion are implemented by the single template shown in Figure 5.7. In short, this code deletes the information stored in the target object This and replaces it by a copy of Source. That would be disastrous if these were the same object, so the assignment operator does nothing in that case. Should the copy process fail because memory is exhausted, you'd probably want to retain access to the original object This. So the assignment operator makes the copy *first*. If that fails, the copy constructor throws an exception, and the assignment operator catches this object, signs the error report, and rethrows. If the copy succeeds, the assignment operator deletes the target Array. (Unless it's already NULL: You'd think applying the delete[] operator to a NULL pointer would have no effect at all, but with Borland C++ that's not necessarily so.) The assignment operator copies the new index bounds and Array pointer from the Source copy. Then it assigns NULL to that copy's Array pointer and returns This to permit chaining. As the return is executed, control leaves the block where the Source copy was declared. That invokes the Vector destructor to delete the Source copy. Had its Array pointer not been made NULL it would coincide with This.Array, which you *don't* want to delete!

The assignment operator that converts a double vector Source to a complex target vector differs in several subtle ways from the one just described. Its code is shown in Figure 5.8. It first constructs a Vector object Copy with the same index bounds as Source and allocates the memory. Should this fail because memory is exhausted, constructor would throw an exception, and the assignment operator would catch it, sign the error report, and rethrow. If allocation

```
template<class Scalar>                              // Assignment:
  Vector<Scalar>& Vector<Scalar>::                  // copy source
      operator=(const Vector<Scalar>& Source) {     // vector.
    try {                                           // Avoid self-
      if (this == &Source) return This;             // assignment.
      Vector<Scalar> Copy = Source;                 // Make copy.
      Low  = Copy.Low;                              // This runs on-
      High = Copy.High;         // ly if there's been no exception.
      if (Array != NULL)        // Delete any storage allocated to
        delete[] Array;         // this . Copy the copy's bounds &
      Array = Copy.Array;       // Array  pointer.  Make the copy's
      Copy.Array = NULL;        // own pointer  NULL  so the  Array
      return This; }            // won't be destroyed with the copy.
    catch(...) {
      cerr << "\nwhile assigning from a  Vector  with  "
              "High,Low = " << Source.High << "," << Source.Low;
      throw; }}
```

Figure 5.7. Vector assignment operators, without type conversion

```
Vector<complex>& Vector<complex>::                  // Assignment:
    operator=(const Vector<double>& Source) {       // copy source
  try {                                             // vector,
    Vector<complex> Copy(Source.HighIndex(),        // converting
                        Source.LowIndex());         // entries to
    for (int k  = Copy.Low;                         // complex .
             k <= Copy.High; ++k)                   // Copy, convert
      Copy[k] = Source[k];                          // each entry.
    Low  = Copy.Low;          // This part runs only if there's
    High = Copy.High;         // been no exception.  Delete any
    if (Array != NULL)        // storage allocated to  this .
      delete[] Array;         // Copy the copy's index bounds and
    Array = Copy.Array;       // Array  pointer.  Make the copy's
    Copy.Array = NULL;        // own pointer  NULL  so the  Array
    return This; }            // won't be destroyed with the copy.
  catch(...) {
    cerr << "\nwhile assigning from a  double  Vector  with  "
            "High,Low = " << Source.HighIndex() << ","
         << Source.LowIndex() << "\nto a  complex  Vector";
    throw; }}
```

Figure 5.8. Vector assignment operator, with conversion from double to complex

succeeds, the `double` entries of `Source` are copied one by one to the target by the scalar assignment operator, which converts `double` entries to `complex` automatically. From here on, the code for this assignment operator agrees with that of the operator described in the previous paragraph.

Equality

A software package that enables you to copy data structures and thus *make* them equal should also enable you to ascertain whether two structures *are* in fact equal. Deciding what *equal* means is sometimes a problem, since measurements in nature—even of the same phenomenon—are rarely exactly equal. Vector algebra provides tools for dealing with that sort of problem. MSP features overloaded equality and inequality operators `==` and `!=` that test whether two vectors are *exactly* equal. The first is declared in `Vector.H`; here's its code, from `VectorL.CPP`:

```
template<class Scalar>                              // Equality
  Boolean Vector<Scalar>::                          // --not if
    operator==(const Vector<Scalar>& W ) const {    // bounds
  if (Low != W.Low || High != W.High )              // dis-
    return False;                                   //   agree.
  for (int i = Low; i <= High; ++i)                 // But if they
    if (This[i] != W[i]) return False;              // agree, check
  return True; }                                    // the entries.
```

The operator checks first whether the index bounds agree and if so compares entries. The corresponding inequality operator `V != W` is defined by a template in MSP header file `General.H`, described in Section 3.1; it simply returns `!(V == W)`.

Keyboard Input and Display Output

An important first step in numerical software development is to construct convenient and reliable keyboard input and display output routines for scalars, vectors, matrices, and so forth. If you try to make do with poorly conceived input/output, you'll find yourself mistaking bugs in that design for bugs in what really counts—the computational software.

The `cin` input stream provides appropriate keyboard input for scalars. The `Scalar` module, described in Section 3.9, includes function `Show` for standard display output. The latter is particularly flexible: One of its arguments specifies the output format, and a default is provided. The default format provides the minimum precision necessary for debugging and for tabulating approximation errors.

For vector input/output there are a few more considerations. Entering several scalar entries in succession on a keyboard is awkward: A prompt should be displayed before each one. Moreover, an output routine should be able to display a vector as a row or as a column. In the former case, the scalar format length is important, to control wrapping at the end of a screen row. In the latter, the routine should provide optional labels for the output vector entries, since the client program can't intervene to do so.

The simple prompted keyboard input routines `Vector& KeyIn(char* Name)` are member functions of the `Vector` classes. You'll find this prototype in header file `Vector.H`, the last part of which is displayed in Figure 5.9. The source code, from `VectorL.CPP`, is shown in Figure 5.10. For example, if `V` is a vector with entries `V[0]`, ..., `V[n]`, then executing `V.KeyIn("V")` will display `V[0] :`, with two spaces after the colon `:`, and await input of that vector entry. Users can backspace if necessary to edit the input and then accept it by pressing the <Enter> key. The process repeats, to input each entry. Borland C++ provides this rudimentary input editing facility via its input stream `cin` and its stream extraction operator `>>`. If any input operation fails, `cin` will convert to the `Boolean` value `False`, and `KeyIn` will classify the failure, display an error

```
        // Input/output  -----------------------------------

        Vector& KeyIn(char* Name);            // Prompted keyboard
                                              // input.
        const Vector& Show(                   // Standard output to
          char*   Name       = EmptyStr,      // display.
          Boolean RowFormat  = True,
          char*   Format     = NULL) const;

        friend istream& operator>>(           // Stream input.
                istream& Source,
          Vector<Scalar>& Target);
        :                                     // End of the definition
    };                                        // of class Vector .

// Unfriendly non-member logical function  ---------------------

template<class Stream, class Scalar>          // Stream
  Stream& operator<<(Stream& Target,          // output.
            const Vector<Scalar>& Source);
```

Figure 5.9. `Vector.H` logical components
Part 3 of 3; continued from Figure 5.1 and 5.3.

```
template<class Scalar>                  // Prompted keyboard
  Vector<Scalar>& Vector<Scalar>::      // input. A prompt
      KeyIn(char* Name) {               // consists of the
    for (int i = Low; i <= High; ++i) { // specified vector
      cout << Name << '[' << i          // name and corre-
        << "] : ";                      // sponding entry
      cin >> This[i];                   // index.
      if (!cin) break; }
    if (cin) return This;
    cerr << "\nException ";             // If unsuccessful,
    if (cin.eof()) {                    // report & throw an
      cin.seekg(ios::end);              // MSP exception. If
      cin.clear();                      // the user entered
      cerr << "EndOfFile";              // <Ctrl-Z> , flush
      throw(Exception(EndOfFile)); }    // the cin buffer &
    cerr << "IOError";                  // clear its end-of-
    throw(Exception(IOError)); }        // file condition.
```

Figure 5.10. Vector function KeyIn

message, and throw an MSP Exception object, which specifies the type of error. If the user pressed <Ctrl-Z> to terminate input, then Library function eof returns True, and KeyIn invokes Library functions seekg and clear, before it throws the exception, to flush the cin input buffer and reset its end-of-file alarm. These three routines are member functions of the Borland C++ IOStream.H class ios to which cin belongs. If input succeeds, KeyIn returns This to permit member function concatenation.

The standard MSP Vector display output function is

```
const Vector& Show(                    // Standard output
    char*   Name       = EmptyStr,     // to display.
    Boolean RowFormat  = True,
    char*   Format     = NULL) const;
```

A member function of each Vector class, it calls Scalar function Show to display each entry of its principal argument vector, using the printf string argument Format. But if you use the NULL default for that, Show will execute MSP Scalar.H function DefaultFormat, described in Section 3.9, to get a format string to produce the shortest output appropriate for debugging. With Boolean parameter RowFormat you can select row or column format; the former is the default. With row format, the specified Name is displayed at the left. Unless it's the default empty string EmptyStr (an MSP macro defined in General.H), it will be followed automatically by the four-character string " : ". Row entries are separated by two blanks. With column format, every output line is prefixed with Name. Unless it's empty, it will be followed by the entry index enclosed by brackets and the string " : ". Here are sample row and column outputs in the default scalar format, with the statements that produced them:

	Row format	Column format
		Name &
Name	---------.............	index ------------
Solution :	1.2e+03 −4.5e−67	D[8] : −1.2
		D[9] : 0.3
		D[10] : −4.0
Show("Solution",True);		Show("D",False);

As usual for functions of this sort, the source code for Vector function Show is uninstructively complicated. In fact, it took longer to write than any other function in this book. It's displayed in Appendix A.1 for your convenience.

Stream Input/Output

The last two logical functions of the Vector module are stream input/output operators intended for use with files. Their implementation requires a file format for a Vector object V. MSP uses a very simple record, reflecting the data structure definition: the int values V.Low and V.High, followed by the entries of V.Array, in order. For the individual items, MSP outputs ASCII numerals. It uses the minimum number of characters for the integers; for the scalars, it uses scientific notation with maximum—15 significant digits—precision. All items are followed by newline characters.

These operators, like all C++ Version 2 stream input/output routines, overload the >> and << operators. Their declarations are shown in Figure 5.9. The input operator >> must be able to change the index bounds of the Target vector, so it must be a member function or a friend of the Vector class. But it can't be a member function, because its left-hand argument must be an Istream, not a Vector, so it must be a friend. The output operator << has no need for friendship.

Source code for the stream input operator >> is shown in Figure 5.11. It shares features with that of the Vector assignment operators and KeyIn function shown in Figures 5.7, 5.8, and 5.10. In short, it deletes the information stored in Target and replaces that with Source stream data. Should this fail due to an input error or because memory is exhausted, you'd probably want to retain access to the original Target object. So the input operator makes the copy *first*, into a

```
template<class Scalar>                        // Input the target
  istream& operator>>(istream& Source,        // vector from the
                Vector<Scalar>& Target) {     // source stream.
    try {
      Vector<Scalar> Input;                   // Read index bounds
      int Lo,Hi;                              // into a temporary
      Source >> Lo >> Hi;                     // Input vector.
      if (Source) {                           // If successful,
        Input.SetUp(Hi,Lo);                   // allocate memory
        for (int i = Lo; i <= Hi; ++i) {      // for its entries,
          Source >> Input[i];                 // and read them.
          if (!Source) break; }}
      if (!Source) {                          // If unsuccessful,
        cerr << "\nException  ";              // report and throw
        if (Source.eof()) {                   // an MSP exception.
          cerr << "EndOfFile";
          throw(Exception(EndOfFile)); }
        cerr << "IOError";
        throw(Exception(IOError)); }
      Target.Low  = Lo;                       // This part runs only if
      Target.High = Hi;                       // there was no exception.
      if (Target.Array != NULL)               // Replace Target Vector
        delete[] Target.Array;                // by Input . Detach the
      Target.Array = Input.Array;             // Array pointer so it
      Input.Array  = NULL;                    // won't be destroyed with
      return Source; }                        // Input . Return Source
    catch(...) {                              // for chaining.
      cerr << "\nwhile extracting a"
           "  Vector  from an input stream";
      throw; }}
```

Figure 5.11. Vector stream input operator >>

temporary Vector called Input. If that fails, Source will convert to False, or Vector function SetUp will throw an exception. The stream input operator will detect either situation, display an error message, and throw or rethrow the exception to a higher-level handler, specifying the type of error. If the copy succeeds, the input operator deletes the target Array. (Unless it's already NULL: You'd think applying the delete[] operator to a NULL pointer would have no effect at all, but with Borland C++ that's not necessarily so.) The input operator copies the new index bounds and Array pointer from the temporary Input vector. Then it assigns NULL to that object's Array pointer and returns Source to permit chaining. As the return is executed, control leaves the

```
template<class Stream, class Scalar>         // Output the source
  Stream& operator<<(Stream& Target,         // vector to the
      const Vector<Scalar>& Source) {        // target stream.
    int OldFormat = Target.setf(             // Save the old
      ios::scientific,ios::floatfield);      // format settings.
    int OldPrecision = Target.precision();   // Select 15
    Target.precision(14);                    // digit scientific
    int Low  = Source.LowIndex(),            // notation.
        High = Source.HighIndex();
    (ostream&)Target << Low  << endl;           // Write index
    (ostream&)Target << High << endl;           // bounds, then
    for (int i = Low; i <= High; ++i)           // the entries,
      (ostream&)Target << Source[i] << endl;    // followed by
    Target.precision(OldPrecision);             // newlines.
    Target.setf(OldFormat,ios::floatfield);     // Restore old
    if (!Target) {                              // settings.
      cerr << "\nException IOError"
              "\nwhile inserting into an output stream a  Vector"
           "  with Low,High = " << Low << ',' << High;
      throw(Exception(IOError)); }
    return Target; }
```

Figure 5.12. Vector stream output operator <<

block where Input was declared. That invokes the Vector destructor to delete it. Had its Array pointer not been made NULL it would coincide with Target.Array, which you *don't* want to delete!

The Vector stream output operator is simpler; its source code is shown in Figure 5.12. First, it invokes IOStream.H Library functions setf and precision to set the output format to "scientific 14". These two routines are member functions of the Borland C++ IOStream.H class ios to which Source belongs. Each returns the previously established setting; the stream output operator stores those values in temporary variables and then invokes precision and setf again when output is finished, to restore the previous settings. Once the format is established, the operator outputs the Source vector's index bounds and entries. If that process encounters an error situation, the Target stream will convert to False; the stream output operator will detect that, display an error message, and throw an MSP exception, specifying the type of error.

Numerical analysis software that seriously supports file input/output needs more consideration of potential error situations than this book includes. That's really a systems programming task, and you should consider sources like Lippman [29, Appendix A] and Stroustrup [52, Chapter 10] which gives it more attention.

Coda

Section 3.9 described the coda portion of source code file Scalar.CPP, which forces the compiler to generate all supported instances of template functions declared in header file Scalar.H but defined in Scalar.CPP. The coda defined

```
template<class Scalar>
  void UseScalar(Scalar t)
     { ... invoke each  Scalar.H  template function ... ; }

void UseScalar()
  { ... invoke  UseScalar(t)  with an argument  t  of  each supported
      Scalar  type ... ; }
```

(These functions are not meant for execution.) The other source code files discussed earlier have similar codas. VectorL.CPP has a coda too, shown in Figure 5.13. But it differs slightly from the earlier ones. Many Vector.H template functions defined in the source code file are Vector class member or friend functions. The declaration Vector<Scalar> V in the UseVectorL(Scalar t) template forces the compiler to generate the corresponding instances of member and friend functions. Therefore, that routine only needs to invoke the *unfriendly non-member* Vector.H template functions defined in VectorL.CPP. Similar considerations hold for the codas of all source code files described later in this book.

```
//*******************************************************************
//                 Generate all supported unfriendly non-member
//                 VectorL.CPP  template function instances.

template<class Scalar>                  // Invoke those functions.
  void UseVectorL(Scalar t) {           // Chained output is need-
    Vector<Scalar> V;                   // ed because the tempo-
    AllEntries(V,t);                    //   rary stream
    AllEntries(V,1.);                   //   variables'
    ofstream f;            ifstream g;  //   types differ
    f    << V << EmptyStr << V;  g  >> V;   //   from those of
    cout << V << EmptyStr << V;  cin >> V; } //  cout & f .

void UseVectorL() {                     // Invoke
  double  x = 0;   UseVectorL(x);       // UseVectorL
  complex z = 0;   UseVectorL(z);       // for each
  int     k = 0;   UseVectorL(k); }     // Scalar  type.
```

Figure 5.13. VectorL.CPP coda

5.2 VECTOR ALGEBRA FUNCTIONS

Concepts

Zero and unit vectors
Random vectors
Max, city, and Euclidean (l_∞, l_1, and l_2) vector norms
Plus and minus
Conjugation
Addition and subtraction
Scalar multiplication
Scalar and inner products
Cross and triple products
Replacement operators

This section details the MSP functions that implement vector algebra. They constitute the mathematical part of the `Vector` module. Its logical functions were discussed in the previous section. Individually, the vector algebra functions aren't very complicated or interesting. But their overall organization is important, because so much of this book's numerical analysis software, especially the matrix algebra functions, is based on them. The vector algebra functions are listed in Figures 5.14 and 5.15, which display part of the MSP `Vector.H` header file. Their source code constitutes the MSP file `VectorM.CPP` on the optional diskette, except for the first four functions described in this section; those are so useful for testing that they're included in file `VectorL.CPP` with the functions discussed in the previous section. The source code for some of the vector algebra functions will be discussed in detail here. For others, which are generally very similar, you may consult the diskette.

This book uses a standard notation for vectors. When no confusion with matrix notation occurs, they'll be denoted by *UPPERCASE* Latin italic letters, and their entries by the corresponding *lowercase* Latin italic letters, usually with subscripts. Sometimes this is emphasized by writing an equation such as $V = [v_k]$. `Monospaced boldface` is used instead of italic in C++ expressions and in reproducing screen displays. Unless explicitly stated otherwise, the lower index bound is assumed to be 1.

In a matrix algebra context, vectors are generally regarded as column matrices and denoted instead by lowercase Greek italic letters. (See Section 7.2.)

```
                                         ⋮
template<class Scalar>
  class Vector {
    private:
      ⋮                    For omitted parts see Section 5.1.
    public:
      ⋮
      // Member algebra functions  ------------------------------

      Vector& MakeRandom();            // Make all entries
      Vector& MakeZero();              // random, zero.
      Vector& MakeUnit(int k);         // Set  =  the  kth
                                       // unit vector.

      Vector operator-() const;        // Negative.
      Vector operator+(                // Sum.
        const Vector& V) const;
      Vector operator-(                // Difference.
        const Vector& V) const;
      Vector operator*(                // Right scalar
        const Scalar& t) const;        // multiple.
      Scalar operator*(                // Scalar product.
        const Vector& V) const;
      double operator|(                // Inner product:
        const Vector& V) const;        // V * conj(W) .
      Vector operator%(                // 3D cross product.
        const Vector& V) const;
                                       // End of the definition
    };                                 // of class  Vector .
```

Figure 5.14. Vector.H mathematical components
Part 1 of 2; continued in Figure 5.15.

Constructing Special Vectors

The Vector module includes functions

```
    template<class Lscalar,              MakeZero()
            class RScalar>
    Vector<LScalar>& AllEntries(         MakeUnit(int I)
      Vector<LScalar>& V,
        const RScalar&  t)               MakeRandom()
```

for constructing special vectors. Executing AllEntries(V,t) sets V[k] = t for each appropriate index k. It's implemented with two template parameters to permit type conversions such as

```
Vector<complex> V(7);   AllEntries(V,11);
```

Without this feature, you'd need to write AllEntries(V,complex(11)). The other three are member functions of the Vector classes. MakeZero makes its principal argument a zero vector by executing AllEntries(This,0). If k lies within bounds, MakeUnit(k) initializes its principal argument This as the kth unit vector; otherwise, it leaves This unchanged, but throws an MSP Exception object, which specifies the error type. MakeRandom initializes its principal argument to a pseudorandom vector by executing MakeRandom(This[k]) for each appropriate index k. The MSP Scalar module defines function MakeRandom for each supported scalar type—see Section 3.9.

These four functions could have been designed as constructors. However, that would involve overloading and complicated syntax conventions to specify or default the various parameters. It seemed more natural to separate the construction and initialization operations and to implement

```
// Unfriendly non-member vector algebra functions   -------------

template<class LScalar, class RScalar>            // Set all   V
  Vector<LScalar>& AllEntries(                    // entries
    Vector<LScalar>& V,                           // = t .
    const RScalar& t);
template<class Scalar>                            // L-infinity
  double MaxNorm (const Vector<Scalar>& V);       // norm.
template<class Scalar>                            // L2  norm.
  double Euclid  (const Vector<Scalar>& V);
template<class Scalar>                            // L1  norm.
  double CityNorm(const Vector<Scalar>& V);
template<class Scalar>                            // Conjugate.
  Vector<Scalar> conj(const Vector<Scalar>& V);
template<class Scalar>              // Left scalar multiple. Were
  Vector<Scalar> operator*(         // it a friend, this would be
    const         Scalar& t,        // ambiguous with the analogous
    const Vector<Scalar>& V);       // Polynom  operator.
template<class Scalar>                            // 3D  triple
  Scalar Triple(const Vector<Scalar>& U,          // product:
                const Vector<Scalar>& V,          // U * (V % W) .
                const Vector<Scalar>& W);
```

Figure 5.15. Vector.H mathematical components
Part 2 of 2; continued from 5.14.

the latter as member functions. They can be invoked quite naturally—for example, this code constructs unit vector U = [0,0,1,0,0]:

```
Vector U(5);  U.MakeUnit(3);
```

The three member functions all return their principal argument, so they permit concatenation. In the previous example, for instance, you could display U by substituting U.MakeUnit(3).Show() for the second statement.

Vector Norms

One fundamental vector algebra operation is computing the magnitude of a vector V. There are several definitions of this concept. When it's used in a mathematical book, it's denoted by $\|V\|$, called the *norm* of V; somewhere in the book you'll find which type of norm, or magnitude, is intended. Here are the three norms of the vector $V = [v_1, \ldots, v_n]$ most common in numerical analysis:

$$max\ (l_\infty)\ \text{norm} = \|V\| = \max_{i=1}^{n} |v_i|$$

$$Euclidean\ (l_2)\ \text{norm} = \|V\| = \sqrt{\sum_{i=1}^{n} v_i^2}$$

$$city\ (l_1)\ \text{norm} = \|V\| = \sum_{i=1}^{n} |v_i|$$

The *max* norm of vector V is simply the largest of the absolute values $|v_k|$ of its scalar entries. That's certainly the easiest vector norm to compute, and it is the one most frequently used in numerical analysis. The *Euclidean* norm is defined by the formula commonly used in calculus and analytic geometry for the distance to a point V from the origin. Because the square root takes relatively long to compute, this norm is generally used only when it's necessary to maintain a theoretical connection with the standard geometric distance concept. The least used of the three is the *city* norm, the sum of the absolute values of the scalar entries. (Its name was chosen because in a city with numbered east-west streets and north-south avenues the city norm of vector V = [7, 11] is the number of blocks you walk from the origin to the intersection of 7th Street and 11th Avenue.) It's listed here mainly because it's closely related to a matrix norm discussed later in Section 7.2. The max, Euclidean, and city norms also occur in higher mathematics in the study of normed linear spaces, where they're called the $l_\infty, l_2,$ and l_1 norms. See [5, Section 7.1] for a discussion of vector norms in general.

These vector norms all satisfy three rules that underlie computations and theoretical considerations in this book:

1. $\|V\| = 0 \Leftrightarrow V = 0$
2. $\|tV\| = |t|\|V\|$
3. $\|V + W\| \leq \|V\| + \|W\|$ (the *triangle* inequality)

The name of inequality (3) stems from the fact that if V and W are points in a Cartesian coordinate system, and Δ is the triangle with vertices O, V, and $-W$, then the Euclidean norm $\|V + W\|$ is the length of the side of Δ opposite O.

The max, Euclidean, and city norms are implemented in MSP via functions `MaxNorm`, `Euclid`, and `CityNorm`. You'll find their declarations in Figure 5.15. They're not member functions, because principal argument notation is inappropriate. Here's the source code for `MaxNorm` :

```
template<class Scalar>                            // L-infinity
  double MaxNorm(const Vector<Scalar>& V) {       // norm.
    int Low  = V.LowIndex();
    int High = V.HighIndex();                     // Treat an empty
    if (Low > High) return 0.;                    // vector like  0 .
    double N = abs(V[Low]);                       // Compute the
    for (int k = Low+1; k <= High; ++k)           // largest of the
      N = max(N,double(abs(V[k])));               // absolute values
    return N; }                                   // of the entries.
```

`max` is defined by a function template in Borland C++ Library header file `StdLib.H`, which is included by MSP header file `Vector.H`. Its two arguments must have the same type. For `double`, `complex`, and `int` scalars, respectively, `abs` is defined in MSP header file `General.H` and in Borland C++ header files `Complex.H` and `StdLib.H`.

MSP function `Euclid(V)` merely returns the square root of the inner product of V with itself; the product is discussed later in this section. For function `CityNorm` you may consult the optional diskette.

Plus, Minus, and Conjugate

MSP overloads singulary C++ operators + and − to implement the corresponding vector operations. The former is implemented (it does nothing) by a template in `General.H`. A member function of the `Vector` class, the negation operator − is shown in Figure 5.16. It constructs a new vector V with the same index range as its principal argument `This`, computes the entries, and then returns V. It's declared `const`, since it doesn't change `This`. The operator must return V by value, since that will disappear immediately after the `return`—before a caller could use a returned address.

```
template<class Scalar>                       // Negation.
  Vector<Scalar> Vector<Scalar>::
      operator-() const {
    try {
      Vector V(High,Low);                    // Construct the
      for (int k = Low; k <= High; ++k)      // negative, en-
        V[k] = -This[k];                     // try by entry;
      return V; }                            // return it.
    catch(...) {
      cerr << "\nwhile computing the negative of a Vector";
      throw; }}
```

Figure 5.16. Vector negation operator

MSP uses a similar technique to implement the complex conjugation operation on vectors: It overloads Complex.H Library function conj. The source code is very similar to that of the negation operator: Instead of computing the negative of each principal argument entry, it computes the conjugate.

Addition and Subtraction

Implementing vector addition requires a design decision: What should you do if the addends' index limits disagree? In some cases you might regard that as an error. But when you use vectors to represent polynomials, you'll *often* want to add vectors with different upper limits: They just stand for polynomials of different degree. In other cases—for example, initial segments of Laurent series—you might want to allow different lower limits, too. It seems reasonable, therefore, not to signal an error when index limits differ, but to pad the vectors with zeros to make their limits match. For example, consider the sum $U = [u_k]$ of these two vectors: $V = [v_k]$ and $W = [w_k]$, whose lower and upper index limits both differ:

k:	-2	-1	0	1	2
v_k:	6	8	1	1	
w_k:			2	7	4	5
$u_k = v_k + w_k$:	6	10	8	5	5

MSP implements this operation as a member function of the Vector classes by overloading the C++ binary operator +. The source code is shown in Figure 5.17. It just constructs a new vector Sum with the appropriate index range, computes its entries, and returns it. For efficiency, the

```
template<class Scalar>                          // Vector
  Vector<Scalar> Vector<Scalar>::               // addition.
      operator+(const Vector& V) const {
  int N = V.High + 1;                           // Regard each
  int M = High + 1;                             // addend as
  int L = min(M,N);                             // padded with
  int K = V.Low;                                // zeros, if
  int J = min(K,M);                             // necessary, so
  int H = Low;                                  // they begin
  int G = min(H,N);                             // and end with
  int i = min(H,K);                             // the same
  try {                                         // indices.
    Vector Sum(max(M,N)-1,i);
    for (; i < G; ++i)  Sum[i] = V[i];          // Compute
    for (; i < H; ++i)  Sum[i] = 0;             // the sum,
    for (; i < J; ++i)  Sum[i] = This[i];       // entry by
    for (; i < K; ++i)  Sum[i] = 0;             // entry.
    for (; i < L; ++i)  Sum[i] = This[i] + V[i];
    for (; i < M; ++i)  Sum[i] = This[i];
    for (; i < N; ++i)  Sum[i] = V[i];
    return Sum; }
  catch(...) {
    cerr << "\nwhile adding  Vectors  with  Low,High = "
        << H << ',' << M << "  and  " << K << ',' << N;
    throw; }}
```

Figure 5.17. Vector addition operator

right-hand addend is passed by reference. Sum must be returned by value, however, because it disappears after the return—before a caller could use a returned address.

Vector subtraction could be implemented in the same way as vector addition. However, it seems best to deal with the index range problem only once. Therefore, the MSP vector subtraction operator

```
template<class Scalar>                          // Vector
  Vector<Scalar> Vector<Scalar>::               // subtraction.
      operator-(const Vector& V) const {
```

merely computes and returns This + (-V).

Scalar Multiplication

There's no complication at all in MSP's implementation of its right scalar multiplication operator as a member function of the Vector classes:

```
template<class Scalar>                  // Right scalar
  Vector<Scalar> Vector<Scalar>::        // multiplication.
    operator* (const Scalar& t) const;
```

It's structured the same as the negation operator already discussed: It just constructs a vector P with the same index range as its principal argument This, computes the entries P[k] = This[k] * t, and returns P. The *left* scalar multiplication operator

```
template<class Scalar>                          // Left scalar
  Vector<Scalar> operator*(const Scalar&  t,    // multiple.
                  const Vector<Scalar>& V) {
    return V * t; }
```

can't be a member function, because its left-hand argument must be a scalar, not a vector.

Scalar and Inner Products

There are two common methods for multiplying vectors $V = [v_1, ..., v_n]$ and $W = [w_1, ..., w_n]$ to get a product p that's a scalar. The simpler one yields the *scalar* product, which has the same type as the vector entries:

$$p = V \star W = \sum_{i=1}^{n} v_i w_i .$$

MSP implements this operator as a member function of the Vector classes by overloading the * operator:

```
template<class Scalar>                          // Scalar
  Scalar Vector<Scalar>::                        // product.
    operator*(const Vector<Scalar>& V) const;
```

If the index bounds agree, it computes and returns the product. If not, it displays an error message and throws an MSP `Exception` object. The source code is straightforward; you'll find it on the optional diskette.

The scalar product satisfies several *linearity* rules often used in numerical analysis computations: for all vectors V, W, X and scalars t:

1. $V * W = W * V$
2. $V * (W + X) = V * W + V * X$
3. $(tV) * W = t(W * V)$
4. $0 * V = 0$

Many other similar rules can be derived from these.

The second kind of product of two vectors is the *inner* product $V|W = V * \overline{W}$. It's used in place of the scalar product in many aspects of complex linear algebra, especially when there's a theoretical connection with the concept of distance. If W is a real vector, conjugation has no effect—$\overline{W} = W$—so the inner product is the same as the scalar product. Inner products satisfy *semilinearity* rules somewhat like the linearity rules for scalar products: for all vectors V, W, X and scalars t,

1. $V|W = \overline{W|V}$
2. $V|(W + X) = (V|W) + (V|X)$
3. $(tV)|W = t(V|W)$
4. $V|(Wt) = (V|W)\bar{t}$
5. $0|V = 0$
6. $V|V \geq 0$
7. $V|V = 0 \Leftrightarrow V = 0$

MSP implements the inner product operator as a member function of the `Vector` classes by overloading C++ operator `|`:

```
template<class Scalar>                       // Inner product.
  double Vector<Scalar>::
    operator|(const Vector<Scalar>& V) const {
  try {
    return real(This * conj(V)); }
  catch(...) {
    cerr << "\nwhile computing an inner product";
    throw; }}
```

The double and int instances of this template coincide with the scalar product, because in those cases conj and real—defined in the MSP Scalar module—have no effect.

The Euclidean norm, described earlier, is closely related to the inner product. In fact, executing Euclid(V) simply computes sqrt(V|V).

Cross and Triple Products

Two further kinds of products are commonly used in three-dimensional vector algebra: the *cross product* U % V and *triple product* U * (V % W) of vectors U, V, and W. If the vector indices range from 1 to 3, then the cross product is the vector $P = [p_1, p_2, p_3]$ defined by equations

$$p_1 = v_2 w_3 - v_3 w_2$$
$$p_2 = v_3 w_1 - v_1 w_3$$
$$p_3 = v_1 w_2 - v_2 w_1.$$

You'll find various notations for these products in the literature—none is standard. The use of % here is bizarre, but it seemed the only feasible choice for MSP. In vector calculus texts you can find a number of computational rules governing these products—for example, the triple product can also be expressed as the determinant of the matrix consisting of the three columns U, V, and W [59, Chapter 13].

MSP implements the cross product as a member function of the Vector classes by overloading C++ operator %:

```
template<class Scalar>                        // 3D cross
  Vector<Scalar> Vector<Scalar>::             // product.
    operator%(const Vector<Scalar>& V) const;
```

This function returns the vector described in the previous paragraph if the indices range from 1 to 3 or an analogous one if they range from 0 to 2. In other cases, it displays an error message and throws an MSP Exception object specifying the error type. The triple product function

```
template<class Scalar>                        // 3D triple
  Scalar Triple(const Vector<Scalar>& U,      // product.
                const Vector<Scalar>& V,
                const Vector<Scalar>& W);
```

simply returns U * (V % W). It can't be a member function, because the principal argument notation is inappropriate. You'll find the source code for these functions on the optional diskette.

Replacement Operators

Via templates in header file `General.H` MSP implements four vector algebra replacement operators: `+=`, `-=`, `%=`, and `*=`. If `V` and `W` are `Vector` objects of the same type, then `V += W` adds `W` to `V`, `V -= W` subtracts `W` from `V`, and `V %= W` cross-multiplies `V` by `W`. If `t` is a scalar, then `V *= t` multiplies `V` by `t`.

5.3 Vector MODULE DEMONSTRATION

> **Concepts**
> *Contours and surfaces*
> *Example: the wedding band*
> *Vector fields*
> *Three-dimensional contour integrals*
> *Surface integrals*
> *Curl of a vector field*
> *Stokes' theorem*
> *Computational verification of Stokes' theorem*

This section presents a demonstration program, which exercises many features of the `Vector` module. It computes related three-dimensional contour and surface integrals and compares them. The computed approximations are extremely close, as predicted by Stokes' theorem in three-dimensional vector analysis.

Contours and Surfaces

You can regard the points of a *curve* in two- or three-dimensional space as the successive locations $X(t)$ of a moving object, where $X(t)$ is a function that maps real numbers to points—that is, two- or three-dimensional vectors. This *parameterization function* is defined for t between initial and final arguments $t = t_0$ and $t = t_*$. Either $t_0 \leq t_*$ or $t_* \leq t_0$ is permissible. The *reverse* of a curve has the same parameterization function, but the initial and final arguments are switched. A *contour* is an indexed family of n curves $C_0, \ldots, C_{n-1}$ of the same dimension, called its *legs*. It's usually denoted by

$$\sum_{k=0}^{n-1} C_k$$

You can regard a curve as a one-legged contour.

Consider, for example, the plane rectangular region R consisting of points [x,y] for which $0 \leq x \leq 2\pi$ and $-1 \leq y \leq 1$. Its boundary ∂R is a closed plane contour with four legs—line segments

$$\left. \begin{array}{l} x = t \\ y = -1 \end{array} \right\} \text{ from } t = 0 \text{ to } 2\pi \qquad \left. \begin{array}{l} x = t \\ y = -1 \end{array} \right\} \text{ from } t = 2\pi \text{ to } 0$$

$$\left. \begin{array}{l} x = 2\pi \\ y = u \end{array} \right\} \text{ from } u = -1 \text{ to } +1 \qquad \left. \begin{array}{l} x = 2\pi \\ y = u \end{array} \right\} \text{ from } u = +1 \text{ to } -1$$

Following the four legs in succession and staying just within R, you'll keep ∂R on your *right*.

A three-dimensional *surface* S is often represented by a *parameterization function* $W(t,u)$, which maps points $[t,u]$ from a plane region R to points (vectors) in S. The region R is usually the interior of a simple closed contour in two-dimensional space. (A *simple closed* contour in the plane is one that intersects itself just once, at the endpoints.) The partial derivative functions $\partial W/\partial t$ and $\partial W/\partial u$ must be defined and continuous in R; for some analyses, one or both must be nonzero.

For example, consider the function $W(t,u)$ given by these equations and the corresponding C++ code:

$$W(t,u) = \begin{bmatrix} x \\ y \\ z \end{bmatrix} \quad \begin{array}{l} x = \cos t \\ y = \sin t \\ z = u \end{array}$$

```
Vector<double> W(double t,
                 double u) {
  Vector<double> V(3);
  V[1] = cos(t);
  V[2] = sin(t);
  V[3] = u;
  return V; }
```

It's defined and appropriately smooth on R and parameterizes the three-dimensional *wedding band* surface S consisting of the points W, such that $x^2 + y^2 = 1$ and $|z| \leq 1$. The boundary ∂S of the band is a contour with two legs:

$$x(t) = W(t,-1) \quad \text{from } t = 0 \text{ to } 2\pi$$
$$x(t) = W(t,+1) \quad \text{from } t = 2\pi \text{ to } 0 .$$

Following the legs in succession and staying just within S on the side facing the origin, you'll keep ∂W on your *right*. The boundary ∂S is contained in, but does not coincide with, the image of ∂R under the parameterization $W(t,u)$.

A *vector field* is a function F(X) that maps three-dimensional vectors X into corresponding vectors. For example, consider these formulas and corresponding C++ code:

$$X = \begin{bmatrix} x \\ y \\ z \end{bmatrix}$$

$$F(X) = \begin{bmatrix} f \\ g \\ h \end{bmatrix} = \begin{bmatrix} z\tan(x+y) \\ x + y\cos z \\ xe^{y+z} \end{bmatrix}$$

```
Vector<double> F(
    const Vector<double>& X) {
    double x = X[1];
    double y = X[2];
    double z = X[3];
    Vector<double> V(3);
    double f = tan(x+y)* z;  V[1] = f;
    double g = x+y*cos(z);   V[2] = g;
    double h = x*exp(y+z);   V[3] = h;
    return V; }
```

(Redundant double variables are used to enhance readability.) You can integrate F over a three-dimensional contour or a surface. The resulting values are called *contour* (line) integrals and *surface* integrals. Those concepts are explained in the following text. Contour integrals, the simpler, are considered first.

Three Dimensional Contour Integrals

The integral of a vector field F(X) over a three-dimensional curve C parameterized by a function X(t) from $t = t_0$ to $t = t_*$ can be defined using Riemann sums, as follows. Under rather general conditions on F(X) and X(t), you can demonstrate existence of a certain limit of sums like

$$\sum_{k=0}^{n-1} F(X(\tau_k)) * (X(t_{k+1}) - X(t_k))$$

The scalar product discussed in Section 5.2 is used here. You choose points t_k and τ_k for indices k as required and define t_n so that if $t_0 < t_*$ then $t_0 < t_1 < \ldots < t_{n-1} < t_n = t_*$ and $t_k \leq \tau_k \leq t_{k+1}$ for each k; if $t_0 > t_*$, reverse all these inequalities. If you require that X(t) be continuously differentiable and $X'(t) \neq 0$ except possibly at $t = t_0$ and t_*, then you can show that the limit depends only on F, on the set of all points X(t), and on the selection of t_0 or t_* as starting point—not on X itself. If you have two parameterizations X(t) satisfying these smoothness requirements, with the same range and starting point, then the two corresponding limits will coincide. Thus, it's reasonable to call such a limit the *integral* $\int_C F$. For a contour concatenated from simpler curves $C_0, \ldots, C_{n-1}$ define

$$\int_{\sum_{k=0}^{n-1} C_k} F = \sum_{k=0}^{n-1} \int_{C_k} F$$

The next paragraph shows how to convert this to a sum of definite integrals, approximated by Riemann sums in the usual sense.

According to the mean-value theorem, each component $x_j(t_{k+1}) - x_j(t_k)$ of a factor $X(t_{k+1}) - X(t_k)$ in the Riemann sum displayed in the previous paragraph can be replaced by a term $x_j'(u_{jk})(t_{k+1} - t_k)$ —where u_{jk} lies between t_{k+1} and t_k. Using delicate analysis, you can show further that the limit of the Riemann sums remains unchanged even if you replace the arguments τ_k and u_{jk} by t_k. Thus, the contour integral $\int_C F$ is the limit of sums

$$\sum_{k=0}^{n-1} F(X(t_k)) * X'(t_k)(t_{k+1} - t_k).$$

These are just the Riemann sums you'd use to approximate the definite integral of $F(X(t)) * X'(t)$ from $t = t_0$ to $t = t_*$. Therefore,

$$\int_C F = \int_{t_0}^{t_*} F(X(t)) * X'(t)\, dt.$$

Since $F(X(t)) * X'(t)$ is $\|X'(t)\|$ times the component of $F(X(t))$ in the direction of $X'(t)$, you can interpret the contour integral as the integral of the speed of the moving point X times the tangential component of $F(X)$. If $F(X)$ is a force field acting on an object at X, then the integral is the work it does to move X along the curve.

```
double Riemann(double    a,              //          ⎡b
               double    b,              // Approximate ⎮ f by a
               double    f(double),      //          ⎣a
               unsigned n) {             // right Riemann sum with
  if (a == b) return 0;                  // n  subdivisions.
  if (n <= 0) n = 1;                     // Use at least  1   step.
  double dx = (b - a)/n;                 // Step size  dx .
  double Sum = 0;                        // Return
  for (unsigned k = 0; k < n; ++k) {     //         n-1
    double xk = a + k*dx;                //         ∑ f(xk)*dx .
    Sum += f(xk); }                      //        k=0
  return Sum*dx; }
```

Figure 5.18. Function Riemann

Figure 5.18 shows function Riemann(a,b,f,n), which computes an *n* term Riemann sum approximation

$$\sum_{k=0}^{n-1} f(t_k)\Delta t \approx \int_a^b f(t)\,dt$$

where $t_k = a + k\Delta t$ for $k = 0, \ldots, n-1$ and $\Delta t = (b-a)/n$. (This is one of several functions that constituted the numerical integration module of an earlier version of MSP [49, Chapter 4]. It's included with the other code for this section in file Stokes.CPP on the accompanying diskette.)

In order to use Riemann to integrate *F* over the boundary ∂S of the wedding band, you need these C++ functions for a derivative of the ∂S parameterization function $W(t, u)$ and for the integrand:

```
Vector<double> dWdt(double t,         // Partial
                    double u) {       // derivative.
  Vector<double> V(3);
  V[1] = -sin(t);                     // ∂x/∂t .
  V[2] =  cos(t);                     // ∂y/∂t .
  V[3] =  0;                          // ∂z/∂t .
  return V; }

double u;                             // Global!
double F_dXdt(double t) {             // Integrand for con-
  return F(W(t,u))*dWdt(t,u); }       // tour integrals.
```

The intermingling of W and X in the code stems from the parameterizations of the legs of ∂W by the functions $X(t) = W(t,-1)$ from $t = 0$ to 2π and $X(t) = W(t,+1)$ from $t = 2\pi$ to 0. The integrand function F_dXdt does double duty: With $u = \pm 1$ it represents the integrand for both legs.

Using 35-term Riemann sums, you can now approximate the integral:

```
int n = 35;                           // Global!
   ⋮
double BdryIntF;
u = -1; BdryIntF  = Riemann(0,2*M_PI,F_dXdt,n);
u =  1; BdryIntF += Riemann(2*M_PI,0,F_dXdt,n);
```

Its value is discussed later in this section. Starting points 0 and 2*M_PI are selected to keep ∂W on your *right* if you follow the legs in succession and stay just within *S* on the side facing the origin. Borland C++ header file Math.H defines macro M_PI equal to π to double precision.

Surface Integrals

The integral of a vector field $G(W)$ over a surface S parameterized by a function $W(t, u)$ over a plane rectangular region R can be defined using Riemann sums somewhat like those for contour integrals. Under rather general conditions on $G(W)$ and $W(t, u)$, you can demonstrate existence of a certain limit of sums like

$$\sum_{j=0}^{m-1} \sum_{k=0}^{n-1} \left[G(W(\tau_j, v_k)), D_{jk}, E_{jk} \right],$$

where $D_{jk} = W(t_{j+1}, u_k) - W(t_j, u_k)$ and $E_{jk} = W(t_j, u_{k+1}) - W(t_j, u_k)$. This formula uses the vector triple product discussed in Section 5.2. If R is the set of points $[t, u]$ satisfying $t_0 \le t \le t_*$ and $u_0 \le u \le u_*$, choose points t_j, τ_j, u_k and v_k for indices k as required, so that

$$t_0 < t_1 < \cdots < t_{m-1} < t_m = t_* \quad \text{and} \quad t_j \le \tau_j \le t_{j+1}$$
$$u_0 < u_1 < \cdots < u_{n-1} < u_n = u_* \quad \text{and} \quad u_k \le v_k \le u_{k+1}$$

for each j and k. Under appropriate requirements on $W(t, u)$ and its partial derivatives, you can show that the limit depends only on G and the set of all points $W(t, u)$ — the range of W — not on W itself. If you have two parameterizations $W(t, u)$ satisfying these requirements, with the same range, then the two corresponding limits will coincide. Thus, it's reasonable to call such a limit the *integral* $\int_C G$. The next paragraph shows how to convert this to a double integral.

According to the mean-value theorem, the components d_{ijk} and e_{ijk} of the factors D_{jk} and E_{jk} of the terms in the Riemann sum displayed in the previous paragraph can be replaced by terms

$$\left. \frac{\partial w_i}{\partial t} \right|_{t_j', u_k} (t_{j+1} - t_j) \quad \text{or} \quad \left. \frac{\partial w_i}{\partial u} \right|_{t_j, u_k'} (u_{k+1} - u_k)$$

where t_j' lies between t_j and t_{j+1} and u_k' between u_k and u_{k+1}. Using delicate analysis, you can show further that the limit of the Riemann sums remains unchanged even if you replace the arguments τ_j and t_j' by t_j and v_k and u_k' by u_k. Thus, the surface integral $\int_C G$ is the limit of sums

$$\sum_{j=0}^{m-1} \sum_{k=0}^{n-1} \left[G(W(t_j, u_k)), \frac{\partial W}{\partial t}, \frac{\partial W}{\partial u} \right] (t_{j+1} - t_j)(u_{k+1} - u_k).$$

These are just the Riemann sums you'd use to approximate the double integral of

$$\left[G(W(t, u)), \frac{\partial W}{\partial t}, \frac{\partial W}{\partial u} \right]$$

over region R. Therefore,

$$\int_S G = \int_R \left[G(W(t,u)), \frac{\partial W}{\partial t}, \frac{\partial W}{\partial u} \right] dt$$

$$= \int_{u_0}^{u_*} \int_{t_0}^{t_*} \left[G(W(t,u)), \frac{\partial W}{\partial t}, \frac{\partial W}{\partial u} \right] dt \, du \, .$$

The triple product integrand of the previous equation is the volume of a parallelepiped with edge vectors $G(W(t,u))$, $\partial W/\partial t$, and $\partial W/\partial u$ emanating from a corner. That's the same as the product of the component of $G(X)$ perpendicular to the surface at $X = W(t, u)$ times the area of a parallelogram that you can regard as the image under the mapping $W(t, u)$ of the rectangle defined by horizontal and vertical unit vectors at the point $[t, u]$. If G is a fluid velocity field, then the integral is the rate at which fluid crosses the surface.

Stokes' Theorem

By Stokes' theorem [59, Section 13.5], if F and Curl F are appropriately smooth and ∂S is traversed as in this example,

$$\int_{\partial S} F = \int_S \text{Curl} F \, .$$

G = Curl F is another vector field closely related to F — you can write

$$F(W) = \begin{bmatrix} f(W) \\ g(W) \\ h(W) \end{bmatrix} \qquad W = \begin{bmatrix} x \\ y \\ z \end{bmatrix} \qquad \text{Curl } F(W) = \begin{bmatrix} \frac{\partial h}{\partial y} - \frac{\partial g}{\partial z} \\ \frac{\partial f}{\partial z} - \frac{\partial h}{\partial x} \\ \frac{\partial g}{\partial x} - \frac{\partial f}{\partial y} \end{bmatrix}$$

In order to use Riemann to integrate the curl of the example vector field F over the wedding band S, you need C++ functions for the remaining derivative of the ∂S parameterization function $W(t, u)$ and for Curl F, as shown in Figure 5.19 Moreover, you need these C++ functions for the integrand $I(t, u)$ of the surface integral, and for its integral $J(u) = \int I(t, u) \, dt$ with a fixed u value:

```
double CurlF_dWdt_dWdu(double t) {      // Integrand for the
    return Triple(CurlF(W(t,u)),        // surface integral.
                  dWdt(t,u),
                  dWdu(t,u)); }
```

```
double InsideInt(double v) {          // Integrand for the
  u = v;                              // outside integral
  return Riemann(0,2*M_PI,            // of the double
    CurlF_dWdt_dWdu,n); }             // integral.
```

The latter serves as the integrand for the outside integral in $\int_S \mathrm{Curl}\, F = \int J(u)\,du = \iint I(t,u)\,dt\,du$. You can use 35-term Riemann sums as before to approximate the surface integral:

```
double SurfIntCurlF = Riemann(-1,1,InsideInt,n);
```

Global variables u and n are used because the Riemann invocation in InsideInt can't transmit that information. You'll find the complete code for this test in file Stokes.CPP on the accompanying diskette.

According to these computations,

$$\int_{\partial S} F \approx 17.545111 \approx \int_S \mathrm{Curl}\, F$$

and the two integral approximations differ by about 1.4×10^{-14}. The number of terms for the Riemann sums was set at n = 35 because beyond that, but not before, the boundary integral

---

```
Vector<double> dWdu(double t,         // Partial derivative.
                    double u) {       // ∂x/∂u = 0 = ∂y/∂u ,
  Vector<double> V(3);                // ∂z/∂u = 1 , forming
  return V.MakeUnit(3); }             // the 3rd unit vector.

Vector<double> CurlF(                 // Curl of the example
    const Vector<double>& W) {        // vector field  F .
  double x = W[1];
  double y = W[2];
  double z = W[3];
  double sec2 = pow(cos(x+y),-2);
  double dfdy = z*sec2;        double dfdz = tan(x+y);
  double dgdz = -y*sin(z);     double dgdx = 1;
  double dhdx = exp(y+z);      double dhdy = x*exp(y+z);
  Vector<double> V(3);
  V[1] = dhdy - dgdz;                 // x component.
  V[2] = dfdz - dhdx;                 // y component.
  V[3] = dgdx - dfdy;                 // z component.
  return V; }
```

Figure 5.19. Code for Stokes' theorem example

approximations agreed to six decimals. Coincidentally, they agreed to *fourteen* decimals with the corresponding surface integral approximations. This coincidence emphasizes the close relationships between the integrals: Corresponding approximations to the boundary and surface integrals don't just approach the same limit, they're closely related algebraically *to each other*. You can verify that by reviewing a proof of Stokes' theorem.

6

Polynomial Classes

Polynomials, such as $a_0 + a_1x + a_2x^2 + \cdots + a_nx^n$, have a dual nature. On one hand, you can regard them as functions $P(x)$. On the other hand, you can subject them to algebraic operations such as addition and subtraction, as though they were vectors with components, $a_0, a_1, a_2, \ldots, a_n$. Designing a data structure for them is complicated by this duality of roles that it must support. Section 6.1 defines classes Polynom for double and complex polynomials. Since they're derived from the corresponding MSP Vector classes, they inherit some structure—in particular, their data components and some logical functions. But new restrictions must be imposed: Subscripts must start at 0 and there may be no trailing zero coefficients. Thus, some Vector operations must be modified when applied to Polynom objects. Some new operations are added—for instance, polynomial multiplication and differentiation—which don't make sense for vectors in general. Finally, to reflect the functional nature of a polynomial, the C++ () operator is overloaded, so that P(x) makes sense when P is a Polynom object and x a scalar. These algebraic parts of the Polynom class definition are described in detail in Section 6.2. That section concludes with a detailed example, which verifies calculus results for Legendre polynomials.

The major application of class Polynom for this book is the computation of polynomial roots. Since the class structure, not the mathematics, is of principal importance, only one mathematical root finding method is considered in detail: Newton-Raphson approximation. The simplest Newton-Raphson routine from Section 4.4 is adapted early in Section 6.3 to handle polynomials. That section continues with discussions of more elaborate techniques for finding *all* real roots and *all* complex roots of a polynomial. The polynomial can have either real or complex coefficients.

This chapter's routines constitute module Polynom. You'll find its header file Polynom.H on the accompanying diskette, and listed in Appendix A.1. Most of its source code is listed later in this chapter. You'll find all of the source code on the optional diskette.

6.1 Polynom MODULE INFRASTRUCTURE

> **Concepts**
> *Polynomial algebras*
> *C++ inheritance features*
> *The* Polynom *classes*
> *The* Degree *selector*
> *Constructors*
> *Function* xTo(n)
> *Copiers*
> *Assignment*

In higher mathematics, polynomials are often represented by vectors. The vector entries are the polynomial coefficients, and most vector algebra operations make sense when applied to polynomials. A special kind of vector space, called a polynomial algebra, is a convenient mathematical framework for studying polynomials. A *vector space* is a mathematical structure like this book's Vector classes. It consists of a set of vectors with entries selected from some scalar field, together with logical features common to any mathematical structure, and the algebraic vector addition, subtraction, and scalar multiplication operators. To accommodate polynomials, this familiar structure is restricted in its membership, but its algebraic features are enhanced. A *polynomial algebra* is a vector space whose vectors all represent polynomials and whose operations include polynomial evaluation and multiplication as well as the vector algebra operations. Within this framework, you can also study the properties of polynomial division, differentiation and integration.

Experience with hierarchical relationships among structures like these led to development of the OOP class *inheritance* paradigm. In C++ you can define a class Special so that it inherits features from a base class Basic. Special is said to be *derived* from Basic and its objects, from Basic objects. Its declaration looks like

```
class Special: public Basic {
    ⋮
    Declarations of additional data components and of member
    functions differing from those of the base class
    ⋮
}
```

The token public specifies that public features of General objects remain public when regarded as features of the derived Special objects. You can also specify inheritance from more than one base class. Generally, a derived class is similar to its base class, except for one or more of these provisions:

- The values of data components of derived objects are more restricted.
- Derived objects have additional data components.
- Some derived class member functions differ from those of the base.
- The derived class has additional member functions.

The first provision is usually carried out by modifying member functions: You restrict those that construct or alter data components.

The Polynom module consists of Polynom classes derived from the double and complex instances of the Vector class template, and some related functions. No new data components are necessary: A polynomial

$$P(x) = p_0 + p_1 x + p_2 x^2 + \cdots + p_n x^n$$

is regarded simply as the vector $P = [p_0, \ldots, p_n]$ of its scalar coefficients. Trailing zero coefficients are banned—for example, the polynomial $P(x) = x^2$ is regarded as the vector $[0,0,1]$ of coefficients of $0 + 0x + 1x^2$, *not* as the vector $[0,0,1,0,0]$ of coefficients of $0 + 0x + 1x^2 + 0x^3 + 0x^4$. The zero polynomial is represented by the *empty* vector. Some new member functions are clearly necessary—for example, one must evaluate a polynomial $P(x)$, given its coefficient vector $[p_0, \ldots, p_n]$ and a scalar argument x. The polynomial class must also include a new operator for multiplying polynomials. Finally, some familiar Vector functions must be specialized to work on polynomials. For example, when you construct a polynomial, the index lower bound will always be zero. When you add or subtract polynomials, you must remove any trailing zero coefficients from the sum or difference vector. Related functions for polynomial differentiation and integration, and for various special aspects of polynomial multiplication and division, are also included in the Polynom module.

The Polynom classes require a definition starting with the phrase

```
template<class Scalar>
  class Polynom: public Vector<Scalar> {
```

as shown in Figure 6.1, which displays part of the Polynom.H header file. The word public specifies that the public features of Vector objects remain public when regarded as features of Polynom objects. The new and modified class components are listed after this phrase.

The rest of this section considers the logical aspects of the Polynom classes: how the logical features of the Vector classes must be specialized for polynomials. The mathematical features are described in Section 6.2.

```
#include "Vector.H"

template<class Scalar>
  class Polynom: public Vector<Scalar> {
    public:

      // Logical functions ------------------------------------

      friend int Degree(              // Return the degree: same
        const Polynom<Scalar>& P );   // as P.HighIndex() .
                                      // Construct a zero poly-
      Polynom();                      // nomial. (Use  Vector
                                      // constructor.)
      Polynom& SetUp(int n);          // Set degree, allocate
                                      // memory.
      Polynom(const double&  t);      // Construct a constant
      Polynom(const complex& t);      // polynomial.

      Polynom(                        // Copy a vector to a
        const Vector<Scalar>& V );    // polynomial.

      Polynom& operator=(const        //
        Polynom<double>& Source);     // Assignment, permitting
      Polynom& operator=(const        // complex = double .
        Polynom<complex>& Source);    //
```

Figure 6.1. Polynom class template
Part 1 of 3; continued in Figure 6.4.

The Degree Selector

The simplest logical feature of the Polynom classes is the Degree selector. It's merely an alternative for the Vector selector HighIndex, syntactically more compatible with mathematical notation:

```
template<class Scalar>            // Return the degree:
  int Degree(                     // conventional notation
    const Polynom<Scalar>& P ) {  // for P.HighIndex() .
      return P.HighIndex(); }
```

It's not a member function, because the principal argument syntax is inappropriate. But some quirk of the C++ implicit type casting criteria requires it to be a friend.

Constructors

C++ requires you to define a default constructor `Polynom()`. This should construct the empty, or *zero* polynomial. (The zero polynomial has no coefficients at all.) This is already done by the default Vector constructor (see Section 5.1), which is invoked automatically whenever an object of a derived class is constructed. Therefore, `Polynom()` needn't do anything at all: Its code is just

```
template<class Scalar>                  // Construct a zero polynomial:
  Polynom<Scalar>::Polynom() {};        // let the default  Vector
                                        // constructor do it.
```

For more serious construction tasks, the `Polynom` classes need a function `SetUp(int n)` to set the index bounds and allocate memory for the coefficients. It's similar to the corresponding Vector function (see Section 5.1), but specialized to ensure that the lower index bound is zero. Here's its code. It merely calls the Vector function and then adjusts the type of the return value.

```
template<class Scalar>                       // Set the degree and
  Polynom<Scalar>&                           // allocate memory:
      Polynom<Scalar>::SetUp(int n) {        // set up a vector
    try {                                    // and call it a
      return (Polynom<Scalar>&)              // polynomial.
        Vector<Scalar>::SetUp(n,0); }
    catch(...) {
      cerr << "\nwhile setting up a polynomial with degree   "
        << n;
      throw; }}
```

The `SetUp` function is invoked, for example, when you construct a *constant* polynomial. For that, MSP provides *three* constructors

```
Polynom<double> ::Polynom(const double&  t),
Polynom<complex>::Polynom(const complex& t),
Polynom<complex>::Polynom(const double&  t).
```

Actually, the `Polynom` class template *declares* in Figure 6.1 the fourth double/complex combination, too, but you wouldn't ever use it. Just these three are implemented in `Polynom.CPP`. The first two are instances of the function template shown in Figure 6.2. The third is implemented separately; its code is identical, except for template features and type specifications. For example, consider these declarations:

```
Polynom<double> O(0);      Polynom<complex> Q(complex(3,4));
Polynom<double> P(2);      Polynom<complex> R(5);
```

```
template<class Scalar>                  // Construct a constant
  Polynom<Scalar>::Polynom(             // polynomial.  The zero
    const Scalar& t) {                  // polynomial is the empty
  try {                                 // vector.  Any other
    if (Scalar(t) == Scalar(0))         // constant has degree  0
      SetUp(-1);                        // and one entry.
    else {
      SetUp(0);
      This[0] = t; }}
  catch(...) {
    cerr << "\nwhile constructing a constant polynomial";
    throw; }}
```

Figure 6.2. Constant `Polynom` constructor template

You could write the first as just `Polynom<double> 0`, which invokes the default constructor, not the `SetUp` function; but that's cryptic. In all four, C++ automatically converts the `int` constants to `double` before it invokes the constructor. (Without *specifically* defining `Polynom<complex>::Polynom(const double& t)`, the fourth example won't work: C++ *doesn't* automatically convert 5 to `complex`. Instead, in the presence of the `Vector` to `Polynom` copier described later in this section, C++ gets this declaration confused with `Vector<complex> R(5)`. That constructs a vector with five components, but doesn't initialize them.)

More complicated polynomials are constructed in stages, as in mathematics: First construct some powers x^n, then perform scalar multiplications to get the terms $p_n x^n$, then add and subtract these and constant polynomials. The multiplication, addition, and subtraction operators are discussed in Section 6.2, with examples. To construct polynomial $P(x) = x^n$, call MSP function `xTo(int n)` like this: `Polynom P = xTo(n)`. Here's its source code:

```
  Polynom<double> xTo(int n) {          // Return the polynomial
    try {                               // x^n .
      Polynom<double> xn;
      return xn.SetUp(n).MakeUnit(n); }
    catch(...) {
      cerr << "\nwhile constructing polynomial  x^" << n;
      throw; }}
```

This function calls `Polynom<double>::SetUp`, too. It's not a member function, because the principal argument syntax is inappropriate.

Copiers, Assignment, and Type Conversion

When `Polynom` objects are manipulated by a `Vector` operator, the result has a `Vector` type; it must be converted to a `Polynom` type. That requires copying the coefficients to a `Polynom` object, ignoring trailing zeros. This process is implemented by the `Vector` to `Polynom` copier, shown in Figure 6.3. If you ask this function to convert a vector whose lower index bound is not zero, it throws an MSP exception.

You need overloaded `Polynom` assignment operators, because the inherited `Vector` assignment operators return `Vector`, not `Polynom`, objects. Overloaded operators for all four double/complex combinations are declared within the `Polynom` class template:

```
template<class Scalar>
  class Polynom: public Vector<Scalar> {
      ⋮
      Polynom& operator=(const              //
        Polynom<double>& Source);           // Assignment, permitting
      Polynom& operator=(const              // complex = double .
        Polynom<complex>& Source);          //
```

```
template<class Scalar>                     // Copy a vector to a
  Polynom<Scalar>::Polynom(                // polynomial.
      const Vector<Scalar>& V) {
    try {                                  // Its lower index
      if (V.LowIndex() != 0) {             // must be  0 . If
        cerr << "Exception  IndexError";   // not, report,
        Exception E(IndexError);           // build, and throw
        throw(E); }                        // an MSP exception.
      int n = V.HighIndex();               // If so, find the
      while (n >= 0 &&                     // degree  n .
        V[n] == Scalar(0))  --n;
      SetUp(n);                            // Allocate memory.
      for (int i = 0; i <= n; ++i)         // Copy the entries,
        This[i] = V[i]; }                  // one by one.
    catch(...) {
      cerr << "\nwhile copying a polynomial to a vector";
      throw; }}
```

Figure 6.3. Vector to Polynom copier

You wouldn't want to convert complex polynomials to double this way, so that combination isn't implemented. The two combinations with no conversion are implemented by a template in source code file Polynom.CPP:

```
template<class Scalar>                              // Assignment,
  Polynom<Scalar>&                                  // with no
    Polynom<Scalar>::operator=(                     // Scalar type
            const Polynom<Scalar>& Source) {        // conversion.
    return (Polynom<Scalar>&)(
      Vector<Scalar>::operator=(Source)); }
```

The assignment operator that converts double to complex is almost identical:

```
Polynom<complex>&                                   // Assignment,
  Polynom<complex>::operator=(                      // with double
    const Polynom<double>& Source) {                // to complex
    return (Polynom<complex>&)(                     // conversion.
      Vector<complex>::operator=(Source)); }
```

The Vector<complex> assignment operator performs the double to complex conversion.

You can use an assignment to construct a complex polynomial x^n: Just execute

```
Polynom<complex> P;  P = xTo(n);
```

Function xTo constructs a Polynom<double> object, which the assignment operator converts to Polynom<complex>. (This technique is a design compromise. MSP has no *copier* with declaration Polynom<complex>::Polynom(const Polynom<double>&), so executing Polynom<complex> P = xTo(n) won't work. Unfortunately, such a copier would conflict with other features of the Vector class. The successful code isn't ideally concise, but now it seems the best solution.)

6.2 POLYNOMIAL ALGEBRA FUNCTIONS

> **Concepts**
> *Negation, addition and subtraction*
> *Multiplication and powers*
> *Division*
> *Replacement operators*
> *Differentiation and integration*
> *Horner's algorithm and evaluation*
> *Testing with Legendre polynomials*

Figure 6.1 showed part of the definition of the `Polynom` classes. But it included only logical features: those with counterparts in any class with a data component allocated dynamically on the heap. This section discusses the mathematical functions in module `Polynom`, which implement polynomial algebra. Declarations of the mathematical member functions are shown in Figures 6.4.

Negation, Addition, and Subtraction

Negation is the simplest polynomial algebra operator. The `Vector` negation operator applies to polynomials, but returns a `Vector` value. The `Polynom` negation operator, a member function of each `Polynom` class, converts that value to a `Polynom` object:

```
template<class Scalar>                            // Polynomial
  Polynom<Scalar> Polynom<Scalar>::               // negation.
    operator-() const {                           // Use the
  try {                                           // analogous
    return Vector<Scalar>::operator-(); }         // Vector
  catch(...) {                                    // operator and
    cerr << "\nwhile negating a polynomial";      // convert back
    throw; }}                                     // to Polynom .
```

The addition operator

```
Polynom operator+(const Polynom& Q) const;
```

is also a member function of the `Polynom` classes. It's implemented by a similar template. The only difference is its `return` statement:

```
return Vector<Scalar>::operator+(Q);
```

You'll find the full template, of course, in file `Polynom.CPP`.
Executing code fragments

```
double t;                    complex t;
Polynom<double> P,Q;         Polynom<complex> P,Q;
  ⋮                            ⋮
Q = P + t;                   Q = P + t;
```

automatically converts the scalars `t` to `Polynom` objects and performs the required additions. But that doesn't occur for the analogous assignments `Q = t + P`. Changing a polynomial's

```
    Polynom operator-() const;        // Polynomial negation.

    Polynom operator+(                // Polynomial addition.
        const Polynom& Q) const;      // The general case
                                      // doesn't cover left sca-
    friend Polynom<Scalar>            // lar addition.  That's a
        operator+(                    // friend, to maintain
            const         Scalar& t,  // consistency with left
            const Polynom<Scalar>& P);// scalar multiplication.

    Polynom operator-(                // Polynomial subtraction.
        const Polynom& Q) const;      // The general case
                                      // doesn't cover left sca-
    friend Polynom<Scalar>            // lar subtraction.  That
        operator-(                    // is a friend, to main-
            const         Scalar& t,  // tain consistency with
            const Polynom<Scalar>& P);// left scalar addition.

    Polynom operator*(                // Polynomial multiplica-
        const Polynom& P)             // tion. The general case
        const;                        // doesn't cover left
                                      // scalar multiplication.
    friend Polynom<Scalar>            // That must be a friend,
        operator* (                   // lest the compiler con-
            const         Scalar& t,  // fuse it with the analo-
            const Polynom<Scalar>& P);// gous  Vector  operator.

    Polynom operator^(int n) const;   // Return This^n .
```

Figure 6.4. Polynom class template
Part 2 of 3; see also Figures 6.1 and 6.7.

constant term is a common operation; no MSP client should have to remember whether to add a scalar on the left or right. So MSP implements the *left* scalar addition operator, too:

```
template<class Scalar>                  // Left scalar addition:  not
    Polynom<Scalar> operator+(          // covered by the general case.
        const         Scalar& t,        // It's a friend, to maintain
        const Polynom<Scalar>& P) { // consistency with left scalar
    return P + t; }                     // multiplication.
```

This operator can't be a member function, because its left operand is a scalar, not a polynomial. In Figure 6.4 it's declared a friend of the Polynom classes—not for any direct reason, but because it's designed exactly like the left scalar multiplication operator described later, which *must* be a friend.

MSP handles subtraction as it does addition. The templates differ from the analogous addition templates only in their return statements. For the polynomial subtraction member function template and the left scalar subtraction template, these are

```
return This + (-Q);        return -P + t;
```

Multiplication and Powers

Multiplication of two polynomials

$$P(x) = \sum_{i=0}^{m} p_i x^i \qquad Q(x) = \sum_{j=0}^{n} q_j x^j$$

is based on the formula

$$P(x)Q(x) = \sum_{i+j=k=0}^{m+n} p_i q_j x^k = \sum_{k=0}^{m+n} \sum_{i=\max(0,k-n)}^{\min(m,k)} p_i q_{k-i} x^k .$$

Its straightforward MSP implementation, as a member function of the Polynom classes, is shown in Figure 6.5.

Mathematics texts usually regard multiplication of a polynomial by a scalar as a *simpler* operation than multiplication of two polynomials, inherited from the underlying vector algebra. For MSP there's no need to treat scalar multiplication separately if the scalar's on the *right*. Executing code fragments

```
double t;                  complex t;
Polynom<double> P,Q;       Polynom<complex> P,Q;
   ⋮                          ⋮
Q = P * t;                 Q = P * t;
```

automatically converts the scalars t to Polynom objects and performs the required multiplications. But that doesn't occur for the analogous assignments Q = t * P. So MSP implements the *left* scalar multiplication operator with a template that's just like those for its left scalar addition and subtraction operators, except for its statement return P * t. Due to a technicality in its implicit type casting criteria, C++ confuses the left scalar multiplication operator with the analogous Vector operator unless it's declared a friend of the Polynom classes, as in Figure 6.5.

```
template<class Scalar>                            // Polynomial
  Polynom<Scalar> Polynom<Scalar>::operator* (    // multiplica-
      const Polynom<Scalar>& P) const {           // tion.
    int m = Degree(This);
    int n = Degree(P);
    try {
      if (m == -1 || n == -1) return 0;           // A nonzero
      Polynom<Scalar> Product;                    // product has
      Product.SetUp(m+n);                         // degree
      for (int k = 0; k <= m+n; ++k) {            // m + n .
        Product[k] = 0;
        for (int i = max(0,k-n);                  // Compute the
                 i <= min(m,k); ++i)              // coefficients,
          Product[k] += This[i] * P[k-i]; }       // one by one.
      return Product; }
    catch(...) {
      cerr << "\nwhile multiplying polynomials with degrees  "
           << m << "  and  " << n;
      throw; }}
```

Figure 6.5. Polynom multiplication operator template

To achieve uniformity in the left scalar addition, subtraction, and multiplication implementations, MSP declares the first two friends also.

MSP overloads C++ operator ^ to compute nonnegative integral powers of polynomials. *Danger:* Operator ^ has lower precedence than that accorded exponentiation in conventional mathematical notation—for example, to get (P^2) + 1, you *must* use the parentheses! The conventional algorithm $P(x)^n = P(x) \cdot P(x) \cdot \cdots \cdot P(x)$ with $n - 1$ multiplications is inefficient, so MSP uses a faster one, which is implemented recursively. The source code is shown in Figure 6.6.

If an exception is thrown in a try block during recursive execution of a function f, the corresponding catch block will handle it. According to MSP exception handling strategy, the catch block should rethrow the exception so the function that called f can identify itself. But that function is likely to be f itself. If the exception occurred during the fifteenth recursive invocation of f, you'd see fifteen error messages from f. That's not helpful, just intimidating. To maintain the MSP strategy, the recursive power function _Power does *not* include try/catch blocks. It can throw an exception, but it can't catch one. Because it's not complete in this sense, it's not declared in any MSP header file, so it's unavailable to MSP clients. Instead, _Power is invoked inside a try block by the ^ operator. If an exception is thrown during execution of _Power, the ^ operator's catch block will handle it in the usual way, and you'll see just one error message identifying that operator.

```
template<class Scalar>                  // Return P(x)^n : to be
  Polynom<Scalar> _Power(                // executed only by
      Polynom<Scalar> P,                 // Polynom::operator^ .
               int n) {
    if (n < 0) {                         // n  must be  >= 0 .
      cerr << "Exception DivideError";   // Recursion:
      throw(Exception(DivideError)); }   // P^0 = 1 ;
    if (n   == 0) return 1;              // for odd  n ,
    if (n%2 != 0) return P*_Power(P,n-1);// P^n = P·P^(n-1) ;
    Polynom<Scalar> Q = _Power(P,n/2);   // for even  n ,
    return  Q*Q; }                       // P^n = (P^½n)^2 .

template<class Scalar>                   // Return This^n . This
  Polynom<Scalar> Polynom<Scalar>::      // shell lets  try/catch
      operator^(int n) const {           // avoid reporting an ex-
    try {                                // ception at each return.
      return _Power(This,n); }
    catch(...) {
      cerr << "\nwhile raising to power   " << n
           << "  a polynomial with degree  " << Degree(This);
      throw; }}
```

Figure 6.6. Polynom power operator

You can follow the source code to see, for example, that for n = 15 MSP computes

$$P(x)^{15} = P(x) \cdot P(x)^{14}$$
$$= P(x) \cdot (P(x)^7)^2$$
$$= P(x) \cdot (P(x) \cdot (P(x)^6)^2$$
$$= P(x) \cdot (P(x) \cdot (P(x)^3)^2)^2$$
$$= P(x) \cdot (P(x) \cdot (P(x) \cdot P(x)^2)^2)^2$$

using 6, rather than $n - 1 = 14$ multiplications. This method is certainly more efficient than the conventional algorithm, but it's *not* the fastest. In fact, you can compute $P(x)^{15}$ with only 5 multiplications: First compute $y = P(x) \cdot P(x)^2$ and then $P(x)^{15} = y \cdot (y^2)^2$. An algorithm that computes every integral power the fastest way would be more complicated, so MSP uses a compromise. There's a large literature on efficient power algorithms: Knuth [28, Section 4.6.3] gives it twenty-five pages!

Division

When you divide a polynomial $F(x) = a_0 + a_1 x + \cdots + a_m x^m$ of degree m by a nonzero polynomial $G(x) = b_0 + b_1 x + \cdots + b_n x^n$ of degree n, you get *quotient* and *remainder* polynomials $Q(x)$ and $R(x)$ such that

$$\frac{F(x)}{G(x)} = Q(x) + \frac{R(x)}{G(x)}, \qquad F(x) = Q(x)G(x) + R(x),$$

and degree(R) < degree(G). If $G(x)$ is just the constant b_0, this is scalar multiplication by $1/b_0$, and $R(x) = 0$. If degree(F) < degree(G), then $Q(x) = 0$ and $R(x) = F(x)$. In other cases, you can compute $Q(x)$ and $R(x)$ by the algorithm you learned in algebra class:

$$
\begin{array}{r}
\dfrac{a_m}{b_n} x^{m-n} + \cdots = Q_1(x) + \cdots \\
G(x) = b_n x^n + \cdots \overline{\bigg) \begin{array}{l} a_m x^m + \cdots \quad = F(x) \\ a_m x^m + \cdots \quad = Q_1(x)G(x) \\ \hline \cdots \quad \begin{array}{l} = F(x) - Q_1(x)G(x) = \\ R(x) \text{ or new } F(x) \end{array} \end{array}}
\end{array}
$$

The first term $Q_1(x)$ of the quotient is $(a_m/b_n)\, x^{m-n}$; you multiply the divisor $G(x)$ by that term and subtract the product from the dividend $F(x)$. If the remainder $R(x) = F(x) - Q_1(x)G(x)$ has degree less than n, you can stop with $Q(x) = Q_1(x)$; otherwise, make $R(x)$ the new dividend, divide by $G(x)$, and add the quotient to the term you've already computed. This familiar algorithm is implemented by MSP function template `Divide(G,Q,R)`; it's declared in the part of header file `Polynom.H` shown in Figure 6.7, and its source code is displayed in Figure 6.8. To take advantage of C++ type casting features, the template defines member functions of the `Polynom` class, and the principal argument, which corresponds to $F(x)$, is suppressed.

MSP also implements as `Polynom` class member functions the operators `/` and `%` for dividing a polynomial F by a nonzero polynomial G and reducing F modulo G. Executing `F/G` or `F % G` invokes `Divide` and then returns the quotient or remainder. Here's the code for the `/` operator:

```
template<class Scalar>                      // Divide a copy of
  Polynom<Scalar> Polynom<Scalar>::          // This by G and
    operator/(                               // return the
      const Polynom<Scalar>& G) const {      // quotient.
    Polynom<Scalar> Q,R;
    Divide(G,Q,R);                           // Divide makes the
    return Q; }                              // copy.
```

The % operator is similar.

The concept of *greatest common divisor* (GCD) of two polynomials is used in Section 6.3 to expedite solution of polynomial equations. It's only slightly more involved than the corresponding concept for integers. A polynomial $D(x)$ is a GCD of polynomials $P(x)$ and $Q(x)$ if $D(x)$ divides them both (with remainder zero) and any other divisor has degree not exceeding that of $D(x)$. Thus, the GCDs of $P(x)$ and $Q(x)$ consist of any single GCD and all its scalar multiples. Under this definition, $P(x)$ is a GCD of $P(x)$ and the zero polynomial. But if $P(x)$ and $Q(x)$ are both zero, their GCD must be defined specifically as the zero polynomial. It's easy to find a GCD $D(x)$ if you consider *prime decompositions* of $P(x)$ and $Q(x)$—that is, write them as products of powers of polynomials $p_0(x), \ldots, p_n(x)$ that can't be factored further, as shown in Figure 6.9. For example, consider

$$P(x) = (x^2 + x)^3(x - 1)^2$$
$$Q(x) = (x^2 + x)^2(x - 1)^3$$
$$D(x) = (x^2 + x)^2(x - 1)^2 = x^2(x^2 - 1)^2 = x^2 - 2x^4 + x^6.$$

However, factoring polynomials, given their *coefficients*, is not a practical way to find a GCD.

```
void Divide(                    // Divide a copy of  This
    const Polynom& G,           // by  G ,  getting
            Polynom& Q,         // quotient  Q  and
            Polynom& R) const;  // remainder  R .

Polynom operator/(              // Divide copies of  This
    const Polynom& G) const;    // by  G ;  return the
Polynom operator%(              // quotient and the
    const Polynom& G) const;    // remainder.

Scalar  Horner(Polynom& Q,      // Divide  This  by  x - t
            const Scalar& t)    // to get quotient  Q(x) ;
    const;                      // return the remainder.

Scalar operator()(              // Evaluate  This
    const Scalar& t) const;     // polynomial at  t .

};              // End of the definition of class  Polynom .
```

Figure 6.7. Polynom class template
Part 3 of 3; continued from Figure 6.4.

```
template<class Scalar>                        // Divide a copy  F
  void Polynom<Scalar>::Divide(               // of  This  by  G ,
    const Polynom<Scalar>& G,                 // getting quotient
        Polynom<Scalar>& Q,                   // Q  and remainder
        Polynom<Scalar>& R) const {           // R .
  try {
    Polynom<Scalar> F = This;                 // If  G = 0 ,
    int n = Degree(G);                        // report,
    if (n < 0) {                              // build, and
      cerr << "\nException  DivideError";     // throw an MSP
      Exception E(DivideError);               // exception.
      throw(E); }                             // Otherwise,
    Q.MakeZero();                             // execute the
    for (;;) {                                // standard
      int m = Degree(F);                      // algorithm.
      if (m < n) break;
      Polynom<Scalar> P;   P  = xTo(m−n);
      Polynom<Scalar> Q1;  Q1 = (F[m]/G[n]) * P;
      F −= Q1 * G;
      if (Degree(F) == m) {
        P = xTo(m);                           // Ensure that
        F −= F[m] * P; }                      // the subtrac-
      Q += Q1; }                              // tion is
    R = F; }                                  // exact.
  catch(...) {
    cerr << "\nwhile dividing polynomials";
    throw; }}
```

Figure 6.8. Polynom function Divide

$$P(x) = \prod_{k=1}^{n} p_j(x)^{k_j} \qquad Q(x) = \prod_{j=1}^{n} p_j(x)^{l_j}$$

$$GCD(x) = \prod_{j=1}^{n} p_j(x)^{m_j} \qquad m_j = \min(k_j, l_j)$$

Figure 6.9. Calculating a GCD of polynomials P and Q by factoring

Instead, you can use an analog of *Euclid's algorithm* for the GCD of integers:

1. Call the polynomials P and Q with degree(P) ≥ degree(Q).
2. P is a GCD if Q is the zero polynomial.
3. Otherwise, find R = P mod Q.
4. Repeat from step 2 with P and Q replaced by Q and R.

Step 4 always decreases the degree of the divisor for the next step 3 so this process will eventually terminate at step 2. The result of this version of the algorithm is often inappropriate in practice, because round-off error makes you work with *approximations* of zero: polynomials whose coefficients are all very small. The process continues too long, because it doesn't reach an exact zero remainder when it should. Therefore, the algorithm must be modified slightly. Replace step 2 by:

2'. P is a GCD if the max norm of Q doesn't exceed a specified tolerance T.

The appropriate tolerance might depend on the problem at hand, so the client is responsible for specifying it.

The modified algorithm is implemented by MSP function template GCD(P,Q,T), shown in Figure 6.10. In header file Polynom.H, parameter T is initialized with default value 0. Here's a

```
template<class Scalar>                      // Return a greatest
  Polynom<Scalar> GCD(Polynom<Scalar> P,    // common divisor of
                      Polynom<Scalar> Q,    // P and Q . Treat
                      double T) {           // a polynomial as 0
    try {                                   // if its max norm
      Polynom<Scalar> R;                    // is  <= T .
      if (Degree(P) < Degree(Q)) {          // Switch  P and  Q
        R = Q;  Q = P;  P = R; }            // if necessary, to
      while (Degree(Q) >= 0                 // ensure  Degree(P)
          && MaxNorm(Q) > T) {              // >= Degree(Q) .
        R = P % Q;                          // Use Euclid's
        P = Q;  Q = R; }                    // algorithm.
      return P; }
    catch(...) {
      cerr << "\nwhile computing the GCD of two polynomials";
      throw; }}
```

Figure 6.10. Polynom function GCD

program fragment, which constructs the polynomials $P(x)$ and $Q(x)$ of degrees 8 and 7 discussed previously, and executes this function to compute a GCD:

```
Polynom<double> x;  x = xTo(1);
Polynom<double> P = (x^2) + x;
Polynom<double> Q = x - 1;
Polynom<double> F = (P^3) * (Q^2);
Polynom<double> G = (P^2) * (Q^3);
GCD(F,G).Show();
```

The program output

```
   0.0      0.0      2.0      0.0     -4.0      0.0      2.0
```

represents the polynomial $2x^2 - 4x^4 + 2x^6 = 2(x^2 - 2x^4 + x^6)$. As in this example, Euclid's algorithm usually computes a scalar multiple of the GCD you'd get by factoring. GCD could have divided the result by its leading coefficient, but there seemed no compelling reason to do so, and the division could add to round-off error, so it was omitted. The default tolerance $T = 0$ succeeded in this example because there was no division round-off error. That was due to the unusually simple coefficients of P and Q; for most applications, you'll need to specify a small positive T, and you'll have to experiment to decide how small this should be.

MSP includes function template LCM(P,Q,T) for computing a *least common multiple* $L(x)$ of polynomials $P(x)$ and $Q(x)$: It's a nonzero multiple of both, but no polynomial of smaller degree is. (If both $P(x)$ and $Q(x)$ are zero, $L(x)$ is defined specifically to be zero.) The least common multiples of $P(x)$ and $Q(x)$ consist of all scalar multiples of any single LCM. If you factor $P(x)$ and $Q(x)$ into prime factors, you can easily construct a least common multiple just as $D(x)$ was built in Figure 6.9: let $m_j = \max(k_j, l_j)$ instead of $\min(k_j, l_j)$. From this you can see how LCD was implemented: It just divides the product $P(x)Q(x)$ by a GCD.

Differentiation and Integration

MSP includes function templates Deriv and Integ for differentiating and integrating a polynomial

$$P(x) = \sum_{x=0}^{n} p_k x^k .$$

Here are the differentiation and integration formulas and the corresponding implementations.

$$P'(x) = \sum_{k=1}^{n} k\, p_k x^{k-1}$$

```
template<class Scalar>                  // Return the
  Polynom<Scalar> Deriv(                // derivative
      const Polynom<Scalar>& P) {       // D(x) = P'(x) .
    try {
      int n = Degree(P);
      Polynom<Scalar> D;  D.SetUp(n-1);
      for (int k = 1; k <= n; ++k) D[k-1] = k * P[k];
      return D; }
    catch(...) {
      cerr << "\nwhile differentiating a polynomial";
      throw; }}
```

$$\int_0^x P(t)dt = \sum_{k=0}^{n} \frac{p_k}{k+1} x^{k+1} = \sum_{k=1}^{n+1} \frac{p_{k-1}}{k} x^k$$

```
template<class Scalar>                  // Return the anti-
  Polynom<Scalar> Integ(                // derivative  J(x)
      const Polynom<Scalar>& P) {       // of P(x) for
    try {                               // which J(0) = 0 .
      int n = Degree(P);
      Polynom<Scalar> J;  J.SetUp(n+1);
      J[0] = 0;
      for (int k = 1; k <= n+1; ++k)
        J[k] = P[k-1] / k;
      return J; }
    catch(...) {
      cerr << "\nwhile antidifferentiating a polynomial";
      throw; }}
```

Horner's Algorithm and Evaluation

The naive way to evaluate $P(t) = p_0 + p_1 t + p_2 t^2 + \ldots + p_n t^n$ is to compute the powers $t^2, \ldots, t^n$ and the terms $p_1 t, p_2 t^2, \ldots, p_n t^n$ and then add. That requires n additions and $2n - 1$ multiplications. There's a much more efficient method, however, which is known as *Horner's algorithm*:

$$q_{n-1} = p_n;$$
$$\text{for } k = n - 1 \text{ down to } 1 \text{ do}$$
$$\quad q_{k-1} = p_k + t q_k;$$
$$P(t) = p_0 + t q_0.$$

For $n = 3$, this amounts to the computation

$$P(t) = p_0 + t(p_1 + t(p_2 + tp_3)).$$

$$\begin{array}{l} q_2 \underline{|} \\ q_1 \underline{||} \\ q_0 \underline{||} \end{array}$$

Horner's algorithm requires only n additions and n multiplications: *Count them*!

Horner's algorithm is also known as *synthetic division*, for the following reason. With its intermediate results $q_0, \ldots, q_{n-1}$ compute

$$\begin{aligned} &P(t) + (x-t)(q_0 + q_1 x + \cdots + q_{n-1} x^{n-1}) \\ &= P(t) - tq_0 - tq_1 x - \cdots - tq_{n-1} x^{n-1} + q_0 x + \cdots + q_{n-2} x^{n-1} + q_{n-1} x^n \\ &= [P(t) - tq_0] + [q_0 - tp_1]x + \cdots + [q_{n-2} - tq_{n-1}]x^{n-1} \\ &= p_0 + p_1 x + \cdots + p_{n-1} x^{n-1} + p_n x^n \\ &= P(x) . \end{aligned}$$

Thus, $q_0, \ldots, q_{n-1}$ are the coefficients of the quotient polynomial $Q(x)$ in the long division

$$\frac{P(x)}{x-1} = Q(x) + \frac{P(t)}{x-t}$$

and $P(t)$ is the remainder—that is, $P(x) = (x-t)Q(x) + P(t)$.

By using the product rule to differentiate the previous equation, you get $P'(x) = Q(x) + (x-t)Q'(x)$; hence, $Q(t) = P'(t)$. Thus, a polynomial $P(x)$ and its derivative can be evaluated easily for an argument t by *two* applications of Horner's algorithm: First apply it to $P(x)$, obtaining $P(t)$ and the coefficients of $Q(x)$ and then apply it to $Q(x)$, obtaining $Q(t) = P'(t)$. This technique will be used for Newton-Raphson iteration in Section 6.3.

MSP implements Horner's algorithm by the member function template Horner. Its declaration is displayed in Figure 6.7, and its source code is shown in Figure 6.11. The code is straightforward, except for the presence of two Scalar type casts. Without them, the compiler would balk, noticing that it's allowed to invoke the intended

```
complex::operator*(complex&)
```

to compute t * Q[i], or to use the constructor

```
Polynom<complex>::Polynom(const complex&)
```

implicitly to convert Q[i] to Polynom<complex>, and then invoke

```
operator*(complex&,const Polynom<complex>&).
```

It doesn't make any mathematical difference *here*, but the compiler doesn't know that. This problem surfaces occasionally and constitutes a failure to achieve the MSP goal of making the C++ code look like the algorithm. I have not been able to find a suitable correction.

Function Horner, in turn, is the basis of the overloaded () operator, which is used to evaluate a Polynom at a specified Scalar argument t:

```
template<class Scalar>                          // Evaluate  This
  Scalar Polynom<Scalar>::operator()(           // polynomial at  t .
      const Scalar& t) const {                  // Use Horner's algo-
    try {                                       // rithm, but ignore
      Polynom Q;                                // the quotient that
      return Horner(Q,t); }                     // it returns in  Q .
    catch(...) {
      cerr << "\nwhile evaluating a polynomial";
      throw; }}
```

This overloading makes it possible to use the expression P(t) for the value of Polynom P at a Scalar t. You'll find several examples in the rest of this section.

Polynomial evaluation is such an important numerical software feature that much effort has been spent analyzing its efficiency. It's known that any algorithm that computes $P(x) = p_0 + p_1 x + \cdots + p_n x^n$ for all possible values of $p_0, \cdots, p_n$ and x must use at least n additions and n multiplications, and any one that achieves this optimality is essentially the same as Horner's algorithm [28, Section 4.6.4].

```
template<class Scalar>                          // Divide  This  by  x - t
  Scalar Polynom<Scalar>::Horner(               // to get quotient   Q(x) .
      Polynom<Scalar>& Q,                       // return the remainder.
      const Scalar&  t) const {
    try {                                       // Why won't this work if
      int n = Degree(This);                     // Degree  isn't a friend?
      if (n <  0) return 0;                     // In case  This  is zero.
      if (n == 0) return This[0];               // Constant  P(x) != 0 .
      Q.SetUp(n-1);
      Q[n-1] = This[n];                                 // Horner's
      for (int I = n-1; I >= 1; --I)                    // algorithm.
        Q[i-1] = This[i] + Scalar(t)*Q[i];              // The  Scalar
      return This[0] + Scalar(t)*Q[0]; }                // casts resolve
    catch(...) {                                // complex * complex
      cerr << "\nwhile executing "              // vs.
        "Horner's algorithm";                   // complex * Polynom
      throw; }}                                 // ambiguity.
```

Figure 6.11. Polynom function Horner

Testing with Legendre polynomials

Legendre polynomials play important roles in numerical analysis. In this section, they're introduced to test and demonstrate the power and convenience of the `Polynom` module. The test routines discussed here are typical of those actually used while developing the MSP modules.

The Legendre polynomials $L_0(x), L_1(x), L_2(x), \ldots$ are defined recursively by equations $L_0(x) = 1$, $L_1(x) = x$, and

$$L_n = \frac{2n-1}{n} x L_{n-1}(x) - \frac{n-1}{n} L_{n-2}(x).$$

The degree of $L_n(x)$ increases by one with each recursion, so $L_n(x)$ has degree n in general. Here are the next three polynomials in the sequence:

$$L_2(x) = \tfrac{3}{2}x^2 - \tfrac{1}{2} \qquad L_3(x) = \tfrac{5}{2}x^3 - \tfrac{3}{2}x \qquad L_4(x) = \tfrac{35}{8}x^4 - \tfrac{15}{4}x^2 + \tfrac{3}{8}.$$

It can be shown that $L_n(x)$ is always even if n is even and odd if n is odd.

MSP includes function `Legendre` for generating these polynomials. Its implementation, shown in Figure 6.12, closely follows the definition. Allocating memory for $n + 1$ Legendre

```
Polynom<double> Legendre(int n) {       // Return the    nth
  try {                                 // Legendre polynomial
    if (n <= 0) return 1;               // Ln .  Special cases
    Polynom<double> x = xTo(1);         // L0 = 1   and   L1 = x .
    if (n == 1) return x;               // L1 = x .
    Polynom<double> LmMinus2 = 1;       // For   n > 1 , use the
    Polynom<double> LmMinus1 = x;       // recursion formula
    for (int m = 2;; ++m) {             // Lm = a*x*LmMinus1
      double M = m;                     //         - b*LmMinus2
      double a = (2*M - 1)/M;           // for   m = 2  to   n .
      double b = (M - 1)/M;
      Polynom<double> Lm = a*x*LmMinus1 - b*LmMinus2;
      if (m == n)   return Lm;
      LmMinus2 = LmMinus1;
      LmMinus1 = Lm; }}
  catch(...) {
    cerr << "\nwhile constructing Legendre polynomial   L" << n;
    throw; }}
```

Figure 6.12. `Polynom` function `Legendre`

polynomials is not strictly necessary, since you only need the two most current ones to construct the next. But restricting the allocation to accommodate just those three would make the program more complicated.

The Legendre polynomials have many complicated (and important) properties that are easily adapted for checking numerical software. The most important one is *orthogonality*:

$$\int_{-1}^{1} L_m(x)L_n(x)\,dx = \begin{cases} \dfrac{2}{2n+1} & \text{if } m = n \\ 0 & \text{if } m \ne n. \end{cases}$$

Figure 6.13 displays a routine employing this property to test Polynom functions Legendre and Integ and the polynomial multiplication and evaluation operators. The figure includes sample test output.

```
double LegInt(int m, int n) {           // Polynom  module  test
  Polynom<double> K = Legendre(m);      // routine. LegInt inte-
  Polynom<double> L = Legendre(n);      // grates the product of
  Polynom<double> J = Integ(K*L);       // Legendre polynomials:
  return  J(1) − J(−1); }               //           ┌1
                                        // Jmn =     |   Lm(x) Ln(x) dx
void Test1() {                          //           ┘-1
  cout << "Enter   m   n = ";
  int m,n;  cin >> m >> n;                            cout <<    "\n";
  cout << "Jmm       = ";   Show(LegInt(m,m),F); cout <<    "\n"
          "2/(2m+1)  = "  << 2.0/(2*m+1);        cout <<    "\n"
          "Jmn       = ";   Show(LegInt(m,n),F); cout <<    "\n"
          "Jnn       = ";   Show(LegInt(n,n),F); cout <<    "\n"
          "2/(2n+1)  = "  << 2.0/(2*n+1);        cout <<    "\n\n"; }
```

Sample output

```
Enter   m   n = 14 15
Jmm       = 0.0689655
2/(2m+1)  = 0.0689655
Jmn       = 0
Jnn       = 0.0645161
2/(2n+1)  = 0.0645161
```

Figure 6.13. Integrating products of Legendre polynomials

```
void Test2() {                                 // Polynom module test
  cout << "Enter  n = ";                       // routine.  Compare
  int n;   cin >> n;  cout << "\n";            // Legendre polynomial
  Polynom<double> Kn = (xTo(2) - 1)^n;         // Ln  with  Kn =
  for (int k = 1; k <= n; ++k)                 //
  Kn = Deriv(Kn);                              //     1    d^n
  Kn *= 1/(pow(2,n)*Factorial(n));             //    ───  ───  (x^2 - 1)^n
  Polynom<double> Ln = Legendre(n);            //    n!2^n dx^n                 .
  cout << "|Kn - Ln| = "
     << MaxNorm(Kn - Ln);  cout << "\n"; }
```

Sample output

```
Enter  n = 16
|Kn - Ln| = 7.27596e-12
```

Figure 6.14. Verifying an alternative definition of Legendre polynomials

A second test uses an alternate definition often given for Legendre polynomials:

$$L_n(x) = \frac{1}{n!2^n} \frac{d^n}{dx^n}(x^2 - 1)^n.$$

The test routine in Figure 6.14 computes a Legendre polynomial according to this definition, but calls it $K_n(x)$. Then it calls function Legendre to compute $L_n(x)$ as usual and outputs the max norm of the difference of the corresponding coefficient vectors. Sample test output is included in the figure. This routine tests functions Legendre and Deriv and the polynomial subtraction, multiplication, and power operators. You'll find the source code for the routines shown in Figures 6.13 and 6.14 in file Legendre.CPP on the accompanying diskette.

For further information on Legendre polynomials, see Wylie [59, Section 9.8].

6.3 POLYNOMIAL ROOTS

Concepts

Implementing Newton-Raphson iteration

Polishing root approximations

Cauchy's bound

Taylor's theorem

Multiplicities of roots

Eliminating multiplicities
Finding a real root
Finding all real roots through deflation
Testing with Legendre polynomials
Finding complex roots

One of the most highly developed areas of numerical analysis is the solution of equations $F(x) = 0$, where $F(x)$ is a polynomial. Solutions x are called *roots* of $F(x)$. Since polynomials can be real or complex, so can their roots. You can regard any real polynomial as complex, and real polynomials may have complex roots—for example, $F(x) = x^2 + 1$ has roots $x = \pm i$. Chapter 4 considered general methods for computing roots of functions; this section specializes them for polynomials. Some of these methods apply equally well to real and complex problems; others, such as the bisection method, are intended only for finding real roots. Since a nonzero polynomial has only finitely many roots, you may reasonably want to find them *all*. That problem usually occurs because an application needs to use one of the roots of some polynomial, and you can't tell which one until you look at them all—for example, it might need a root with minimum absolute value or maximum imaginary part. This section first considers adaptations of the Chapter 4 methods that apply to both real and complex polynomial roots, then discusses special techniques for real roots, and finally, special techniques for complex roots. None of these treatments is exhaustive—you can't even regard them as deep. The mathematical aspects of this subject would fill a monograph. This book's goal, though, is to show that polynomial root approximation algorithms can be expressed gracefully and usefully in C++, so it only needs to consider a selection of common techniques.

Implementing Newton-Raphson Iteration

The problem of finding a root of a polynomial $F(x)$ is common enough that you'll want to modify the MSP root finding functions already considered, using a parameter F of a Polynom type instead of defining a function specifically to evaluate F and perhaps its derivative. It's a simple task. Here are the declarations of the original and modified Newton-Raphson functions, from MSP header files Equate1.H and Polynom.H:

```
—— Equate1.H ——
template<class Scalar>
  Scalar NR(Scalar FdF(Scalar  x,
                       Scalar& dF),
           Scalar x,
           double T,
           int&   Code);
```

```
—— Polynom.H ——
template<class Scalar>
  Scalar NR(
      const Polynom<Scalar>& F,
              Scalar  x,
              double  T,
              int&    Code);
```

Here's the only source code change beyond this, alongside the original version:

```
    ⋮                                       Polynom<Scalar> Q;
Fx = FdF(x,dF);                             Fx = F.Horner(Q,x);
    ⋮                                       dF = Q(x);
```

On the left, Equate1 function NR executes function FdF, supplied by the caller, to obtain Scalar values Fx and dF, representing $F(x)$ and $F'(x)$. On the right, Polynom function NR executes Horner to obtain the value Fx and the quotient polynomial Q that results when you divide $F(t)$ by $t - x$. (Regard t as the variable and x as a constant.) As mentioned in Section 6.2, NR can then evaluate Q at x to get dF. You'll find the complete source code for Polynom function NR in file Polynom.CPP. You're invited to modify Equate1 functions FixPoint and Secant the same way; they're easier, since they don't use derivatives. Analogous modifications to other Equate1 functions are described later in this section.

Repeating nearly all the NR code to get the Polynom version is inelegant programming. Ideally, NR should be implemented by a template such as this:

```
template<class Scalar,
        class Function1>
  Scalar NR(const Function1& FdF,
            Scalar    x,
            double    T,
            int&      Code);
```

The first parameters of both NR functions would be objects of type Function1, instead of Scalar(Scalar,Scalar&) and Polynom<Scalar>. A future stage in MSP development, that design hasn't been completed for this book. It's complicated by the goal to make it apply to multivariable functions, especially for use in integration and solution of differential equations.

Later parts of this section consider methods for computing approximations $r_1, \ldots, r_n$ to roots of a polynomial $F(x)$, which can possibly be improved—or *polished*—by individual applications of Newton-Raphson iteration. MSP implements the polishing process with function template Polish, shown in Figure 6.15. Invoke it as you would NR, except with a Vector of initial estimates, not just one. Use the Code parameter as in NR to opt for intermediate output. Polish returns Code = 0 if this process succeeds. If NR encounters an exception while improving an approximation r_k, it leaves that one unchanged, but—after processing *all* r_k—returns Code = 1.

Cauchy's Bound

Computing a root of a polynomial by Newton-Raphson iteration or some other method usually involves finding some initial estimate. Computing *all* roots, you'll face the problem of determin-

```
template<class Scalar>                  // Root contains
   void Polish(Vector<Scalar>& Root,    // approximations to
         const Polynom<Scalar>& F,      // roots of F . Use
                  double   T,           // NR to improve
                  int&     Code) {      // them. If NR
   int C = Code, C1 = 0;                // encounters an
   for (int k  = Root.LowIndex();       // exception, skip
             k <= Root.HighIndex(); ++k) {  // that entry and
      try {                             // set  Code = 1
         C = Code;                      // to report failure.
         Root[k] = NR(F,Root[k],T,C); }
      catch(...) { C1 = 1; }}
   Code = C1; }
```

Figure 6.15. Polynom function Polish

ing when you've found them all. The following theorem provides a tool you can use in both situations: Any real or complex root r of an nth degree polynomial $F(x) = a_0 + a_1 x + \cdots + a_n x^n$ satisfies the inequality

$$r \leq 1 + \max_{k=0}^{n} \left| \frac{a_k}{a_n} \right|.$$

You can find a simple proof in Householder [20, Section 2.2]; it's based on the triangle inequality and geometric series. The expression on the right is called *Cauchy's bound* for the roots of F. Figure 6.16 shows the MSP function template CauchyBound for computing it.

Multiplicity of a Root

Finding roots of a polynomial $F(x)$ becomes more complicated when it has a root of multiplicity greater than one. To determine whether you've found *all* roots, you often need to count them according to multiplicity. That concept is easiest to understand after you've considered *Taylor's theorem* for polynomials: If $F(x)$ is a polynomial, n = degree(F), and t is a scalar, then there are unique scalars $a_0, \ldots, a_n$ such that

$$F(x) = \sum_{k=0}^{n} a_k (x-t)^k ;$$

in fact, $a_k = F^{(k)}(t)/k!$ for $k = 0, \ldots, n$, where $F^{(k)}(x)$ stands for the kth derivative of $F(x)$. From Taylor's theorem you can see that the following three conditions on a positive integer $m \leq n$ are equivalent:

- $F(x)$ is divisible by the polynomial $(x - t)^m$

- $F(x) = \sum_{k=m}^{n} a_k (x - t)^k$

- $F^{(k)}(t) = 0$ for $k < m$.

When $(x - t)^m$ divides $F(x)$ but $(x - t)^{m+1}$ does not, t is called a root of *multiplicity m*. Thus, t is a root of multiplicity m if the values of F and its first $m - 1$ derivatives at t are zero but $F^{(m)}(t) \neq 0$. Roots of multiplicity 1,2,3,... are called *single, double, triple*, ... It's useful to assign *multiplicity* 0 to numbers that *aren't* roots; if $F(x)$ is the zero polynomial, assign *every* number multiplicity –1.

If t is a root of $F(x)$ of multiplicity m, then

$$F(x) = (x - t)^m (a_m + a_{m+1}(x - t) + \cdots)$$

so $F(x)$ has the same sign as $(x - t)^m a_m$ for x close enough to t. Thus, $F(x)$ changes sign at t if, and only if, m is odd.

MSP provides function template Vector<int> Multiplicities(x,P,E) to compute the multiplicity, with respect to polynomial P, of each entry x[j] of vector x. It returns these integers as the corresponding entries of a Vector<int> object. You'll find the source code in Figure 6.17. It's straightforward, although you'll have to ponder its nested for loop to see that it actually implements the definitions in the previous text.

```
template<class Scalar>                      // Return Cauchy's
  double CauchyBound(                       // bound for the
      const Polynom<Scalar>& F) {           // absolute values of
    try {                                   // the roots of  F .
      int n = Degree(F);
      if (n <= 0) return 1;
      double M = abs(F[0]);
      for (int k = 1; k < n; ++k)
        M = max(M,abs(F[k]));
      return 1 + M/abs(F[n]); }
    catch(...) {
      cerr << "\nwhile computing  CauchyBound  for a polynomial";
      throw; }}
```

Figure 6.16. Polynom function CauchyBound

```
template<class Scalar>                  // Return a vector  M
  Vector<int> Multiplicities(           // such that for each
       const   Vector<Scalar>& x,       // j ,  M[j]  is the
       const Polynom<Scalar>& P,        // multiplicity of
                  double    E) {        // x[j]  as a root
    try {                               // of  P .  Regard as
      int High = x.HighIndex(),         // zero any scalar
          Low  = x.LowIndex();          // with  abs < E .
      Vector<int> M(High,Low);
      if (Degree(P) == -1) {
        return AllEntries(M,-1); }
      int Mj; Polynom<Scalar> Q;
      for (int j = Low; j <= High; ++j) {
        for (Mj = 0, Q = P; abs(Q(x[j])) < E; ++Mj) Q = Deriv(Q);
        M[j] = Mj; }
      return M; }
    catch(...) {
      cerr << "\nwhile computing the  Multiplicities  of a "
              "polynomial's roots";
      throw; }}
```

Figure 6.17. Polynom function Multiplicities

Let $t_1, \ldots, t_l$ denote all the roots of $F(x)$, with multiplicities $m_1, \ldots, m_l$, and consider a prime decomposition

$$F(x) = Q(x) \prod_{j=1}^{l} (x-t)^{m_j} ,$$

where $Q(x)$ is a product of prime factors of $F(x)$ with no roots. From this equation it's apparent that

$$\sum_{j=1}^{l} m_j \leq n .$$

That is, $F(x)$ has *at most* n roots, counted by multiplicity. The fundamental theorem of algebra states that a real or complex polynomial of degree n has *exactly* n *complex* roots, counted by multiplicity. Thus, if you're looking for *all* the roots, you can stop when you've found n of them, counted this way; and you never need to look outside the circle centered at the origin whose radius is Cauchy's bound.

Finding Real Roots

The considerations under previous headings in this section apply to both real and complex polynomial roots. Chapter 4 discussed some techniques that lead just to real roots. It's easy to adapt them to apply to the particular case of polynomials. MSP's `Polynom` module includes functions

```
Boolean Bracket(double&   xL,       // Find a bracketing interval
                double&   xR,       // [xL,xR]  of length  <= T
   const Polynom<double>& F,        // within the specified one.
                double    T);       // Return  True  if successful.

double NRBisect(                    // Return an approximation
   const Polynom<double>& F,        // to a root of  F ,
                double    xL,       // computed by the hybrid
                double    xR,       // Newton-Raphson
                double    T,        // Bisection method.
                int&      Code);
```

In fact, the *only* change in function `Bracket` is its calling sequence. The changes required for function `NRBisect` are analogous to those for `NR`, discussed earlier under the heading "Implementing Newton-Raphson Iteration." The lines on the left in the following code segment are from Figure 4.18; on the right are their adaptations:

```
 —— Equate1.H ——                     —— Polynom.H ——
    ⋮                                    ⋮
double x,h,yL,yLp,y,yp;             double x,h,yL,y,yp;
x = (xL + xR)/2;                    Polynom<double> Q;
h = x - xL;                         x  = (xL + xR)/2;
yL = FdF(xL,yLp);                   h  = x - xL;
do {                                yL = F.Horner(Q,xL);
    y = FdF(x,yp);                  do {
    ⋮                                   y  = F.Horner(Q,x);
                                        yp = Q(x);
                                        ⋮
```

Because only these few changes were required, there's no need to display the full source code for these versions of `Bracket` or `NRBisect`. You'll find it in file `Polynom.CPP` on the optional diskette.

Some polynomials, such as $x^2 + 1$, have no real roots at all. There are theorems that indicate when polynomials must have real roots. The only simple one states that any polynomial $F(x)$ of odd degree must have at least one root, because it's continuous and its values for sufficiently large positive and negative x have opposite signs.

Function `NRBisect` is the basic MSP tool for solving polynomial equations. But it requires an initial bracketing interval. A polynomial whose roots all have even multiplicity never changes sign, so it has *no* bracketing inverval. Even when one exists, how do you find it? To use function `NRBisect` to approximate a root t of a polynomial $F(x)$, for which there may be no bracketing interval, you must replace $F(x)$ by a related function that does change sign at t. You can accomplish that by dividing $F(x)$ by a greatest common divisor $G(x)$ of $F(x)$ and $F'(x)$. Such a $G(x)$ is computed by the function `GCD` described in Section 6.2. Why does this work? In a prime decomposition of $F(x)$, each linear factor $x - t_j$ has exponent m_j. In the decomposition of $F'(x)$, this factor has exponent $m_j - 1$. Therefore, each factor $x - t_j$ also occurs in the decomposition of $G(x)$ with exponent $m_j - 1$. When you divide $F(x)$ by $G(x)$, you get a quotient $Q(x)$ with linear factors $x - t_j$, each with exponent 1. Finally, $Q(x)$ has no further roots: Since $F(x) = G(x)Q(x)$, any root of $Q(x)$ is a root of $F(x)$. Thus, $Q(x)$ has the same roots as $F(x)$, but they all have multiplicity 1.

Using the GCD idea in the last paragraph, you can convert any polynomial into one for which all roots have bracketing intervals. Function `Bracket` can then help you find the intervals. However, it will only find a bracketing interval that's inside an initially specified interval. How can you select an initial interval that's guaranteed to be large enough? Compute the Cauchy bound C and then apply function `Bracket` to the initial interval $[-C,C]$.

Finding *All* Real Roots

The technique outlined under the previous heading finds only one real root at a time. How do you find several in succession? How do you know when you've found them all? Here's the most common method:

1. Handle constant polynomials first as special cases.
2. If 0 is a root, record that; find its multiplicity m; and divide $F(x)$ by x^m.
3. Divide $F(x)$ by the GCD of $F(x)$ and $F'(x)$ to replace $F(x)$ by a polynomial with the same roots, all single.
4. If the new polynomial $F(x)$ is linear or quadratic, use elementary algebra to find a root or decide there's none.
5. Use `CauchyBound` and `Bracket` to find a bracketing interval. `Bracket` may report that there's none. Otherwise, use `NRBisect` to find a root t.
6. Use `Horner` to divide $F(x)$ by $x - t$, obtaining a quotient polynomial $Q(x)$.
7. Apply steps 4 through 7 recursively to find the roots of $Q(x)$ — that is, replace $F(x)$ by $Q(x)$ and repeat from step 4.

Step 6 is called *deflation*. *Caution:* Subtraction of nearly equal quantities in the quadratic formula numerator can cause loss of accuracy. You may want to incorporate some safeguards [15].

```
Vector<double> RealRoots(            // Return a vector of all
    Polynom<double> F,               // real roots of  F ,  ap-
          double  S,                 // proximated by
          double  T,                 // NRBisect . Ignore any
          int&    Code) {            // two separated by  < S .
  try {                              // Use parameters  T  and
    int n = Degree(F);               // Code  like  NRBisect .
    if (n < 0) {
      Code = 0;                      // Return one root  0  for
      Vector<double> V(1); V[1] = 0; // the zero polynomial.
      return V; }
    if (n == 0) {                    // A nonzero constant
      Code = 0;                      // polynomial has no root:
      Vector<double> V(0);           // return an empty vector.
      return V; }
    int k; for (k = 0; F[k] == 0; ++k);// Deflate if  0  is a
    F /= xTo(k);                     // root.
    Polynom<double> D = Deriv(F),    // Divide by  GCD(F,F')
                    G = GCD(F,D,T);  // to discard duplicate
    F /= G;                          // roots.
    n = Degree(F);                   // There are at most  n+1
    Vector<double> Root(n+1);        // roots. Allocate their
    if (k > 0) {                     // storage. If  0  is a
      k = 1;  Root[k] = 0; }         // root, report it. From
    Boolean Done = False;            // now on,  k  counts
    int C, C1 = 0;                   // roots.  C , C1  handle
    double B  = CauchyBound(F);      // the  Code  parameter.
```

Figure 6.18. Polynom function RealRoots
Part 1 of 2; continued in Figure 6.19.

Figures 6.18 and 6.19 display MSP function RealRoots, which implements this deflation technique to approximate all real roots of a double polynomial $F(x)$. The corresponding parameter F is passed by value because it's used to store successive deflations of F. Since it uses NRBisect to calculate the approximations, RealRoots has a parameter T for specifying the tolerance and a reference parameter Code for requesting intermediate output and reporting success or failure. (See Appendix A.3 for intermediate output details. The dual nature of the Code parameter requires its awkward manipulation with RealRoots variables C and C1.) Since it uses Bracket to find a bracketing interval, RealRoots also has a parameter S for specifying the root separation: It won't search intervals shorter than S for a sign change. There's one further awkward programming point: When it can make a reasonable estimate of the number of roots, RealRoots allocates a vector Root for storing them and uses a variable k to count them. When it's finished, Root may not be full, so its first k entries are copied to a new vector for return to

```
    while (!Done) {                            // B is a bound for
      if (n > 2) {                             // the roots. Use it
        double xL = -B, xR = B;                // to find a bracket-
        if (n%2 == 0 &&                        // ing interval.
          !Bracket(xL,xR,F,S))                 // Quit if F never
            Done = True;                       // changes sign.
        if (!Done) {
          C = Code;                            // Find, record,
          Root[++k] = NRBisect(F,xL,xR,T,C);   // and count a
          C1 |= C;                             // root, then
          Polynom<double> Q;                   // deflate.
          F.Horner(Q,Root[k]);                 // Deflation may
          F = Q;                               // decrease the
          B = min(B,CauchyBound(F));           // Cauchy bound.
          --n; }}
      else {                                   // Elementary
        Done = True;                           // algebra.
        if (n == 2) {
          double D =
            F[1]*F[1] - 4*F[2]*F[0];           // Discriminant.
          if (D >= 0) {                        // No root if
            double S = sqrt(D),                // D < 0 .
                   T = 2*F[2];                 // Quadratic
            Root[++k] = (-F[1] + S)/T;         // formula. k
            Root[++k] = (-F[1] - S)/T; }}      // counts roots!
        else                                   // Linear
          Root[++k] = -F[0]/F[1]; }}           // equation.
    Code = C1;
    Vector<double> V(k);                       // Copy the roots into a
    V.Copy(1,Root,1,k);                        // vector of length k
    return V; }                                // and return it.
  catch(...) {
    cerr << "\nwhile computing the real roots of a polynomial";
    throw; }}
```

Figure 6.19. Polynom function RealRoots
Part 2 of 2; continued from Figure 6.18.

the caller. This awkwardness stems from the fact that an MSP Vector object isn't quite the appropriate data structure for accumulating roots.

The mathematical aspects of RealRoots are straightforward. It follows the outline of the algorithm except in one small detail. With each recursive step the polynomial F changes; hence, its Cauchy bound changes. But the roots don't change, so RealRoots always uses the smallest Cauchy bound computed so far, to keep Bracket from wasting time.

Testing with Legendre Polynomials

The Legendre polynomials L_n, introduced in Section 6.2, provide a doubly useful test of functions RealRoots and Polish, because their roots play a significant role in approximating integrals. L_n has n distinct roots, which all lie in the open interval (–1,1). Since L_n is either odd or even, the roots are located symmetrically about zero, and zero *is* a root if n is odd. At right in the following chart are the root approximations that result from executing the code at left:

`Separation = 0.01;`	–.973907	+.973907
`Tolerance   = 1e-3;`	–.865063	+.865063
`RealRoots(Legendre(10),`	–.679410	+.679412
`          Separation,`	–.433393	+.433385
`          Tolerance,`	–.148884	+.148889
`          Code)`		

Root estimates $t_1, \ldots, t_m$ computed later in the deflation process tend to be less accurate than those produced at early stages, because the coefficients of the deflated polynomials are the results of long sequences of calculations depending on all earlier steps. Round-off error can grow noticeably. Therefore, it's usually prudent to *polish* the estimates after deflation. The simplest method is to compute new root approximations using Newton-Raphson iteration with the original (not deflated) polynomial and the initial approximations $t_1, \ldots, t_m$. This process is implemented by MSP function Polish, described earlier in this section.

Polish was tested by applying it to the vector $[r_1, \ldots, r_{10}]$ of root approximations computed earlier for Legendre polynomial L_{10}. The largest value $|L_{10}(r_j)|$ was 4.0×10^{-5}. With the same tolerance 10^{-3}, polishing required one Newton-Raphson iteration; the polished approximations are denoted by $[s_1, \ldots, s_{10}]$. The largest value $|L_{10}(s_j)|$ was 1.9×10^{-10}, which represents a slight improvement. The largest adjustment in a single root was $|r_j - s_j| \approx 1.5 \times 10^{-5}$. Here are the original and polished root estimates, side by side:

Original	*Polished*	*Original*	*Polished*
–.973907	–.973907	+.973907	+.973907
–.865063	–.865063	+.865063	+.865063
–.679410	–.679410	+.679412	+.679410
–.433393	–.433395	+.433385	+.433395
–.148884	–.148874	+.148889	+.148874

Finding Complex Roots

You've probably seen many application problems whose solutions are real roots of some polynomial. What kind of problem leads to *complex* roots? One of the more familiar examples is an nth-order linear homogeneous ordinary differential equation (ODE) $a_0 y + a_1 y' + a_2 y'' + \cdots + a_n y^{(n)} = 0$ with constant real coefficients $a_0, \ldots, a_n$. For what values λ does the function $y = e^{\lambda x}$ satisfy the ODE? With that function, $y^{(k)} = \lambda^k y$ for $k = 0, \ldots, n$, so

$$(a_0 + a_1 \lambda + a_2 \lambda^2 + \cdots + a_n \lambda^n) y = 0,$$
$$a_0 + a_1 \lambda + a_2 \lambda^2 + \cdots + a_n \lambda^n = 0.$$

The real solutions $y = e^{\lambda x}$ occur just when λ is a real root of the polynomial with coefficients $a_0, \ldots, a_n$. When $\lambda = a + bi$ is a complex root, $y = e^{\lambda x} = u(x) + v(x)i$ is a complex solution of the ODE, where $u(x) = e^{ax}\cos bx$ and $v(x) = e^{ax}\sin bx$. But then

$$0 = a_0 y + a_1 y' + a_2 y'' + \cdots + a_n y^{(n)}$$
$$= (a_0 u + a_1 u' + a_2 u'' + \cdots + a_n u^{(n)}) + (a_0 v + a_1 v' + a_2 v'' + \cdots + a_n v^{(n)}) i$$
$$a_0 u + a_1 u' + a_2 u'' + \cdots + a_n u^{(n)} = 0 = a_0 v + a_1 v' + a_2 v'' + \cdots + a_n v^{(n)}.$$

Thus, the single complex root λ yields *two* real ODE solutions: $y = e^{ax}\cos bx$ and $y = e^{ax}\sin bx$.

Some of the root finding techniques considered in Chapter 4 and earlier in this section work for complex equations; some do not. Those that involve the bisection method are based on the *ordering* of real scalars; they cannot be directly extended to complex equations.

A method known as *Lehmer's* could be regarded as a kind of extension of the bisection method for complex polynomials. It uses a theoretical criterion for the existence of a root within a circle of given center and radius. One certainly exists within the circle centered at 0 with the Cauchy bound as radius. When a circle is determined to contain a root, it's covered by seven circles with smaller radii, and the same criterion is applied repeatedly until a smaller circle is found to contain a root. This process continues recursively until the radius of the enclosing circle is smaller than a specified tolerance. As you see, the idea is simple enough to describe in a paragraph. But the details of the algorithm are too involved to present here [20, Section 2.7].

The secant and Newton-Raphson methods do apply to complex equations, because the mathematics that underlies their convergence properties generalizes to complex functions. (If you want either method to converge to a nonreal root, you'll have to start with a nonreal initial estimate; otherwise, the algebra will result only in successive real approximations.) The MSP adaptation of Newton-Raphson iteration for double or complex polynomials was described in earlier in this section. The secant method isn't considered further because it presents the same problems as Newton-Raphson.

Numerical analysts have learned from experience that in complex applications, Newton-Raphson iteration frequently *overshoots*—that is, the derivative used in correcting an approximation x_n to a desired root x of a function f can be so small that the next approximation

$$x_{n+1} = x_n - \delta x_n = x_n \frac{f(x_n)}{f'(x_n)}$$

becomes huge, relative to the size of the root. This results in overflow or at least requires many iterations to get back to an approximation near x. A possible remedy is to ensure that $|\delta x_n|$ doesn't exceed $|x|$, by dividing the correction by $|x|$ if necessary. This idea is implemented by MSP function NRStep, which is exactly like function NR except for these code fragments:

Code from function NR *described earlier*

```
Fx = F.Horner(Q,x);
dF = Q(x);
if (Fx == Scalar(0)) dx = 0;
   else dx = Fx/dF;



x -= dx; }
```

Corresponding NRStep *code*

```
Fx = F.Horner(Q,x);
dF = Q(x);
if (Fx == Scalar(0)) dx = 0;
   else dx = Fx/dF;
double N = abs(x);
if (x != Scalar(0)
      && abs(dx) > N) dx /= N;
x -= dx; }
```

Function NRStep underlies the MSP routine for computing *all* roots of a double or complex polynomial. Here's its template, from header file Polynom.H:

```
template<class Scalar>          // Return a vector of all
   Vector<complex> Roots(       // the roots of  F ,  as
      Polynom<Scalar> F,        // approximated by
               double T,        // NRStep .
                 int& Code);
```

Roots operates much like the routine RealRoots described earlier in this section which computes all *real* roots of a *real* polynomial. However, to begin approximating each root *r*, the earlier routine used the Cauchy bound and function Bracket to find an interval containing *r*; it then employed a hybrid Newton-Raphson/bisection algorithm to ensure that *r* is ultimately captured inside an interval whose length is smaller than the tolerance *T*. The bisection hybrid is not applicable to complex polynomials, so Roots uses another, less reliable, starting device. First, an initial estimate $x_0 + y_0 i$ for Newton-Raphson iteration is chosen *randomly* so that $0 \leq x_0, y_0 < 1$. If NRStep doesn't achieve convergence within the specified tolerance after 1000 iterations, Roots tries another random initial estimate. It makes up to ten such attempts before returning the last approximation—which is probably incorrect—and setting parameter Code = 1. The source code for Roots is shown in Figure 6.20. Compare it with the RealRoots code shown in Figures 6.18 and 6.19. You'll see that the code differences stem from

- The randomized initial estimates
- The possible need to convert double to complex data
- Possible complex square roots in the quadratic formula
- The fact that as soon as you've eliminated multiple roots, you know *exactly* how much memory to allocate for the returned vector (if a complex polynomial of degree n has only *single* roots, it has *exactly* n roots)

```
template<class Scalar>                  // Return a vector  of all
  Vector<complex> Roots(                // roots of  F ,  approxi-
      Polynom<Scalar> F,                // mated by  NRStep .
              double  T,                // Use parameters  T  and
              int&    Code) {           // Code  like  NRStep .
    try {
      int n = Degree(F);
      if (n <  0) {
        Code = 0;                       // Return one root  0  for
        Vector<complex> V(1); V[1] = 0; // the zero polynomial.
        return V; }
      if (n == 0) {                     // A nonzero constant
        Code = 0;                       // polynomial has no root:
        Vector<complex> V(0);           // return an empty vector.
        return V; }
      int k; for (k = 0;                // Deflate if  0  is
        F[k] == Scalar(0); ++k);        // a root.
      Polynom<Scalar> xk;  xk = xTo(k);
      F /= xk;                          // Divide by this GCD
      Polynom<Scalar>  D = Deriv(F),    // to discard dupli-
                       G = GCD(F,D,T);  // cate roots. Con-
      Polynom<complex> FF; FF = F/G;    // vert to  complex .
      n = Degree(FF);                   // There are  n  or
      Vector<complex>                   // n+1  roots. Allocate
        Root(k == 0 ? n : n + 1);       // their storage. If  0
      if (k > 0) {                      // is a root, report it.
        k = 1;  Root[k] = 0; }          // From now on,  k  counts
      int C, C1 = 0;                    // roots.  C , C1  handle
```

Figure 6.20. Polynom function Roots
Continues on the following page.

```
      while (n > 2) {                    // the Code parameter.
        int m = 0; complex x;            // Deflation loop, for
        do {                             // nontrivial cases. Do
          MakeRandom(x);                 // NRStep m <= 10 times
          C = Code;                      // with random initial
          x = NRStep(FF,x,T,C); }        // estimates.
          while (C != 0 && ++m < 10);
        Root[++k] = x;                   // Record the root, set
        if (C != 0) C1 = 1;              // C1 = 1 if it's
        Polynom<complex> Q;              // unreliable.
        FF.Horner(Q,x);                  // Deflate.
        FF = Q; n = Degree(FF); }
      if (n == 2) {
        complex D = FF[1]*FF[1] - 4*FF[2]*FF[0],  // Quadratic
                S = sqrt(D),                      // formula.
                U = 2*FF[2];
        Root[++k] = (-FF[1] + S)/U;      // k counts
        Root[++k] = (-FF[1] - S)/U; }    // roots.
      if (n == 1)                        // Linear
        Root[++k] = -FF[0]/FF[1];        // equation.
      Code = C1;
      return Root; }
    catch(...) {
      cerr << "\nwhile computing a polynomial's Roots .";
      throw; }}
```

Figure 6.20. Polynom function Roots
Continued from previous page.

Before beginning Newton-Raphson iteration to approximate a root of a polynomial f RealRoots divided f by the GCD of f and f'. This ensured that the root has multiplicity 1, so that Bracket could detect a sign change to determine the initial interval. Although that step is unnecessary for complex polynomials, Roots includes it anyway, because it eliminates the problem of extremely slow convergence to multiple roots. It's conceivable that the extra computations required to compute the GCD could cause accuracy problems with some polynomials f. You may want to modify Roots slightly, so that, for example, an input value 2 for the Code parameter would cause it to bypass the GCD computation.

MSP functions Polish and Multiplicities, discussed earlier in this section, can be used unaltered with vectors returned by Roots to polish approximate roots of complex polynomials and determine their multiplicities.

NRStep and Roots were suggested for use with complex scalars, but they work as well with double. Executing Roots on the Section 4.4 "bouncing" polynomial

$$f(x) = x^5 - 8x^4 + 17x^3 + 8x^2 - 14x - 20$$

with tolerance $T = 0.01$ produces these results, accurate to three decimal places:

Initial Estimate	Iterations	Root
−12	21	1.465
10 + 10i	10	3.969 + 1.430i
10 − 10i	10	3.969 − 1.430i
−10 + 10i	12	−0.701 + 0.524i
−10 − 10i	12	−0.701 − 0.524i

This illustrates the theorem that the complex roots of a real polynomial occur in conjugate pairs. Section 4.4 reported that the unmodified Newton-Raphson routine NR required more than 50 iterations to reach root 1.465 from the initial estimate −12, and the hybrid Newton-Raphson/bisection routine NRBisect required 14. Overstep limitation seems reasonably successful in this case.

7

Matrix Classes

The MSP `Matrix` class implements an abstract data type for matrix algebra. It's based on the `Vector` class and provides for

- scalars of any type supported by the MSP `Scalar` module (`double`, `complex`, or `int`)
- matrices with varying starting indices
- matrices of varying dimensions
- bounds checking.

The first provision allows you to perform real, complex, and integer matrix algebra in the same program using formulas that closely resemble conventional mathematics. The second provision means that you can use matrices

$$A = \begin{bmatrix} a_{00} & a_{01} & \cdots \\ a_{10} & a_{11} & \cdots \\ \vdots & \vdots & \end{bmatrix} \qquad B = \begin{bmatrix} b_{11} & b_{12} & \cdots \\ b_{21} & b_{22} & \cdots \\ \vdots & \vdots & \end{bmatrix}$$

in the same context, too. The `Matrix` class provides complete but flexible bounds checking: Matrix operations check their operands for consistency of lower and upper index bounds. References to entries a_{ij} of matrix A are implemented by the `Vector` operator that selects an entry of the vector representing row i of A. As noted in Section 5.1, you can easily turn off that operator's error checking to increase your program's run-time efficiency.

MSP functions that manipulate matrices are classified as *logical* or *mathematical*. The logical `Matrix` functions, described in Section 7.1, almost exactly parallel the logical `Vector` functions described in Section 5.1. The mathematical functions implement matrix algebra; they're discussed in Section 7.2. You'll find many examples of the use of these features in Chapters 8 and 9.

7.1 Matrix MODULE INFRASTRUCTURE

> **Concepts**
> *Design goals*
> *Files* Matrix.H, MatrixL.CPP, *and* MatrixM.CPP
> Matrix *classes*
> *Selectors, subscript operator*
> *Constructors and destructor*
> *Converting row and column matrices to and from* Vector *types*
> *Copiers and assignment*
> *Equality and inequality*
> *Keyboard input and display output*
> *Stream input/output*

As with other MSP modules, the Matrix module has a single header file Matrix.H. Its source code is so long, however, that it's split into two files MatrixL.CPP and MatrixM.CPP for logical and mathematical functions. You'll find the header file on the accompanying diskette and in Appendix A.2, and the entire source code on the optional diskette. The logical part of the header file is shown in four figures, starting with Figure 7.1. It automatically includes the MSP header file Vector.H and other MSP and Borland C++ header files that Vector.H loads. Parts of MatrixL.CPP are displayed later in this section. The mathematical part of the header file and its corresponding source code will be discussed in Section 7.2.

The Matrix Data Structure

The Matrix module is designed to handle matrices of scalars of any type supported by the MSP Scalar module, so the Matrix classes are implemented via a template, as shown in Figure 7.1. The design must permit matrices of varying dimensions. To facilitate ascertaining the lower index bound and upper row and column index bounds, they form data components Low, HighRow, and HighCol of Matrix objects. Moreover, it was decided that a matrix should be stored as an array of vectors, so that the extensive MSP vector handling features could be used for matrices, too. Since the array size is indeterminate, it must be located in free memory. The Row component of a Matrix object is a *pointer* to its first entry in free memory. If m = HighRow − Low + 1, then vectors Row[0], ..., Row[m] represent the rows of the matrix. A row's memory image consists of its lower and upper index bounds and a pointer to another array in free memory that stores its Scalar entries. Duplicating the index bounds Low and HighRow in each row vector is inefficient use of memory. In view of the overall goal of making numerical analysis programs easier to read,

```
//*******************************************************************
//                        Matrix.H

#ifndef   Matrix_H
#define   Matrix_H

#include "Vector.H"

template<class Scalar>
  class Matrix {                        // The private parts of
                                        // class  Matrix  are its
     private:                           // data components.
        int               Low;          // Index bounds for the
        int               HighRow;      // matrix entries.
        int               HighCol;
        Vector<Scalar>*   Row;          // Row vector storage.
```

Figure 7.1. Matrix.H logical components
Part 1 of 4; continued in Figure 7.2.

however, this was judged an acceptable cost. In order to enforce bounds checking, the index bounds and row array must be private components, and member functions must be provided to manipulate them as needed. In designing the class, no other private components proved necessary.

MSP uses a standard representation for *empty* matrices: Low = 0, HighRow = HighCol = −1, and Row = NULL.

It's sometimes more convenient mathematically to regard a matrix as an array of *column* vectors instead of row vectors. Unfortunately, in defining the structure, you have to choose one or the other, and C++ array indexing conventions make the row vector array, chosen for MSP, easier and more efficient to program. As you'll see, operations with column vectors are consequently somewhat clumsy and inefficient.

An alternative MSP design strategy would make *matrices* basic, rather than vectors. The matrix data structure would consist of the three index bounds and a pointer to a two-dimensional array of scalars in free memory. Vectors could be defined and manipulated as matrices with a single row or a single column, as appropriate for an application. That would save storage space, and some Vector module code could be eliminated. The cost is the disappearance of the data structure corresponding to a *row of a matrix*. The row structures wouldn't be available in that format in the matrix structure and would have to be created whenever an application needed them.

According to the overview presented in Chapter 2, the functions necessary for logical manipulation of matrices include

- *Selectors*, to provide limited access to the data components
- *Constructors*, to initialize the index bounds and the row array pointer and—usually—to set up the rows in free memory

- A *destructor*, to deallocate the row vectors when a matrix is deleted or passes out of scope
- *Converters*, to convert row and column matrices to and from vector format
- A *copy constructor*, for the compiler to use during initialization or function calls
- An *assignment* operator, overloaded to copy the row vectors
- A general *copier*, for manipulating pieces of matrices
- Overloaded *equality* and *inequality* operators, to compare the index bounds and row vectors
- *Keyboard input* and *display output* functions
- *Stream input/output* functions

These features are described in detail in the rest of this section. Sometimes, source code is given here also. For the display output function, you're referred to Appendix A.1. Three (slightly) mathematical functions, whose names begin with Make..., are included in Figure 7.2, which continues the class template started in Figure 7.1, because you may need them to test this sections's other features. They're described in Section 7.2. You'll find all the code on the optional diskette.

Selectors

Selectors are public member functions that provide limited access to private data components—for example, the Matrix classes provide functions LowIndex, HighRowIndex, and HighColIndex to return the Low, HighRow, and HighCol components. They're declared in Figure 7.2.

Since the row vectors of a matrix are stored in a private array in free memory, you can't access them directly with the subscript operator [ ]. Instead, this must be overloaded so that if A is a Matrix object and k is an int value within the A row index bounds, then A[k] is a reference to the appropriate Row entry in free memory. Since the A and Row indices range over intervals starting with Low and zero, respectively, the appropriate Row entry has index k − Low. Here's the source code:

```
template<class Scalar>                          // Row selector.  If
   Vector<Scalar>& Matrix<Scalar>::             // j  is in bounds,
      operator [](int j) const {                // adjust for a non-
   if (Low <= j && j <= HighRow)                // zero low index.
      return Row[j-Low];                        //   Other-
   cerr << "\nException   IndexError"           //   wise
        "\nwhile selecting the  jth row of a  " //   report,
            "Matrix  with  Low,j,HighRow = "    //   throw an
         << Low << ',' << j << ',' << HighRow;  //   MSP ex-
   throw(Exception(IndexError)); }              //   ception.
```

```
public:

  // Selectors  ------------------------------------------

  int LowIndex()             const;     // Selectors for
  int HighRowIndex()         const;     // the index
  int HighColIndex()         const;     // bounds.

  Vector<Scalar>& operator [](          // Row selector.
                    int j) const;
  Vector<Scalar> Col(int k) const;      // Column selector.

  // Constructors and destructor  -------------------------

  Matrix();                             // Default constructor.

  Matrix& SetUp(int HiRow,              // Set index bounds and
                int HiCol,              // allocate storage.
                int Lo = 1);

  Matrix(int HiRow,                     // Construct a matrix with
         int HiCol,                     // specified index bounds.
         int Lo = 1);

  Matrix& MakeRandom();                 // Make entries random.
  Matrix& MakeZero();                   // Make all entries zero.
  Matrix& MakeIdentity();               // Set = identity matrix.

  ~Matrix();                            // Destructor.

  // Converters to and from type Vector  -----------------

  operator Vector<Scalar>();            // Row, column to vector.

  Matrix(const Vector<Scalar>& V);  // Vector to row matrix.
```

Figure 7.2. Matrix.H logical components
Part 2 of 4; see also Figures 7.1 and 7.7.

The matrix row selector lets you use the standard C++ double subscripting technique to access matrix entries: A[j][k] is the kth entry of the jth row of A, provided j and k fall within bounds.

Unfortunately, in C++ you can't use the more conventional double subscripting notation A[j,k]. The reason is that the *comma* is a standard C++ operator: Executing j,k evaluates j,

```
template<class Scalar>                         // Column selector.
  Vector<Scalar> Matrix<Scalar>::              // If  k  is in
    Col(int k) const {                         // bounds, construct
    try {                                      // a vector, fill it
      if (Low <= k && k <= HighCol) {          //   with the col-
        Vector<Scalar> V(HighRow,Low);         //   umn entries
        for (int j = Low; j <= HighRow; ++j)   //   and return it.
          V[j] = This[j][k];                   // Otherwise, re-
        return V ; }                           // port and throw
      cerr << "\nException IndexError";        // an MSP
      throw(Exception(IndexError)); }          // exception.
    catch(...) {
      cerr << "\nwhile selecting the  kth column of a  Matrix  "
              "with  Low,k,HighCol = "
           << Low << ',' << k << ',' << HighCol;
      throw; }}
```

Figure 7.3. Matrix column selector Col

then k, and returns k. That's not what you want. If you overloaded the comma to implement conventional double subscript notation, you'd disable its standard use.

The matrix *column selector* is a member function Col with one parameter, the column index. Thus, the vector A.Col(k) is the kth column of matrix A, provided k falls within bounds. The column selector is less efficient than the row selector, since it must build a new vector. Its source code is displayed in Figure 7.3.

Constructors

As discussed in Section 2.6, each Matrix class needs a *default constructor*, which the compiler will call n times whenever you declare an array of n matrices. It needn't do much; the function just initializes an empty matrix by setting Low = 0, HighRow = HighCol = -1, and Row = NULL.

When you declare a matrix, however, you usually want to set up specific index bounds and allocate its row storage arrays in free memory. This process needs to be carried out in several different situations, so it's implemented as a separate public member function SetUp(int HiRow, int HiCol, int Low = 1). The Matrix.H header file specifies the default parameter. Figure 7.4 contains the source code. SetUp will simply initialize an empty matrix if HighRow = Low - 1. If HighRow < Low - 1, it displays an error message and throws an MSP Exception object, which specifies an index error. Otherwise, SetUp asks operator new to allocate space for an array of

```
template<class Scalar>                        // Set the index
  Matrix<Scalar>& Matrix<Scalar>::            // bounds and
    SetUp(int HiRow,                          // allocate storage.
          int HiCol,
          int Lo) {
    int j = Lo - 1;
    try { try {
      int m = HiRow - Lo + 1;                 // m rows.  Index
      if (m < 0) {                            // error if HighRow
        cerr << "\nException IndexError";     // < Low - 1 :
        throw(Exception(IndexError)); }       // report and throw
      if (m == 0) {                           // an MSP exception.
         HighRow = -1;      Low = 0;          // If  HighRow =
         HighCol = -1;      Row = NULL; }     // Low - 1 , set up
       else {                                 // an empty matrix.
         HighRow = HiRow;   Low = Lo;         // Otherwise, set the
         HighCol = HiCol;                     // index bounds and
         Row = new Vector<Scalar>[m];         // allocate memory
         for (j = Lo; j <= HiRow; ++j)        // for the rows.
           This[j].SetUp(HiCol,Lo); }         // Return This for
      return This; }                          // chaining.
    catch(xalloc) {                           // If memory is ex-
      cerr << "\nException OutOfMemory";      // hausted, throw an
      throw(Exception(OutOfMemory)); }}       // MSP exception.
    catch(...) {
      if (j >= Lo) { delete[] Row; }            // Release allo-
      cerr << "\nwhile allocating memory "      // cated memory.
               "for a  Matrix  with  "           // Report excep-
               "HighRow,HighCol,Low = "          // tion details.
            << HiRow << ','                      // Set up an
            << HiCol << ',' << Lo;               // empty matrix,
      HighRow = -1;  Low =  0;                   // and pass on
      HighCol = -1;  Row = NULL;                 // the same
      throw; }}                                  // exception.
```

Figure 7.4. Matrix function SetUp

HighRow - Low + 1 row vectors, invoking the default Vector constructor to initialize each one as an empty row. If new succeeds, it returns a pointer to the allocated space, which SetUp assigns to Row. Then SetUp calls its Vector counterpart to set the row index limits properly and allocate free memory storage for the matrix entries. If new fails, it throws an xalloc object. SetUp catches that; its inner catch block displays an error message and throws an Exception object, which specifies an out-of-memory error. All the SetUp code described so far is contained in an outer try block. The outer catch block handles any MSP Exception thrown in either of these error situations or by the Vector function SetUp. It displays a message identifying the location of the error and then throws the same object to the next handler. If it succeeds, SetUp returns the value of its principal argument in order to permit member function concatenation.

Warning: As mentioned in Section 5.1, the MSP exception handling technique may not maintain program control when memory is exhausted.

To construct an array of double matrices (of possibly varying dimensions), execute the default constructor n times by declaring Matrix<double> A[n] and then call SetUp in a for loop n times to set the index bounds and allocate the memory. The matrices are then A[0], ..., A[n−1] and—for example— A[0][1][1] is the top left entry of A[0].

Client programs use the constructor

```
Matrix(int HiRow, int HiCol, int Lo = 1)
```

most often. It merely calls SetUp(HiRow,HighCol,Lo) to set up the index bounds and allocate the memory—for example, you may invoke it with the last parameter defaulted by declaring Matrix<double> A(2,3) to set up matrix A as shown on the left:

$$A = \begin{bmatrix} a_{11} & a_{12} & a_{13} \\ a_{21} & a_{22} & a_{23} \end{bmatrix} \qquad B = \begin{bmatrix} b_{00} & b_{01} & b_{02} \\ b_{10} & b_{11} & b_{12} \end{bmatrix}.$$

Alternatively, you can include the last parameter explicitly: Declaring Matrix<double> B(1,2,0) sets up a matrix B as displayed.

Destructor

When you construct a local Matrix object A by declaring Matrix<int> A, for example, the four components A.Low, A.HighRow, A.HighCol and A.Row are stored on the stack. The last is the address of an array of vectors in free memory. Each of these vectors, in turn, includes a pointer to another array—of scalars—in free memory. When control passes out of the block where A was declared, the first four components are destroyed. But you must provide a *destructor* to release the free memory areas. The compiler will arrange to execute the destructor at the proper time. Its source code is simple:

```
  template<class Scalar>                         // Destructor.
    Matrix<Scalar>::~Matrix() {                  // Destroy all row
      if (Row != NULL) delete[] Row; }           // vectors.
```

This deletes the entire array of HighRow − Low + 1 = *m* row vectors; during that process, it invokes the corresponding Vector destructor *m* times implicitly to release the corresponding scalar arrays. The compiler will also execute this function to destroy temporary Matrix objects it constructed for passing function parameters and return values and to complete the destruction when you execute delete A for a Matrix object A.

Converters

Matrix algebra is often applied to a vector by regarding it as a matrix with a single row or column. Thus, MSP should include functions to convert a vector to a row or column matrix and back again. To convert a row or column matrix to a vector, MSP provides the conversion operator Matrix::Vector(). Its code, shown in Figure 7.5, is straightforward. To convert a vector to a row matrix, use this constructor:

```
template<class Scalar>                              // Convert
  Matrix<Scalar>::Matrix(const Vector<Scalar>& V) { // vector
    try {                                           // V to a
      int Lo   = V.LowIndex(),                      // matrix
          HiCol = V.HighIndex();                    // with one
      SetUp(Lo,HiCol,Lo);                           // row.
      This[Lo] = V; }
    catch(...) {
      cerr << "\nwhile converting a  Vector  to a row  Matrix";
      throw; }}
```

```
template<class Scalar>                              // Convert a row or
  Matrix<Scalar>::                                  // column matrix to
    operator Vector<Scalar>() {                     // a vector.  If
  if (HighCol == Low)                               // there's only one
    return Col(Low);                                // row or column, re-
  else if (HighRow == Low)                          // turn that.  Other-
    return This[Low];                               // wise,
  cerr << "\nException  IndexError"                 // report,
          "\nwhile converting to a  Vector  a  "    // throw an
          "Matrix  with  HighRow,HighCol,Low = "    // MSP  ex-
       << HighRow << "," << HighCol << ","          // ception.
       << Low;
  throw(Exception(IndexError)); }
```

Figure 7.5. Matrix to Vector converter

```
template<class Scalar>                          // Convert a vector
  Matrix<Scalar> operator~(                     // to a column.
      const Vector<Scalar>& V) {
    try {
      int Lo = V.LowIndex(),                    // Allocate its
          Hi = V.HighIndex();                   // storage and insert
      Matrix<Scalar> C(Hi,Lo,Lo);               // its entries, one
      for (int j = Lo; j <= Hi; ++j)            // by one.
        C[j][Lo] = V[j];
      return C; }
    catch(...) {
      cerr << "\nwhile converting a  Vector  to a column  "
              "Matrix";
      throw; }}
```

Figure 7.6. Vector to column matrix converter

As index of the single row of the matrix, the constructor uses the vector's lower index bound. To convert a vector to a column matrix, it seemed appropriate to overload the ~ operator, which is used in another context to represent matrix transposition. Thus, if V is a vector with lower index bound 1, then ~V is a column matrix, and (~V)[j][1] = V[j] for each index i. The parentheses are necessary because operator ~ has lower precedence than []. Figure 7.6 displays the source code. This operator is not a member function, because principal argument syntax would be inappropriate.

Copiers

The Matrix classes' copy constructors use their Vector counterparts to copy each row. They're declared in header file Matrix.H, whose listing is continued in Figure 7.7. Their implementation, by the template shown in Figure 7.8, is straightforward.

You'll need another copier to manipulate parts of matrices—for example, to assemble a matrix for Gauss elimination or disassemble one after LU factorization. The source code documentation for the Copy function template shown in Figure 7.9 explains the details. It copies a specified piece of the source matrix to the target matrix starting at a given location. The function first ensures that the specified piece is consistent with the source matrix bounds and then trims it, if necessary, so that the copy doesn't overflow the target bounds. The copy operation is performed by repeated application of the analogous Vector function.

```
// Copiers, assignment, and equality    --------------------

Matrix(const Matrix& Source);       // Copy constructor.

Matrix& Copy(int    LoRow,          // Copy the source matrix
             int    LoCol,          // rectangle with indicat-
       const Matrix& Source,        // ed corner indices to
             int    SourceLoRow,    // the target matrix so
             int    SourceLoCol,    // that its  Lo   corner
             int    SourceHiRow,    // falls at target posi-
             int    SourceHiCol);   // tion   (LoRow,LoCol) .

Matrix& operator=(const             //
  Matrix<double>& Source);          // Assignment, possibly
                                    // with type conversion,
Matrix& operator=(const             // but only
  Matrix<complex>& Source);         // double-to-complex
                                    // conversion is
Matrix& operator=(const             // implemented.
  Matrix<int>& Source);             //

Boolean operator==(                 // Equality.
  const Matrix& B) const;
```

Figure 7.7. Matrix.H logical components
Part 3 of 4; see also Figures 7.2 and 7.10.

```
template<class Scalar>                    // Copy constructor.
  Matrix<Scalar>::Matrix(
    const Matrix& Source) {
   try {
     int Lo   = Source.Low,               // Copy the
         HiRow = Source.HighRow,          // index bounds,
         HiCol = Source.HighCol;          // allocate storage
     SetUp(HiRow,HiCol,Lo);               // for the rows,
     for (int j = Lo; j <= HiRow; ++j)    // and copy them.
       This[j] = Source[j]; }
   catch(...) {
     cerr << "\nwhile copying a  Matrix";
     throw; }}
```

Figure 7.8. Matrix copy constructor

```
template<class Scalar>                        // Copy the source
  Matrix<Scalar>& Matrix<Scalar>::            // matrix, between
      Copy(          int TargetLoRow,         // indicated corners,
                     int TargetLoCol,         // to This . The
           const Matrix<Scalar>& Source,      // Lo corner should
                     int SourceLoRow,         // fall at
                     int SourceLoCol,         // (LoRow,LoCol) .
                     int SourceHiRow,
                     int SourceHiCol) {
    TargetLoRow = max(TargetLoRow,Low);                 // Ensure
    TargetLoCol = max(TargetLoCol,Low);                 // reason-
    SourceLoRow = max(SourceLoRow,Source.Low);          // able
    SourceLoCol = max(SourceLoCol,Source.Low);          // indices.
    SourceHiRow = min(SourceHiRow,Source.HighRow);
    SourceHiCol = min(SourceHiCol,Source.HighCol);
    int jS = SourceLoRow,
        jT = TargetLoRow;
    while (jS <= SourceHiRow && jT <= HighRow) {
      This[jT].Copy(TargetLoCol,                        // Copy each
        Source[jS],SourceLoCol,SourceHiCol);            // row.
      ++jS;  ++jT; }
    return This; }
```

Figure 7.9. Function to copy part of a Source matrix

Assignment

Just like its Vector module, the MSP Matrix module *declares* overloaded assignment operators for all possible combinations of scalar types:

```
Matrix& operator=(const Matrix<double>&  Source);
Matrix& operator=(const Matrix<complex>& Source);
Matrix& operator=(const Matrix<int>&     Source);
```

However, it implements only those instances where the source and target types agree, as well as the one that assigns Matrix<double> to Matrix<complex>. (Please refer to the justification in Section 5.1.) The operators without type conversion are implemented with

```
template<class Scalar>
  Matrix<Scalar>& Matrix<Scalar>::
    operator=(const Matrix<Scalar>& Source);
```

The assignment operator

```
Matrix<complex>& Matrix<complex>::
  operator=(const Matrix<double>& Source);
```

that converts Matrix<double> to Matrix<complex> is implemented by a separate definition. They're *identical* in structure to the corresponding Vector assignment operators. You'll find their source code in file MatrixL.CPP on the optional diskette.

Equality

As with its Vector module, MSP provides overloaded equality operators ==, which test whether two matrices are *exactly* equal:

```
template<class Scalar>
  Boolean Matrix<Scalar>::operator==(const Matrix<Scalar>& A) const;
```

Its code is almost identical to that of the analogous Vector operator discussed in Section 5.1; there's an additional index bound to check, and, instead of comparing entries, the Matrix operator uses the Vector operator == to compare whole rows. You'll find the code in file MatrixL.CPP on the accompanying diskette. The corresponding inequality operator A != B is defined by a template in MSP header file General.H, described in Section 3.1; it simply returns !(A == B).

Keyboard Input and Display Output

MSP's Matrix module includes a simple keyboard input member function Matrix& KeyIn(char* Name). If you've declared Matrix<double> A(2,2), for example, then executing A.KeyIn("A") prompts the user for input:

```
j  k  A[j,k]
1  1   ...
1  2   ...

2  1   ...
2  2   ...
```

The user, of course, supplies input where the dots appear here. There's room for one- or two-digit indices j and k. This function is declared in header file Matrix.H, the last part of which is

```
        // Input/output  -----------------------------------------

    Matrix& KeyIn(char* Name);        // Prompted keyboard
                                      // input.
    const Matrix& Show(               // Standard display
      char* Name   = "",              // output.
      char* Format = NULL) const;

    friend istream& operator>>(       // Stream input.
           istream& Source,
      Matrix<Scalar>& Target);
      ⋮
    };                // End of the definition of class  Matrix .

// Unfriendly non-member logical functions  --------------------

template<class Scalar>                          // Vector to
  Matrix<Scalar> operator~(                     // column.
    const Vector<Scalar>& V);

template<class Stream, class Scalar>            // Stream
  Stream& operator<<(Stream& Target,            // output.
      const Matrix<Scalar>& Source);
```

Figure 7.10. Matrix.H logical components
Part 4 of 4; continued from Figure 7.7.

displayed in Figure 7.10. You'll find its straightforward source code in Figure 7.11. Users can backspace if necessary to edit the input and then accept it by pressing the <Enter> key. KeyIn handles the <Ctrl-Z> input termination keystroke and other exceptional conditions exactly as its Vector counterpart does; that was described in Section 5.1.

The standard MSP Matrix display output routines

```
    const Matrix& Show(                // Standard display
      char* Name   = EmptyStr,         // output using  printf
      char* Format = NULL) const;      // Format  string.
```

are member functions of the Matrix classes, declared in the part of header file Matrix.H shown in Figure 7.10. They call the analogous Vector function to display each row of the principal argument matrix. Consequently, they use the printf string argument Format in the same way. The effect of the default Format argument was described in Section 3.9. The specified Name is displayed before each row, at the left margin. Unless it's the empty string (the default), it's followed automatically by the row index, enclosed by brackets [], and then the four characters

```
template<class Scalar>                      // Keyboard input,
  Matrix<Scalar>& Matrix<Scalar>::KeyIn(    // prompted by matrix
      char* Name) {                         // name.
    cout << "  j   k   " << Name << "[j,k]";
    for (int j = Low; j <= HighRow; ++j) {
      cout << endl;                                  // Enter each
      for (int k = Low; k <= HighCol; ++k) {         // input on a
        cout << setw(2) << j << setw(3) << k         // new line.
             << "  ";                                // Skip between
        cin >> This[j][k];                           // rows.
        if (!cin) break; }}
    if (cin) return This;
    cerr << "\nException   ";               // If unsuccessful,
    if (cin.eof()) {                        // report & throw an
      cin.seekg(ios::end);                  // MSP exception. If
      cin.clear();                          // the user entered
      cerr << "EndOfFile";                  // <Ctrl-Z> , flush
      throw(Exception(EndOfFile)); }        // the cin buffer &
    cerr << "IOError";                      // clear its end-of-
    throw(Exception(IOError)); }            // file condition.
```

Figure 7.11. Matrix function KeyIn

" : ". Row entries are separated by two blanks. If you've declared Matrix<double> A(2,3), for example, then A.Show("A") might produce output like this:

```
----....------------.------------..---------
A[1]  :    3.0         −0.3           4.7
A[2]  :   −6.7          7.6          −5.1
```

The dots and dashes are not part of the output; they just help you ascertain the spacing specified by the default Format code. Show will align the row indices, using the minimum possible space. Its source code is included in Appendix A.1.

Stream Input/Output

The last two logical functions of the Matrix module are stream input/output operators intended for use with files. Their implementation requires a file format for a Matrix object A. MSP uses a record reflecting the data structure definition: the int values A.Low, A.HighRow, and A.HighCol, followed by vectors A[Low] to A[HighRow], in order. Each vector is stored in the format described in Section 5.1: the int values of its index bounds Low and High, followed by

```
template<class Scalar>                          // Input the target
  istream& operator>>(istream& Source,          // matrix from the
             Matrix<Scalar>& Target) {          // source stream.
    try {
      int Lo,HiRow,HiCol;                       // Read the index
      Matrix<Scalar> Input;                     // bounds into a
      Source >> Lo >> HiRow >> HiCol;           // temporary Input
      if (!Source) {                            // matrix. If
        cerr << "\nException  ";                // unsuccessful,
        if (Source.eof()) {                     // report and throw
          cerr << "EndOfFile";                  // an MSP exception.
          throw(Exception(EndOfFile)); }
        else {
          cerr << "IOError"  ;
          throw(Exception(IOError));    }}
      Input.SetUp(HiRow,HiCol,Lo);              // Otherwise, set up
      for (int j = Lo; j <= HiRow; ++j)         // its rows and read
        Source >> Input[j];                     // them.
      Target.Low     = Lo;                   // This part runs only if
      Target.HighRow = HiRow;                // there was no exception.
      Target.HighCol = HiCol;                // Replace Target Matrix
      if (Target.Row != NULL)                // by Input . Detach the
        delete[] Target.Row;                 // Row  pointer so it won't
      Target.Row = Input.Row;                // be destroyed with
      Input.Row  = NULL;                     // Input . Return Source
      return Source; }                       // for chaining.
    catch(...) {
      cerr << "\nwhile extracting a  Matrix"
           "   from an input stream";
      throw; }}
```

Figure 7.12. Matrix stream input operator >>

its array of Scalar values. For the individual items, MSP outputs ASCII numerals. It uses the minimum number of characters for the integers; for the scalars, it uses scientific notation with maximum, 15 significant digits, precision. All items are followed by newline characters.

The stream input/output operators, like the analogous Vector stream input/output routines, overload the << and >> operators. They return their left-hand arguments to permit concatenation. The declarations are shown in Figure 7.10. As with the analogous Vector features described in Section 5.1, the input operator is a friend of the Matrix class, while the output operator is neither member nor friend. The Matrix operators' source code is shown in Figures 7.12 and 7.13. It's structured rather like that of the analogous Vector operators. Occasional differences are due

```
template<class Stream, class Scalar>         // Output the source
  Stream& operator<<(Stream& Target,         // matrix to the
         const Matrix<Scalar>& Source) {     // target stream.
    int Low     = Source.LowIndex(),         // The  catch  block
        HighRow = Source.HighRowIndex(),     // needs these three
        HighCol = Source.HighColIndex();     // variables.
    try {
      (ostream&)Target << Low     << endl    // Write the index
                      << HighRow << endl     // bounds,
                      << HighCol << endl;
      if (!Target) {
        cerr << "\nException IOError";
        throw(Exception(IOError)); }
      for (int j = Low; j <= HighRow; ++j)   // then each row, in
        Target << Source[j];                 // turn, followed by
      return Target; }                       // newlines.
    catch(...) {
      cerr << "\nwhile inserting into an output stream a  Matrix"
           "  with  HighRow,HighCol,Low = "
          << HighRow << ','
          << HighCol << ',' << Low;
      throw; }}
```

Figure 7.13. Matrix stream output operator <<

to the fact that where the latter use standard C++ stream input/output to handle vector entries, the Matrix routines use their Vector counterparts to handle whole rows. Thus, the Matrix routines must interface with the Vector operators' exception handling system.

7.2 MATRIX ALGEBRA FUNCTIONS

Concepts

Zero and identity matrices

Random matrices

Row and column norms

Plus and minus

Conjugate and transpose

Addition and subtraction

Trace

> *Scalar multiplication*
> *Matrix multiplication*
> *Replacement operators*

This section details the MSP functions that implement matrix algebra. They constitute the mathematical part of module `Matrix`. Its logical functions were discussed in the previous section. Individually, the matrix algebra functions aren't very complicated or interesting. But their overall organization is important, because so much of the this book's numerical analysis software is based on them. The matrix algebra functions are shown in Figures 7.14 and 7.15, which form part of

```
    :
template<class Scalar>
  class Matrix {                            For omitted parts, see Section 5.1.
      :
    public:
      :
      // Member algebra functions ------------------------------

      Matrix& MakeRandom();                 // Make all entries
      Matrix& MakeZero();                   // random, zero.
      Matrix& MakeIdentity();               // Set This = iden-
                                            // tity matrix.
      Matrix operator~() const;             // Transpose, conju-
      Matrix operator!() const;             // gate transpose.
      Matrix operator-() const;             // Negative.
      Matrix operator+(                     // Sum.
         const Matrix& W) const;
      Matrix operator-(                     // Difference.
         const Matrix& W) const;
      Matrix operator*(                     // Right scalar
         const Scalar& t) const;            // multiple.
      Matrix operator*(                     // Matrix * matrix.
         const Matrix& A) const;
      Vector<Scalar> operator*(             // Matrix *
         const Vector<Scalar>& W) const;    // column vector.

    };              // End of the definition of class  Matrix .
```

Figure 7.14. `Matrix.H` mathematical components
Part 1 of 2; continued in Figure 7.15.

```
// Unfriendly non-member matrix algebra functions  -------------

template<class LScalar, class RScalar>           // Set all  A
  Matrix<LScalar>& AllEntries(                   // entries
    Matrix<LScalar>& A,                          // = t .
      const RScalar& t);
template<class Scalar>                           // Row norm.
  double RowNorm(const Matrix<Scalar>& A);
template<class Scalar>                           // Column norm.
  double ColNorm(const Matrix<Scalar>& A);
template<class Scalar>                           // Trace.
  Scalar Trace(const Matrix<Scalar>& A);
template<class Scalar>                           // Conjugate.
  Matrix<Scalar> conj(
    const Matrix<Scalar>& A);
template<class Scalar>                           // Left scalar
  Matrix<Scalar> operator*(                      // multiple.
    const       Scalar&  T,
    const Matrix<Scalar>& A);
template<class Scalar>                           // Row vector
  Vector<Scalar> operator*(                      //  * matrix.
    const Vector<Scalar>& V,
    const Matrix<Scalar>& A);
```

Figure 7.15. Matrix.H mathematical components
Part 2 of 2; continued from Figure 7.14.

the MSP Matrix.H header file. Their source code constitutes the MSP file MatrixM.CPP on the optional diskette, except for the first four functions described in this section. Those are so useful for testing that they're included in file MatrixL.CPP with the functions discussed in the previous section. The source code for some of the matrix algebra functions will be discussed in detail here. For others, which are generally very similar, you may consult the diskette.

This book uses a standard notation for matrices. They'll be denoted by *UPPERCASE* Latin italic letters, and their entries will be denoted by the corresponding *lowercase* letters, usually with subscripts. Sometimes this is emphasized by writing an equation such as $A = [a_{jk}]$. Monospaced boldface is used instead of italic in C++ expressions and in reproducing screen displays. Unless explicitly stated otherwise, the lower row and column index bound is assumed to be 1. When vectors are regarded as *column* matrices, they're denoted by lowercase Greek letters, and their entries are denoted by the corresponding lowercase Latin letters. Sometimes this is emphasized by writing an equation such as $\xi = [x_k]$. The transposition sign $\sim$ is used to indicate the corresponding row matrix $\xi^\sim$.

Constructing Special Matrices

The `Matrix` module includes functions

```
template<class Lscalar,                    MakeZero()
         class RScalar>
  Matrix<LScalar>& AllEntries(             MakeIdentity()
    Matrix<LScalar>& A,
      const RScalar& t);                   MakeRandom()
```

for constructing special matrices. They're designed, implemented, and used just like their `Vector` counterparts, which were described in Section 5.2. Executing `AllEntries(A,t)` sets all entries of `A` equal to `t`. Its implementation as a template with two parameters permits type conversions like

```
Matrix<complex> A(2,3);   AllEntries(A,4);
```

Without this feature, you'd need to write `AllEntries(A,complex(4))`. The other three are member functions of the `Matrix` classes. `MakeZero` makes its principal argument a zero matrix by executing `AllEntries(0)`. `MakeIdentity` sets equal to 1 all diagonal entries `This[j][j]` of its principal argument, and all other entries equal to 0. That makes `This` an identity matrix if it's square. But executing `Matrix<int> I(2,3)` and then `I.MakeIdentity()`, for example, sets

$$I = \begin{bmatrix} 1 & 0 & 0 \\ 0 & 1 & 0 \end{bmatrix}.$$

`MakeRandom` initializes its principal argument to a pseudorandom matrix by executing `This[j].MakeRandom()` (the corresponding `Vector` function) for each appropriate index `j`.

Matrix Norms

Corresponding to each vector norm is a *matrix norm*: a measure $\|A\|$ of the magnitude of a matrix A. It's defined by the equation

$$\|A\| = \inf_{\xi \neq 0} \frac{\|A\xi\|}{\|\xi\|}$$

The right-hand side is the *infimum*, or greatest lower bound, taken over all nonzero vectors ξ. The numerator and denominator of the fraction are norms of vectors.

In general, matrix norms are hard to compute. They're the subject of an extensive theory. However, those corresponding to the l_∞ and l_1 vector norms—the max and city norms—are easy. They're called the *row* and *column norms* of A. Short arguments show that for an $m \times n$ matrix A,

$$\text{Row norm } \|A\| = \max_{j=1}^{m} \sum_{k=1}^{n} |a_{jk}|$$

$$\text{Column norm } \|A\| = \max_{k=1}^{n} \sum_{j=1}^{m} |a_{jk}|.$$

Thus, the row norm of a matrix A is the maximum city norm of its rows, and the column norm is the maximum city norm of its columns. The connection between these corresponding vector and matrix norms is even closer:

- If A is a column matrix, then its row norm is the max norm of the corresponding vector
- If A is a row matrix, then its row norm is the city norm of the corresponding vector.

For the column norm, you must reverse these two statements. Because the row and column norms are so similar, there's rarely a need for both. Therefore, for the remainder of this book, the notation $\|A\|$ *stands for the row norm of matrix* A. For a vector ξ (regarded as a column matrix), $\|\xi\|$ automatically indicates the max norm, and $\|\xi^-\|$ denotes the city norm of the corresponding row vector.

The row norm satisfies four rules that form the basis for the computations and theoretical considerations in this book:

1. $\|A\| = 0 \Leftrightarrow A = 0$
2. $\|tA\| = |t|\,\|A\|$
3. $\|A + B\| \leq \|A\| + \|B\|$ (the *triangle* inequality)
4. $\|AB\| \leq \|A\|\,\|B\|$

The name of inequality (3) stems from the analogous one for vector norms.

The row and column norms are implemented in MSP via functions RowNorm and ColNorm. You'll find their declarations in Figure 7.15. They're not member functions, because principal argument notation is inappropriate. Figure 7.16 displays the RowNorm source code; ColNorm(A) merely returns RowNorm(~A).

```
template<class Scalar>                          // Return the
  double RowNorm(const Matrix<Scalar>& A) {     // row norm:
      int HighRow = A.HighRowIndex();           // the largest
    try {                                       // city norm of
      int     Low = A.LowIndex();               // the matrix
      double  N   = CityNorm(A[Low]);           // rows.
      for (int j = Low+1; j <= HighRow; ++j)
        N = max(N,CityNorm(A[j]));
      return N; }
    catch(...) {
      cerr << "\nwhile computing the  RowNorm  of a  Matrix  "
        "with  HighRow = " << HighRow;
      throw; }}
```

Figure 7.16. Matrix function RowNorm

Plus and Minus

MSP overloads the singular C++ operators + and – to implement the corresponding matrix operations. The former is implemented (it does nothing) by a template in General.H. A member function of the Matrix class, the negation operator – is shown in Figure 7.17. It constructs a new matrix N with the same index ranges as the principal operand, computes the entries, and then returns N. Compare it with the code for the corresponding Vector operator in Figure 5.16; you'll find them nearly identical.

Conjugate and Transpose

MSP uses a similar technique to implement the complex conjugation operation on matrices: It overloads Math.H Library function conj to take a Matrix parameter A. The source code is very similar to that the – operator: Instead of computing the negative of each row of A, it computes the conjugate.

In Section 7.1 the C++ operator ~ was overloaded to convert a vector to a column matrix—that is, ~V was defined for a Vector operand V. MSP further overloads ~ to implement the matrix transposition operator. The operator must be placed *before* its Matrix operand; however, in mathematical formulas, this book will continue to use the more conventional postfix notation:

$$\text{MSP notation } \sim\!A \qquad A^\sim \text{ mathematics.}$$

You'll find the source code in Figure 7.18.

```
template<class Scalar>                              // Negation.
  Matrix<Scalar> Matrix<Scalar>::
    operator-() const {
  try {                                             // Construct the
    Matrix<Scalar> N(HighRow,HighCol,Low);          // negative  N ,
    for (int j = Low; j <= HighRow; ++j)            // row by row,
      N[j] = -This[j];                              // and return
    return N; }                                     // it.
  catch(...) {
    cerr << "\nwhile computing the negative of a  Matrix";
    throw; }}
```

Figure 7.17. Matrix negation operator

In complex matrix algebra, the *conjugate transpose* operator plays a more important role. It's implemented in MSP by overloading the C++ operator !; it merely computes and returns conj(~A). Unfortunately, this notation differs widely from standard mathematics:

$$\text{MSP notation } !A \qquad A* \text{ mathematics.}$$

Nothing can be done about that, because overloading the C++ singulary operator * would interfere with its use in manipulating pointers.

```
template<class Scalar>                              // Transpose.
  Matrix<Scalar> Matrix<Scalar>::
    operator~() const {
  try {
    Matrix<Scalar> T(HighCol,HighRow,Low);          // Construct the
    for (int j = Low; j <= HighCol; ++j) {          // transpose
      for (int k = Low; k <= HighRow; ++k)          // T ,  entry by
        T[j][k] = This[k][j]; }                     // entry, and
    return T; }                                     // return it.
  catch(...) {
    cerr << "\nwhile computing the transpose of a  Matrix";
    throw; }}
```

Figure 7.18. Matrix transposition operator

Addition and Subtraction, Trace

Although MSP permits vector addition when addends' index limits disagree, it regards the analogous situation as an error for *matrix* addition. MSP implements this operation by overloading the C++ binary operator +. The source code is shown in Figure 7.19. It first checks whether the index limits agree. If so, it constructs a new matrix Sum with the appropriate index ranges, computes its entries, and returns it. If not, it sets the error signal and returns an empty matrix.

Matrix subtraction could be implemented like matrix addition. However, it seems better to deal with index compatibility only once. Therefore, the MSP matrix subtraction operator

```
template<class Scalar>
  Matrix<Scalar> Matrix<Scalar>::
    operator-(const Matrix& A) const;
```

merely computes and returns This + (-A).

The *trace* of an $m \times n$ matrix A is the sum of its diagonal entries:

$$\operatorname{tr} A = \sum_{k=1}^{\min(m,n)} a_{kk}.$$

```
template<class Scalar>                          // Matrix
  Matrix<Scalar> Matrix<Scalar>::operator+(     // addition.
      const Matrix<Scalar>& A) const {
    try {
      if (   Low     != A.Low                   // If the index
          || HighRow != A.HighRow               // bounds disagree,
          || HighCol != A.HighCol) {            // report and throw
        cerr << "Exception  IndexError";        // an MSP exception.
        throw(Exception(IndexError)); }         //   Otherwise,
      Matrix<Scalar> Sum = This;                // make  Sum
      for (int j = Low; j <= HighRow; ++j)      // a copy of
        Sum[j] += A[j];                         // This , add
      return Sum; }                             // to its rows
    catch(...) {                                // those of  A ,
      cerr << "\nwhile adding  Matrices"        // & return it.
              "  with  HighRow,HighCol,Low =\n"
        <<   HighRow << ',' <<   HighCol << ',' <<   Low << ';'
        << A.HighRow << ',' << A.HighCol << ',' << A.Low;
      throw; }}
```

Figure 7.19. Matrix addition operator

This concept is implemented by MSP function

```
template<class Scalar>                      // Return
  Scalar Trace(const Matrix<Scalar>& A);    // tr A .
```

It's not a member function, because the principal argument syntax is inappropriate. Its code is straightforward; you'll find it on the diskette.

Scalar Multiplication

There's no complication at all in the MSP implementation of its right scalar multiplication operator as a member function of the Matrix class:

```
template<class Scalar>                      // Right scalar
  Matrix<Scalar> Matrix<Scalar>::           // multiple.
    operator*(const Scalar& t) const;
```

It's structured like the negation operator already discussed: It just constructs a matrix P with the same index range as its principal argument This, computes the entries P[j] = This[j] * t, and returns P. The *left* scalar multiplication operator

```
template<class Scalar>                      // Left scalar
  Matrix<Scalar> operator*(                 // multiple.
      const          Scalar&  t,
      const Matrix<Scalar>& A);
    return A * t; }
```

can't be a member function, because its left-hand argument must be a scalar, not a matrix.

Matrix Multiplication

MSP implements three related forms of matrix multiplication by overloading the C++ binary * operator:

(1 ×m row vector V) * (m ×n matrix   B) = 1 ×n row vector   V * B
(l ×m matrix   A) * (m ×n matrix   B) = l ×n matrix   A * B
(m ×n matrix   A) * (n ×1 column vector W) = m ×1 column vector   A * W.

The first form is regarded as the most basic: for each index k, the kth entry of the product V * B is the scalar product of V by the kth column of B. The second form is computed row by row: for each row index j, the jth row of the product A * B is the product of the jth row of A by the matrix B. You'll find their source codes in Figure 7.20.

```
template<class Scalar>                        // Row vector
  Vector<Scalar> operator*(                   // * Matrix .
      const Vector<Scalar>& V,
      const Matrix<Scalar>& A) {              // Think of  V ,  A
    int n = A.HighColIndex();                 // and their product
    try {                                     // P  as   1xm ,
      int Low = V.LowIndex();                 // mxn ,  and  1xn .
      Vector<Scalar> P(n,Low);                // Construct   P
      for (int k = Low; k <= n; ++k)          // entry by entry.
        P[k] = V * A.Col(k);                  // Scalar product.
      return P; }
    catch(...) {
      cerr << "\nwhile computing the product of a row  Vector"
              "  and a  Matrix  with  HighCol = " << n;
      throw; }}

template<class Scalar>                        // Matrix
  Matrix<Scalar> Matrix<Scalar>::operator*(   // * Matrix .
      const Matrix<Scalar>& A) const {
    int L = HighRow;                          // Think of  This ,
    try {                                     // A ,  and their
      int n = A.HighCol;                      // product  P  as
      Matrix<Scalar> P(L,n,Low);              // Lxm ,  mxn ,  and
      for (int j = Low; j <= L; ++j)          // Lxn .  Construct
        P[j] = This[j] * A;                   // P  row by row.
      return P; }
    catch(...) {
      cerr << "\nwhile computing the product of a  Matrix  with"
              "  HighRow = " << L << "  by another  Matrix";
      throw; }}
```

Figure 7.20. Vector * matrix and matrix * matrix operators

The third form of matrix multiplication—(matrix A) * (column vector W)—is computed as a special case of the second:

```
template<class Scalar>                        // Matrix
  Vector<Scalar> Matrix<Scalar>::operator*(   // * column
      const Vector<Scalar>& W) const {        //       vector.
    try { return This * ~Matrix<Scalar>(W); } // Matrix
    catch(...) {                              // * matrix.
      cerr << "\nwhile computing the product "
              "of a  Matrix  by a  column  Vector";
      throw; }}
```

This routine first uses a converter, described in Section 7.1, to convert vector W to a row matrix, transposes it to form a column matrix, and then uses the matrix * matrix operator in Figure 7.20 to multiply This by the column matrix.

There's no need to check index range compatibility in the matrix multiplication routines: That's done by the Vector scalar multiplication operator.

MSP also overloads the ^ operator to compute integral powers M^n of a square matrix M. Its definition for $n \geq 0$ could be discussed here, but for $n < 0$ that requires the concept of inverse matrix. Therefore, the power operation is considered in Section 8.6.

Replacement Operators

Via templates in header file General.H, MSP implements six matrix algebra replacement operators: +=, -=, and four forms of *=. If A and B are Matrix<Scalar> objects for any supported Scalar type, then A += B adds B to A, A -= B subtracts B from A, and A *= B postmultiplies A by B. If t is a Scalar, then A *= t replaces A by A * t. If V is a Vector<Scalar> object, then V *= B postmultiplies V (regarded as a row vector) by B, and B *= V postmultiplies B by V (regarded as a column vector).

8

Matrix Computations

Systems of linear equations are found everywhere in scientific and engineering applications problems; hence, they constitute one of the most important branches of numerical analysis. This chapter is devoted to two algorithms: solution of a linear system by Gauss elimination and computation of its eigenvalues by Leverrier's method. The Gauss elimination algorithm is implemented first in Section 8.2 in its simplest form. In later sections it's enhanced to include

- Maximal column pivoting
- Computation of the determinant
- LU decomposition
- Matrix inversion
- Gauss-Jordan elimination to compute the reduced echelon form of a nonsquare or singular matrix

Leverrier's algorithm first computes the coefficients of the characteristic polynomial and then finds its roots by Newton-Raphson iteration and deflation. You can then use Gauss-Jordan elimination to study the eigenvectors for each eigenvalue. These algorithms and their implementations work with both real and complex scalars.

8.1 SYSTEMS OF LINEAR EQUATIONS

Concepts

Example: temperature distribution on a metal plate

Matrix notation

How many solutions can a linear system have?

Everywhere in applications of mathematics, you'll find systems of linear equations. Here's just one. Consider a square metal plate with a 7-cm edge. Suppose some heating elements keep the temperatures along its edges at fixed values ranging from 0° to 6°, as shown in Figure 8.1. For measurement, the plate is divided into 1-cm-square cells, and the temperature is regarded as constant within each cell. Edge cells have fixed temperatures, but the interior cells have variable temperatures $x_1, \ldots, x_{25}$ as shown. If no other heat sources are applied, the interior cell temperatures will approach steady state values. To compute these you can assume that in the steady state each of $x_1, \ldots, x_{25}$ is the average of the temperatures in its four neighboring cells. Thus, you can write a system of 25 linear equations, including

$$\begin{cases} x_1 = \tfrac{1}{4} \cdot 1 + \tfrac{1}{4} \cdot 1 + \tfrac{1}{4} x_2 + \tfrac{1}{4} x_6 \\ x_2 = \tfrac{1}{4} \cdot 2 + \tfrac{1}{4} x_1 + \tfrac{1}{4} x_3 + \tfrac{1}{4} x_7 \\ \quad \vdots \\ x_7 = \tfrac{1}{4} x_2 + \tfrac{1}{4} x_6 + \tfrac{1}{4} x_8 + \tfrac{1}{4} x_{12} \\ \quad \vdots \end{cases}$$

These can be rearranged as follows:

$$\begin{cases} x_1 - \tfrac{1}{4} x_2 \quad\quad - \tfrac{1}{4} x_6 \quad\quad\quad\quad\quad\quad\quad\quad = \tfrac{1}{2} \\ -\tfrac{1}{4} x_1 + x_2 - \tfrac{1}{4} x_3 \quad\quad - \tfrac{1}{4} x_7 \quad\quad\quad\quad\quad = \tfrac{1}{2} \\ \quad\quad - \tfrac{1}{4} x_2 \quad\quad - \tfrac{1}{4} x_6 + x_7 - \tfrac{1}{4} x_8 - \tfrac{1}{4} x_{12} = 0 \\ \quad\quad\quad\quad\quad\quad\quad\quad \vdots \end{cases}$$

The solution $x_1, \ldots, x_{25}$ of this system of 25 equations is the steady state temperature distribution. Clearly, this method can be adapted for an arbitrarily shaped plate, and the system would become much larger if smaller cells were required for finer temperature resolution.

Most of this chapter is concerned with the general problem of solving a system of m linear equations

$$\begin{cases} a_{11} x_1 + \cdots + a_{1n} x_n = b_1 \\ \quad \vdots \quad\quad\quad\quad \vdots \quad\quad \vdots \\ a_{m1} x_1 + \cdots + a_{mn} x_n = b_m \end{cases}$$

in n unknowns $x_1, \ldots, x_n$ where the coefficients a_{ij} and b_j are known scalars (m and n can be any positive integers, not necessarily equal). Matrix and vector algebra notation is generally used to abbreviate this system as $A\xi = \beta$, where A is an $m \times n$ matrix, and ξ and β are $n \times 1$ and $m \times 1$ column vectors:

$$A = \begin{bmatrix} a_{11} & a_{12} & \cdots & a_{1n} \\ a_{21} & a_{22} & \cdots & a_{2n} \\ \vdots & \vdots & & \vdots \\ a_{m1} & a_{m2} & \cdots & a_{mn} \end{bmatrix} \quad \xi = \begin{bmatrix} x_1 \\ x_2 \\ \vdots \\ x_n \end{bmatrix} \quad \beta = \begin{bmatrix} b_1 \\ b_2 \\ \vdots \\ b_n \end{bmatrix}.$$

A linear system might have no solution at all, exactly one, or infinitely many. For example, the system

$$\begin{cases} x_1 + 2x_2 = 3 & (1) \\ 4x_1 + 5x_2 = 6 & (2) \end{cases}$$

has a unique solution, determined as follows:

$$\begin{cases} x_1 + 2x_2 = 3 & (3) \\ -3x_2 = -6 & (4) = (2) - 4 \cdot (1) \end{cases}$$

$$\begin{cases} x_2 = 2 & (5) = (4) \text{ solved.} \\ x_1 = -1 & \text{Substitute (5) into (3) and solve.} \end{cases}$$

Systems with unique solutions are called *nonsingular*. Next, the system

$$\begin{cases} x_1 + 2x_2 = 3 & (6) \\ 4x_1 + 8x_2 = 12 & (7) \end{cases}$$

6.	5.	4.	3.	2.	1.	0.
5.	x_{21}	x_{22}	x_{23}	x_{24}	x_{25}	1.
4.	x_{16}	x_{17}	x_{18}	x_{19}	x_{20}	2.
3.	x_{11}	x_{12}	x_{13}	x_{14}	x_{15}	3.
2.	x_6	x_7	x_8	x_9	x_{10}	4.
1.	x_1	x_2	x_3	x_4	x_5	5.
0.	1.	2.	3.	4.	5.	6.

Figure 8.1. Temperature distribution on a metal plate

has more than one solution—for example, $x_1, x_2 = 0, 3/2$ or $1, 1$. In fact, any solution of (6) also satisfies (7), so there are infinitely many solutions. Finally, this system obviously has no solution:

$$\begin{cases} x_1 + 2x_2 = 3 & (8) \\ 4x_1 + 8x_2 = 13 & (9). \end{cases}$$

Section 8.2 presents the Gauss elimination method for solving nonsingular square systems. Section 8.4 implements that method using this book's MSP software. Section 8.5 discusses singular or nonsquare systems in detail.

8.2 GAUSS ELIMINATION

> **Concepts**
> *Gauss elimination*
> *Downward pass*
> *Upper triangular systems*
> *Upward pass*
> *Square nonsingular systems have unique solutions*
> *Efficiency: counting arithmetic operations*

The main algorithm considered in this chapter for solving linear systems $A\xi = \beta$ is called *Gauss elimination*. Its basic strategy is to replace the original system step by step by equivalent simpler ones until the resulting system can be analyzed very easily. Two systems are called *equivalent* if they have the same sets of solution vectors ξ. Just two kinds of operations are used to produce the simpler systems:

(I) Interchange two equations;

(II) Subtract from one equation a scalar multiple of another.

Obviously, the operation (I) doesn't change the set of solution vectors: it produces an equivalent system. Here's an application of operation (II), from the example in Section 8.1:

$$\begin{cases} x_1 + 2x_2 = 3 & (1) \\ 4x_1 + 5x_2 = 6 & (2) \end{cases}$$

$$\begin{cases} x_1 + 2x_2 = 3 & (3) \\ -3x_2 = -6 & (4) = (2) - 4 \cdot (1). \end{cases}$$

In general, this operation has the following appearance:

$$\begin{cases} a_{i1}x_1 + \cdots + a_{in}x_n = b_i & (1) \\ a_{j1}x_1 + \cdots + a_{jn}x_n = b_j & (2) \\ (a_{j1} - ca_{i1})x_1 + \cdots + (a_{jn} - ca_{in})x_n = b_j - cb_i & (3) = (2) - c \cdot (1). \end{cases}$$

Clearly, any solution $x_1, \ldots, x_n$ of (1) and (2) satisfies (3). On the other hand, (2) = (3) + $c \cdot$ (1); hence, any solution of (1) and (3) also satisfies (2). Thus, operation (II) doesn't change the solution set; it produces an equivalent system.

Downward Pass

The first steps of Gauss elimination, called the *downward pass*, convert the original system

$$\begin{cases} a_{11}x_1 + \cdots + a_{1n}x_n = b_1 \\ \vdots \qquad \vdots \quad \vdots \\ a_{m1}x_1 + \cdots + a_{mn}x_n = b_m \end{cases}$$

into an equivalent upper triangular system

$$\begin{cases} a_{11}x_1 + a_{12}x_2 + \cdots + a_{1,m-1}x_{m-1} + a_{1,m}x_m + \cdots + a_{1,n}x_n = b_1 \\ \qquad\quad a_{22}x_2 + \cdots + a_{2,m-1}x_{m-1} + a_{2,m}x_m + \cdots + a_{2,n}x_n = b_2 \\ \qquad\qquad\qquad\qquad \vdots \qquad\qquad \vdots \qquad\qquad \vdots \quad \vdots \\ \qquad\qquad\qquad a_{m-1,m-1}x_{m-1} + a_{m-1,m}x_m + \cdots + a_{m-1,n}x_n = b_{m-1} \\ \qquad\qquad\qquad\qquad\qquad\qquad\qquad a_{m,m}x_m + \cdots + a_{m,n}x_n = b_m \end{cases}$$

A linear system is called *upper triangular* if $a_{ij} = 0$ whenever $i > j$.

Figure 8.2 contains pseudocode for the downward pass. The algorithm considers in turn the diagonal coefficients $a_{1,1}, \ldots, a_{n-1,n-1}$, called *pivots*. Lines 2, 3, and 4 ensure that each pivot is nonzero, if possible. That is, if a diagonal coefficient is zero, you search downward for a nonzero coefficient; if you find one, you interchange rows to make it the pivot. If you don't, then you proceed to the next equation. Nonzero pivots a_{kk} are used in lines 6 and 7 to eliminate the unknown x_k from all equations below equation k. This process clearly produces an equivalent upper triangular system.

1. for ($k = 1$; $k < \min(m, n)$; $++k$) {
2.   for ($i = k$; $a_{ik} == 0$ && $i \leq m$; $++i$);
3.   if ($i \leq m$) {
4.   if ($i > k$) { interchange equations (i) and (k); }
5.   for ($i = k + 1$; $i \leq m$; $++i$) {
6. $M = a_{ik}/a_{kk}$;
7.   subtract M times equation (k) from equation (i); }}}

Figure 8.2. Pseudocode for the downward pass

Here's an example downward pass to convert a system of five equations in six unknowns $x_1, \ldots, x_6$ into an equivalent upper triangular system:

$$\begin{cases} 7x_1 + 3x_2 + 8x_3 - 15x_5 - 12x_6 = -18 & (1) \\ 7x_1 + 3x_2 + 8x_3 - 2x_4 - 10x_5 - 9x_6 = -15 & (2) \\ -14x_1 - 6x_2 - 16x_3 - 4x_4 + 30x_5 + 18x_6 = 28 & (3) \\ -7x_1 - 6x_3 - 2x_4 + 15x_5 + 10x_6 = 13 & (4) \\ 14x_1 + 3x_2 + 14x_3 + 4x_4 - 40x_5 - 31x_6 = -41 & (5) \end{cases}$$

$$\begin{cases} 7x_1 + 3x_2 + 8x_3 - 15x_5 - 12x_6 = -18 & (6) = (1) \\ - 2x_4 + 5x_5 + 3x_6 = 3 & (7) = (2) - (1) \\ - 4x_4 - 6x_6 = -8 & (8) = (3) + 2 \cdot (1) \\ 3x_2 + 2x_3 - 2x_4 - 2x_6 = -5 & (9) = (4) + (1) \\ -3x_2 - 2x_3 + 4x_4 - 10x_5 - 7x_6 = -5 & (10) = (5) - 2 \cdot (1) \end{cases}$$

$$\begin{cases} 7x_1 + 3x_2 + 8x_3 - 15x_5 - 12x_6 = -18 & (11) = (6) \\ 3x_2 + 2x_3 - 2x_4 - 2x_6 = -5 & (12) = (9) \\ - 4x_4 - 6x_6 = -8 & (13) = (8) \\ - 2x_4 + 5x_5 + 3x_6 = 3 & (14) = (7) \\ -3x_2 - 2x_3 + 4x_4 - 10x_5 - 7x_6 = -5 & (15) = (10) \end{cases}$$

$$\begin{cases} 7x_1 + 3x_2 + 8x_3 - 15x_5 - 12x_6 = -18 & (16) = (11) \\ 3x_2 + 2x_3 - 2x_4 - 2x_6 = -5 & (17) = (12) \\ - 4x_4 - 6x_6 = -8 & (18) = (13) \\ - 2x_4 + 5x_5 + 3x_6 = 3 & (19) = (14) \\ 2x_4 - 10x_5 - 9x_6 = -10 & (20) = (15) + (12) \end{cases}$$

$$\begin{cases} 7x_1 + 3x_2 + 8x_3 - 15x_5 - 12x_6 = -18 & (21) = (16) \\ 3x_2 + 2x_3 - 2x_4 - 2x_6 = -5 & (22) = (17) \\ - 4x_4 - 6x_6 = -8 & (23) = (18) \\ - 2x_4 + 5x_5 + 3x_6 = 3 & (24) = (19) \\ - 5x_5 - 6x_6 = -7 & (25) = (20) + (19) . \end{cases}$$

You can assess this algorithm's efficiency by counting the number of scalar arithmetic operations it requires. They're all in lines 6 and 7 of the pseudocode shown in Figure 8.2. When you execute these for particular values of k and i, you need one division to evaluate M, then $n - k + 1$ subtractions and as many multiplications to subtract M times equation (k) from equation (i). (While the equations have $n + 1$ coefficients, you know the first k coefficients of the result of the subtraction are zero.) Thus, the total number of scalar operations is

$$\sum_{k=1}^{m-1} \sum_{i=k+1}^{m} [1 + 2(n-k+1)] = \sum_{k=1}^{m-1} (m-k)(2n - 2k + 3)$$

$$= \sum_{k=1}^{m-1} \left[m(2n+3) - (2m+2n+3)k + 2k^2 \right]$$

$$= m(m-1)(2n+3) - (2m+2n+3)\sum_{k=1}^{m-1} k + 2\sum_{k=1}^{m-1} k^2$$

$$= m(m-1)(2n+3) - (2m+2n+3)\frac{m(m-1)}{2} + 2\frac{m(m-1)(2m-1)}{6}$$

$$= m^2 n - \tfrac{1}{3} m^3 + \text{lower order terms.}$$

In particular, for large $m = n$, about $2/3 n^3$ operations are required.

This operation count has two major consequences. First, solving large linear systems can require an excessive amount of time. For example, computing the steady state temperature distribution for the problem at the beginning of this section required solving a system of 25 equations, one for each interior cell. The result will be only a coarse approximation, because each cell is one square centimeter. To double the resolution—to use cells with a 0.5-cm edge—requires four times as many cells and equations, hence $4^3 = 64$ times as many operations. Second, solving large linear systems can require a huge number of scalar operations, each of which depends on the preceding results. This can produce excessive round-off error. For example, computing the temperature distribution just mentioned for a 0.5-cm grid requires about 700,000 operations.

These large numbers justify spending considerable effort to minimize the use of Gauss elimination in numerical analysis applications. Unfortunately, there are few alternatives. Linear systems of special form—where nonzero coefficients are rare and occur in symmetric patterns—can sometimes be solved by special algorithms that are more efficient. For systems of general form, there are algorithms that are somewhat more efficient than Gauss elimination, but only for extremely large systems. They are considerably more difficult to implement in software. Thus, round-off error in solving linear systems is often unavoidable. This subject has been studied in detail, particularly by Wilkinson [56].

Upward Pass for Nonsingular Square Systems

When you apply Gauss elimination to a system of m linear equations in n unknowns $x_1, \ldots, x_n$, the downward pass always yields an equivalent upper triangular system. This system may have a unique solution, infinitely many, or none at all. In one situation, you can easily determine a unique solution: namely, when $m = n$ and none of the diagonal coefficients of the upper triangular system is zero. Such a system is called *nonsingular*. For example, if the upper triangular system has the form

$$\begin{cases} a_{11}x_1 + a_{12}x_2 + a_{13}x_3 + a_{14}x_4 = b_1 & (1) \\ \phantom{a_{11}x_1 + {}} a_{22}x_2 + a_{23}x_3 + a_{24}x_4 = b_2 & (2) \\ \phantom{a_{11}x_1 + a_{12}x_2 + {}} a_{33}x_3 + a_{34}x_4 = b_3 & (3) \\ \phantom{a_{11}x_1 + a_{12}x_2 + a_{13}x_3 + {}} a_{44}x_4 = b_4 & (4) \end{cases}$$

with $a_{11}, a_{22}, a_{33}, a_{44} \neq 0$, then you can solve (4) for x_4, substitute this value into (3), and then solve that equation for x_3. You can substitute the x_3 and x_4 values into equation (2) and solve that for x_2. Finally, equation (1) would yield a value for x_1. This process is called the *upward pass* of Gauss elimination. In pseudocode, it has the form for $(k = n;\ k \geq 1;\ --k)$

$$x_k = \frac{b_k - \sum_{j=k+1}^{n} a_{kj} x_j}{a_{kk}}.$$

For $k = n$ to 1, the assignment statement in the upward pass requires $n - k$ subtractions, $n - k$ multiplications, and one division. Thus, the total number of scalar arithmetic operations required is

$$\sum_{k=1}^{n}[2(n-k)+1] = 2\sum_{k=1}^{n}(n-k) + n = 2\sum_{k=0}^{n-1} k + n = 2\frac{(n-1)n}{2} + n$$

$$= n^2 + \text{lower - order terms}.$$

Notice two facts about nonsingular square systems. First, their solutions are unique: Each iteration of the for loop in the upward pass pseudocode *determines* the value of one entry of the solution vector. Second, the nonsingularity criterion (no zero among the diagonal coefficients of the upper triangular system) doesn't involve the right-hand sides of the equations. Here's a summary of the preceding discussion of the downward and upward passes.

Consider a square linear system

$$\begin{cases} a_{11}x_1 + \cdots + a_{1n}x_n = b_1 \\ \vdots \qquad \vdots \quad \vdots \\ a_{m1}x_1 + \cdots + a_{mn}x_n = b_m, \end{cases}$$

that is, $A\xi = \beta$ with

$$A = \begin{bmatrix} a_{11} & a_{12} & \cdots & a_{1n} \\ a_{21} & a_{22} & \cdots & a_{2n} \\ \vdots & \vdots & & \vdots \\ a_{n1} & a_{n2} & \cdots & a_{nn} \end{bmatrix} \quad \xi = \begin{bmatrix} x_1 \\ x_2 \\ \vdots \\ x_n \end{bmatrix} \quad \beta = \begin{bmatrix} b_1 \\ b_2 \\ \vdots \\ b_n \end{bmatrix}.$$

Suppose the downward pass of Gauss elimination yields an upper triangular system

$$\begin{cases} a'_{11}x_1 + \cdots + a'_{1n}x_n = b'_1 \\ \qquad \ddots \quad \vdots \quad \vdots \\ \qquad \qquad a'_{nn}x_n = b'_n \end{cases}$$

with no zero among the diagonal coefficients a'_{kk}. Then this situation will occur for *any* coefficient vector β on the right-hand side, and each of these systems has a unique solution β. Each solution may be computed by the upward pass, executed after the downward. For large n, the downward pass requires about $2/3 n^3$ scalar arithmetic operations; the upward, requires about n^2.

8.3 DETERMINANTS

> **Concepts**
> *Definition of the determinant*
> *Computing the determinant during the downward pass*
> *Efficiency of the computation*

For every square matrix A there is a corresponding scalar det A: its *determinant*. For a 2×2 matrix A, the definition is simple:

$$\det \begin{bmatrix} a_{11} & a_{12} \\ a_{21} & a_{22} \end{bmatrix} = a_{11}a_{22} - a_{12}a_{21}.$$

You may recall a somewhat more complicated equation for defining the determinant of a 3×3 matrix. For $n \times n$ matrices in general, the definition is significantly more complicated:

$$\det A = \sum_\pi \text{sign}(\pi) a_{1\pi_1} \cdots a_{n\pi_n}.$$

The sum extends over all $n!$ permutations π of the index set $\{1, \ldots, n\}$. For example, if $n = 2$ there are two permutations: The first is the identity, with $<\pi_1, \pi_2> = <1,2>$; the second is the transposition, with $<\pi_1, \pi_2> = <2,1>$. Any permutation can be constructed as a succession of transpositions; while you can do that in many ways, for a given permutation π, you'll always use an even number of transpositions, or always use an odd number. Sign (π) is defined to be 1 or -1 in these two cases. Thus, the $n \times n$ definition agrees with the earlier 2×2 definition.

Determinants find many applications in mathematics, particularly in evaluating the effects of coordinate changes on areas, volumes, and so forth. In n dimensions, they appear as Jacobian determinants in the substitution rule for multiple integrals. For a nonsingular system $A\xi = \beta$, *Cramer's rule* gives an explicit formula for the entries x_i of the solution vector ξ:

$$x_i \frac{\det A_i}{\det A}$$

where A_i is the matrix obtained from A by replacing column i by B. (Although perhaps familiar from elementary algebra, Cramer's rule is an inefficient method for computing solutions, as you'll see later.)

The definition of a determinant provides no practical method for computing $n \times n$ determinants in general, because it requires $n(n! - 1)$ scalar arithmetic operations. For $n = 25$, this amounts to 3.9×10^{26} operations. On my 90-MHz Pentium machine, a `double` operation requires about 1.9×10^{-7} seconds, so a 25×25 determinant would take about 2.3×10^{12} years to compute, following the definition!

Although the theory of determinants is perhaps the hardest part of elementary linear algebra, some of its results yield a simple method for computing the determinant of an $n \times n$ matrix A:

(I) Interchanging two rows of A changes the sign of det A.

(II) Subtracting a scalar multiple of one row of A from another leaves det A unchanged.

(III) The determinant of an upper triangular matrix A is the product of its diagonal entries.

Finkbeiner [12, Chapter 5] discusses this in detail. Since the downward pass of Gauss elimination uses a sequence of steps of types (I) and (II) to convert A to upper triangular form, you can compute det A during that process:

- Initialize a variable $D = 1$.
- Each time you interchange rows of A, execute $D = -D$.
- When the downward pass is complete and A is upper triangular, multiply D by the product of the diagonal entries of A.

The resulting value of D is the determinant of the original matrix.

You saw in Section 8.2 that for large n the downward pass requires about $^2/_3 n^3$ scalar arithmetic operations. Computing the determinant during that process only adds $n - 1$ more multiplications, so the total remains approximately $^2/_3 n^3$. On my machine, the `double` arithmetic operations for the 25×25 determinant considered earlier would require about 0.002 second; with MSP software, the total process requires about 0.04 second.

Even with this fast method for computing determinants, Cramer's rule is inefficient: It computes $n + 1$ determinants, each of which requires about $^2/_3 n^3$ scalar operations—hence, it requires about $^2/_3 n^4$ operations in all. But Gauss elimination only requires about $^2/_3 n^3$.

In Section 8.2, a square matrix was defined as singular if its diagonal contained a zero after the downward pass. Since the determinant is ± the product of these diagonal entries, a square matrix is singular just in case its determinant is zero.

Another property of determinants, used occasionally in later discussions, is its relationship with matrix multiplication: For any $n \times n$ matrices A and B, det AB = det A det B.

8.4 GAUSS ELIMINATION SOFTWARE

> **Concepts**
> *Basic* `GaussElim` *routine*
> *Maximal column pivoting*
> *Using a row finder*
> `RowFinder` *class*
> *Downward pass: function* `Down`
> *Computing the determinant with function* `det`
> *Upward pass: functions* `Up` *and* `Solve`
> *LU factorization: function* `LU`

The Gauss elimination software described in this chapter forms MSP module `GaussEl`. As usual, it consists of a header file `GaussEl.H` and a source code file `GaussEl.CPP`. The former is listed in Appendix A.2, and you'll find it on the accompanying diskette. Much of the source code is displayed in this chapter. Both files, of course, are included on the optional diskette. These routines work with any scalar types supported by the MSP `Scalar` module: at present, `double` or `complex`.

This section presents first a basic Gauss elimination routine `GaussElim`, and then some enhancements. Incorporating these into the software requires considerable modification of `GaussElim`, which splits it into two functions `Down` and `Up` corresponding to the two passes of Gauss elimination. The section concludes with a discussion of the LU factorization of a square matrix, a theoretical result that parallels the software organization.

Basic Gauss Elimination Routine

In Section 8.2 you saw pseudocode for the downward and upward passes of the Gauss elimination method for solving linear systems $A\xi = \beta$. That code is combined in Figure 8.3 and specialized to the square case: A is $m \times m$; β and ξ are $m \times 1$.

This pseudocode is implemented by

```
template<class Scalar>                  // Return the solution X
  Vector<Scalar> GaussElim(             // of the linear system
    Matrix<Scalar> A,                   // AX = B .  Set status
    Vector<Scalar> B,                   // report Code .
            int& Code);
```

GaussElim uses the status report parameter Code as follows:

Code	Status report
-1	X was computed and an interchange was required.
0	X was computed without interchanging any equations.
1	A is singular.

The interchange status is reported because for some considerations—particularly LU factorization—it's useful information. The GaussElim source code is listed in Figures 8.4 and 8.5. It adheres to the pseudocode except for four features:

- It sets Code.
- It accommodates an arbitrary lower index bound (0 and 1 are both common).
- It reports and throws an MSP Exception object if the index bounds of A and B are inconsistent.
- It returns an empty vector X immediately when it detects that A is singular.

Singularity is not regarded as an error: It's *ordinary* to invoke GaussElim to ascertain whether a matrix is singular or not.

---

1. for $(k = 1; k < m; ++k)$ {
2.   for $(i = k; a_{ik} == 0\ \&\&\ i \leq m; ++i)$;
3.   if $(i \leq m)$ {
4.   if $(i > k)$ { interchange equations (i) and (k); }
5.   for $(i = k + 1; i \leq m; ++i)$ {
6. $M = a_{ik}/a_{kk}$;
7.   subtract M times equation (k) from equation (i); }}}
8. for $(k = m; k \geq 1; --k)$
9. $$x_k = \frac{b_k - \sum_{j=k+1}^{m} a_{kj} x_j}{a_{kk}}$$

Figure 8.3. Gauss elimination pseudocode

```
template<class Scalar>                  // Return solution X of
  Vector<Scalar> GaussElim(              // linear system AX = B .
      Matrix<Scalar> A,                  // Set Code =
      Vector<Scalar> B,                  //   1 : A is singular,
              int& Code) {               //   0 : solved without
    try {                                //          swapping,
      int Low = A.LowIndex(),            //  -1 : solved with a
          m   = A.HighRowIndex();        //          swap.
      if (Low != B.LowIndex()       ||   // Return X empty if A
            m != B.HighIndex()      ||   // is singular. If
            m != A.HighColIndex()   ||   // the index ranges
            m  < Low ) {                 // are inconsistent,
        cerr << "\nSetError(IndexError)";// report and throw
        throw(Exception(IndexError)); }  // an MSP exception.
      Vector<Scalar> X;                  // Construct X
      Boolean Swapped = False;           // empty. Assume A
      Code = 1;                          // singular until you
      int k,i,j;                         // prove otherwise.
      for (k = Low; k < m; ++k) {
        for (i  = k;                                  // Downward
             i <= m && A[i][k] == Scalar(0); ++i);    // pass.
        if (i <= m) {
          if (i > k) {
            Vector<Scalar> T;   Scalar t;             // Swap
            T    = A[i];        t = B[i];             // Equa-
            A[i] = A[k];        B[i] = B[k];          // tions
            A[k] = T;           B[k] = t;             // i & k .
            Swapped = True; }
          for (i = k+1; i <= m; ++i) {    // Subtract M
            Scalar M = A[i][k]/A[k][k];   // times Equa-
            A[i] -= M*A[k];               // tion k from
            B[i] -= M*B[k]; }}            // Equation i .
          else
            return X; }
      if (A[m][m] == Scalar(0))           // A is    Return
        return X;                         // singu-  X
      Code = (Swapped ? -1 : 0);          // lar!    empty.
      X.SetUp(m,Low);
      for (k = m; k >= Low; --k) {        // A is non-
        Scalar S = B[k];                  // singular. Do
        for (j = k+1; j <= m; ++j)        // upward pass.
          S -= A[k][j]*X[j];
        X[k] = S/A[k][k]; }
      return X; }
```

Figure 8.4. Basic routine `GaussElim`
Part 1 of 2; continued in Figure 8.5.

```
catch(...) {
  cerr << "\nwhile solving   AX = B   by Gauss elimination,"
          "\nwhere for   A ,   Low,HighRow,HighCol = "
       << A.LowIndex() << ',' << A.HighRowIndex()
                       << ',' << A.HighColIndex() <<
          "\nand     for   B ,   Low,High = "
       << B.LowIndex() << ',' << B.HighIndex() ;
  throw; }}
```

Figure 8.5. Basic routine `GaussElim`
Part 2 of 2; continued from Figure 8.4.

Pivoting

In solving an $m \times m$ linear system $A\xi = \beta$, the basic Gauss elimination function `GaussElim` must sometimes interchange equations to ensure that a pivot a_{kk} is not zero. For $k = 1$ to $m - 1$, it searches for the first nonzero entry on or below the diagonal and interchanges equations, if necessary, to make *that* the diagonal entry. It's generally good practice to search instead for the entry *of largest norm* on or below the diagonal: If the diagonal entry happens to be nonzero but very small, the subsequent division in pseudocode step 6 (Figure 8.3) could greatly magnify any round-off errors already present. Using the pivot of largest possible norm tends to minimize this problem. This technique is called *maximal column pivoting*.

As an example, consider this 2×2 system $A\xi = \beta$:

$$\begin{cases} 10^{-14} x_1 + x_2 = 1 \\ x_1 + x_2 = 2 . \end{cases}$$

Its exact solution is

$$\begin{cases} x_1 = \dfrac{10^{14}}{10^{14} - 1} = \overline{1.000\,000\,000\,000\,01} \\ x_2 = \dfrac{10^{14} - 2}{10^{14} - 1} = \overline{1.999\,999\,999\,999\,98} . \end{cases}$$

(The overbars signify repeating decimals.) Because the pivot $a_{11} = 10^{-14}$ is so small, GaussElim returns the inaccurate solution

$$x_1 = 0.999\,\overline{2}\ldots \qquad x_2 = 0.999\,999\,999\,999\,\overline{99}.$$

If you reverse the order of the equations, implementing maximal column pivoting in this one instance, GaussElim returns the solution

$$x_1 = 1.000\,000\,000\,000\,00 \qquad x_2 = 0.999\,999\,999\,999\,99,$$

which is as accurate as possible. Of course, this example is artificially generated—it's due essentially to George Forsythe [15]. But comparable situations frequently occur in practice, so maximal column pivoting is generally recommended.

Unfortunately, the problem of round-off error in solving linear systems is not this simple. For example, if you multiply the first equation in Forsythe's system by 10^{15}, you get the equivalent system

$$\begin{cases} 10x_1 + 10^{15} x_2 = 10^{15} \\ \quad\quad x_1 + x_2 = 2. \end{cases}$$

Maximal column pivoting doesn't interchange the equations of this system, and GaussElim returns the inaccurate solution described in the previous paragraph. However, if you reverse the order of the equations—*contrary* to maximal pivoting strategy—then GaussElim returns the most accurate solution possible! Thus, maximal column pivoting *could* make results worse. Further consideration of pivoting strategy is beyond the scope of this book; consult Forsythe and Moler [14] for more information.

In view of the preceding two paragraphs, the basic Gauss elimination routine GaussElim should be enhanced to permit, but not to require, maximal column pivoting. This is simple to do: To steps 2 and 3 of the pseudocode shown in Figure 8.3, provide an alternative, steps 2.1 through 2.5 as follows:

2.1 $Max = |a_{kk}|$; $i = k$;
2.2 for ($j = k + 1$; $j \le m$; $++j$) {
2.3 $\quad N = |a_{jk}|$;
2.4 $\quad$ if ($N > Max$) { $Max = N$; $i = j$; }
2.5 if ($Max > 0$) {

These steps will be built into the enhanced routine Down discussed later.

Using a Row Finder

Later, you'll need to solve in succession several systems $A\xi = \beta$ with the same $m \times m$ matrix A, but different vectors β. (Sometimes each β depends on the solution ξ of the previous system.) Most of the downward pass of Gauss elimination doesn't involve β. It would be unfortunate if you had to repeat for each β in succession the computations involving only A. In fact, you can avoid that. In the pseudocode shown in Figure 8.3, all downward pass computations involving β are in step 7: subtracting a multiple of one equation from another. The multipliers M are computed in step 6. If the multipliers—for indices $k = 1, \ldots, m - 1$ and $i = k + 1, \ldots, m$—were saved somewhere, then all computations dependent only on A could be done once and for all, and the β computations could be postponed until the beginning of the upward pass. Where could you save the multipliers? There's an ideal place: After you've computed M in step 6, you'll never again use entry a_{ik} in the lower triangular part of A. Therefore, you could add to step 6 the statement $a_{ik} = M$, and move the β computations from step 7 to the upward pass.

There's still one reference to β in the downward pass. It's involved in step 4: interchanging equations. To move those manipulations to the upward pass, you have to keep a record of the interchanges, in order to perform them later on β. This problem is solved as a byproduct of an efficiency enhancement to be considered next.

Interchanging equations in routine GaussElim is inefficient. Doing it for every pivot would require $n + 1$ Scalar assignments for each of $n - 1$ equations: a total of $n^2 - 1$ assignments. It's much more efficient to use a *row finder*: a vector $[r_1, \ldots, r_m]$ of indices of rows of A. Each entry r_k indicates where the row now regarded as row k is actually stored. You initialize $r_k = k$ for each k. Instead of interchanging equations (i) and (k) in pseudocode step 4, you interchange r_i and r_k. You never actually *move* the rows of A or, later, the entries of B.

To incorporate a row finder into the Gauss elimination routine, you must construct and initialize $R = [r_1, \ldots, r_m]$, replace the statements interchanging equations by statements interchanging row finder entries, and then replace all references to entries a_{kj} by references to $a_{r_k j}$ and all references to b_k by references to b_{r_k}. In the source code, replace all references A[k][j] and B[k] by A[R[k]][j] and B[R[k]].

To handle row finder initialization and entry swapping conveniently, MSP derives in GaussEl.H a class RowFinder from Vector<int>:

```
class RowFinder: public Vector<int> {
  public:
    RowFinder();                        // Default constructor.
    RowFinder& SetUp(int Hi,            // Set the index bounds
                     int Lo = 1);       // and allocate storage.
    void Swap(int i,                    // Interchange entries
              int j); };                // i and j .
```

C++ requires the default constructor, but it does nothing whatever:

```
RowFinder::RowFinder() {};    // Default constructor:  let the
                              // Vector  constructor do it.
```

(C++ calls the `Vector` constructor automatically to set up an empty vector.) `SetUp` calls the analogous `Vector` function to allocate storage for the row finder entries and then performs the special `RowFinder` initialization:

```
RowFinder& RowFinder::SetUp(int Hi,       // Set index bounds,
                            int Lo) {     // allocate entry
  try {                                   // storage. Initial-
    Vector<int>::SetUp(Hi,Lo);            // ize This as the
    for (int i = Lo; i <= Hi; ++i)        // identity permuta-
      This[i] = i;                        // tion.  Return
    return This; }                        // This
  catch(...) {                            // for
    cerr << "\nwhile setting up a RowFinder .";  // chain-
    throw; }}                             // ing.
```

Finally, the `Swap` function is standard:

```
void RowFinder::Swap(int i,        // Interchange
                     int j) {      // entries  i
  int  T  = This[i];               // and  j .
  This[i] = This[j];
  This[j] = T; }
```

Downward Pass

The basic Gauss elimination routine `GaussElim` can now be enhanced according to the previous discussions and split into two parts `Down` and `Up`. Given an $m \times m$ matrix `A` and an $m \times 1$ vector `B`, `GaussElim` returned a solution vector `X` for the linear system `AX = B`. The new downward pass routine `Down` operates only on matrix `A`. It transforms `A` to upper triangular form using a row finder, optionally employing maximal column pivoting. It stores the multipliers in the lower triangular part of `A` and returns `A` and the row finder for later use by the upward pass routine. `Down` also returns the determinant of `A` as function value. Thus, `Down` has template

```
                                  // Downward pass on  A .
template<class Scalar>            // Max. col. pivoting if
Scalar Down(Matrix<Scalar>& A,    // Code = 0 .  Return
                RowFinder& Row,   // det A ,  A  with stored
                     int& Code);  // multipliers, row find-
                                  // er.  Set status  Code .
```

```
template<class Scalar>                      // Downward pass on  A .
  Scalar Down(Matrix<Scalar>& A,            // Use maximal column
              RowFinder& Row,               // pivoting if  Code = 0 .
              int& Code) {                  // Return  det A , A with
    int Low = A.LowIndex(),                 // stored multipliers, and
        m   = A.HighRowIndex(),             // the row finder. Set
        n   = A.HighColIndex();             // Code  like  GaussElim .
    try {                                   // If  A  isn't
      if (m != n || m < Low) {              // square, re-
        cerr << "\nException  IndexError";  // port, throw
        throw(Exception(IndexError)); }     // an MSP excep-
      Row.SetUp(m,Low);                     // tion.  Ini-
      Boolean MCPivot = (Code == 0);        // tialize row
      Boolean Swapped = False;              // finder.  A  is assumed
      Code = 1;                             // singular until proven
      Scalar Mik, Det = 1;                  // not. Multiplier, de-
      double Max,N;                         // terminant, and tempor-
      int     k,i,j;                        // ary storage for norms.
```

Figure 8.6. GaussEl function Down
Part 1 of 2; continued in Figure 8.7.

The source code for Down is listed in Figures 8.6 and 8.7. For your convenience, MSP also includes a function that invokes Down but returns just the determinant:

```
template<class Scalar>
   Scalar det(Matrix<Scalar> A);            // Determinant of  A .
```

Upward Pass

The Gauss elimination process for solving an $m \times m$ linear system $A\xi = \beta$, where ξ and β are $m \times 1$ column vectors, has now been split into two passes. The downward pass routine transforms A into an upper triangular matrix using a row finder $\rho = [r_1, \ldots, r_m]$ and preserves the multipliers and ρ for later use by the upward pass. All computations involving β have been deferred to the upward pass. Here's the resulting pseudocode for the upward pass:

$$\text{for } (k = 1; k < m; ++k)$$
$$\quad \text{for } (i = k + 1; i \le m; ++i)$$
$$\quad\quad b_{r_i} = b_{r_i} - M_{r_i k} b_{r_k}$$

$$\text{for } (k = m; k \geq 1; --k)$$

$$x_k = \frac{b_{r_k} - \sum_{j=k+1}^{m} a_{r_k j} x_j}{a_{r_k k}}$$

The first three lines of pseudocode represent the β computations from the earlier version of the downward pass. $M_{r_i k}$ is the multiplier M stored earlier in $a_{r_i k}$ during the downward pass.

```
    for (k = Low; k < m; ++k) {
      if (MCPivot) {
        Max = abs(A[Row[k]][k]);  i = k;       // Maximal
        for (j = k+1; j <= m; ++j) {            // column
          N = abs(A[Row[j]][k]);                // pivoting.
          if (N > Max) {
             Max = N;  i = j; }}
        if (Max == 0) return 0; }               // Singular!
      else {
        for (i  = k; i <= m &&
           A[Row[i]][k] == Scalar(0); ++i);
        if (i > m) return 0; }                  // Singular!
      if (i > k) {
        Row.Swap(i,k);  Swapped = True;         // Interchange
        Det = -Det; }                           // rows  i , k .
      Det *= A[Row[k]][k];
      for (i = k+1; i <= m; ++i) {
        Mik = A[Row[i]][k]/A[Row[k]][k];        // Subtract  Mik
        for (j = k+1; j <= m; ++j)              // times  row  k
           A[Row[i]][j] -= Mik*A[Row[k]][j];    // from row  i .
        A[Row[i]][k] = Mik; }}                  // Store  Mik .
    Det *= A[Row[m]][m];
    if (Det == Scalar(0)) return 0;             // Singular!
    Code = (Swapped ? -1 : 0);                  // Nonsingular.
    return Det; }
  catch(...) {
    cerr << "\nduring the downward pass on a matrix with  Low,"
            "HighRow,HighCol = " << Low << ','
                    << m   <<  ',' << n;
    throw; }}
```

Figure 8.7. GaussEl function Down
Part 2 of 2; continued from Figure 8.6.

This routine can be adapted easily to solve matrix equations $AX = B$, where A is $m \times m$ and B and X are $m \times n$. Just solve n separate systems $A\xi^{(l)} = \beta^{(l)}$, where $\beta^{(l)}$ is column l of B. Construct X with columns $\xi^{(1)}, \ldots, \xi^{(n)}$. Then $AX = B$ because, for each l

$$(\text{column } l \text{ of } AX) = A(\text{column } l \text{ of } X)$$
$$= A\xi^{(l)} = \beta^{(l)} = \text{column } l \text{ of } B.$$

That amounts to enclosing the previous paragraph's pseudocode in this loop:

$$\text{for } (l = 1;\, l \leq n;\, ++l)\, \{\, \ldots \}\, .$$

Header file GaussEl.H includes the prototype for the enhanced upward pass routine:

```
template<class Scalar>              // Upward pass on   AX =
  Matrix<Scalar> Up(                // B :  return  X . A
    const Matrix<Scalar>& A,        // is  mxm ,  X , B  are
            Matrix<Scalar>  B,      // mxn .  A ,  Row  were
    const RowFinder& Row);          // prepared by  Down .
```

These functions assume that A and Row have been prepared properly by the downward pass routine Down described earlier in this section. They report and throw an MSP Exception object if the dimension of A doesn't agree with the number of rows of B. After the upward pass, they return the solution X. The source code for Up is listed in Figures 8.8 and 8.9.

```
template<class Scalar>                    // Upward pass on
  Matrix<Scalar> Up(                      // AX = B :  return  X .
    const Matrix<Scalar>& A,              // A is  mxm ,  X , B
            Matrix<Scalar>  B,            // are  mxn .  A , Row
    const RowFinder& Row) {               // have been prepared by
    try {                                 // Down .
      int m    = A.HighRowIndex(),
          n    = B.HighColIndex(),
          Low  = A.LowIndex();            // If the index
      if ( m != B.HighRowIndex() ||       // ranges are
           Low != B.LowIndex()   ||       // inconsistent,
           m < Low ) {                    // report and
        cerr << "\nException  IndexError";// throw an MSP
        throw(Exception(IndexError)); }   // exception.
      Matrix<Scalar> X;  X.SetUp(m,n,Low);// Set up solu-
      int L,k,i,j;                        // tion matrix.
      Scalar S;                           // L will index
                                          // its columns.
```

Figure 8.8. GaussEl function Up
Part 1 of 2; continued in Figure 8.9.

```
        for (L = Low; L <= n; ++L) {
          for (k = Low; k < m; ++k)              // This   B  computa-
            for (i = k+1; i <= m; ++i)           // tion was once in
              B[Row[i]][L] -=                    // the downward pass
                A[Row[i]][k]*B[Row[k]][L];       // of  GaussElim .
          for (k = m; k >= Low; --k) {
            S = B[Row[k]][L];                    // Original upward
            for (j = k+1; j <= m; ++j)           // pass from
              S -= A[Row[k]][j]*X[j][L];         // GaussElim .
            X[k][L] = S/A[Row[k]][k]; }}
        return X; }
      catch(...) {
        cerr << "\nduring the upward pass on  AX = B"
                "\nfor  A ,  Low,HighRow,HighCol = "
             << A.LowIndex() << ',' << A.HighRowIndex()
                             << ',' << A.HighColIndex() << "  and"
                "\nfor  B ,  Low,HighRow,HighCol = "
             << B.LowIndex() << ',' << B.HighRowIndex()
                             << ',' << B.HighColIndex() << "  and";
        throw; }}
```

Figure 8.9. GaussEl function Up
Part 2 of 2; continued from Figure 8.8.

You'll usually want to execute Down and Up in immediate succession to solve a single linear system AX = B where A is $m \times m$ and B is $m \times n$. That facility is provided by MSP function

```
                                      // Return  X   such that
     template<class Scalar>           // AX = B ,  where  A  is
     Matrix<Scalar> Solve(            // mxm  and  X ,  B   are
             Matrix<Scalar> A,        // mxn . Use  Code  like
       const Matrix<Scalar>& B,       // Down . Throw an MSP
                    int& Code);       // Exception  object if  A
                                      // is singular.
```

Its straightforward source code is shown in Figure 8.10.

LU Factorization

Previous sections described the downward pass of the Gauss elimination method for converting a linear system $A\xi = \beta$ with matrix A to an equivalent upper triangular system with matrix U. The algorithm was enhanced by storing the multipliers in entries of A not needed for further computations, so that all computations involving the right-hand side B could be done *after* U is

```
template<class Scalar>                      // Return the X such
  Matrix<Scalar> Solve(                     // that  AX = B , where
           Matrix<Scalar>  A,               // A is  mxm  and  X , B
    const Matrix<Scalar>& B,                // are  mxn .  Use Code
                    int& Code) {            // like  Down .
    try {
      RowFinder Row;                                // Perform the
      Down(A,Row,Code);                             // downward pass
      if (Code <= 0)                                // and, if it
        return Up(A,B,Row);                         // succeeds, the
      cerr << "\nException  DivideError";           // upward.  If
      throw(Exception(DivideError)); }              // A  is singu-
    catch(...) {                                    // lar, report
      cerr << "\nwhile solving a linear"            // and throw
              " system by Gauss elimination";       // an MSP
      throw; }}                                     // exception.
```

Figure 8.10. GaussEl function Solve

computed. When the downward pass can be completed without interchanging rows, this process is called *LU factorization*, for the following reason. First, call a matrix *unit lower triangular* if it has diagonal entries 1 and entries 0 above or to the right of the diagonal. If L is the unit lower triangular matrix containing the multipliers, then LU is the original matrix A. In fact, if the downward pass of Gauss elimination on a square matrix A can be completed without interchanging rows, then A has a *unique* factorization as a product of a unit lower triangular matrix L and an upper triangular matrix U. The proof of this result is not difficult [5, Section 6.6].

The condition that the downward pass be completed without row interchanges is often expressed differently. The *principal minors* of a matrix A are the square submatrices in its upper left corner. By considering the effect of the downward pass on larger and larger principal minors of A in succession, you can see that the downward pass on A can be completed without row interchanges, just in case *all principal minors of A are nonsingular.*

LU factorization is implemented by

```
template<class Scalar>                      // Return  det A  and  L,U
  Scalar LU(Matrix<Scalar>& L,              // decomposition of  A
            Matrix<Scalar>& U,              // if all principal minors
            Matrix<Scalar>  A,              // are nonsingular.  Set
                     int& Code);            // Code  like  GaussElim .
```

This is easy to write: Just construct a row finder Row, execute Down(A,Row,Code), and then construct upper triangular and unit lower triangular matrices U and L from the entries of A. The source code is included on the optional diskette.

8.5 GENERAL RECTANGULAR SYSTEMS

> **Concepts**
> Singular or nonsquare systems
> Reduced echelon systems
> Gauss-Jordan elimination
> Inconsistent systems
> MSP function GaussJordan
> Rank of a matrix
> Homogeneous systems

Nonsquare or Singular Systems

The discussion of Gauss elimination in Sections 8.2 and 8.4 applied only to square, nonsingular systems. What about nonsquare systems or square ones with one or more diagonal zeros in the equivalent upper triangular system? Systems of the last type are called *singular*. Instead of proceeding with the upward pass, you can perform further elimination operations (interchanging equations or subtracting a multiple of one equation from another) to convert the upper triangular system to an equivalent *reduced echelon* system. Such a system has two defining properties:

1. If a_{ij} is the first nonzero coefficient in equation (i), then $a_{kl} = 0$ whenever $k > i$ and $l \leq j$, or $k < i$ and $l = j$.
2. Any equations whose a coefficients are all zero come last.

To construct an equivalent system satisfying the first criterion, just use the diagonal coefficient in each equation of the upper triangular system in turn, if it's not zero, to eliminate the corresponding variable from all preceding equations. Here's pseudocode for that process:

```
for (k = 1; k ≤ min(m, n); ++k)
    if (a_kk ≠ 0)
        for (i = 1; i < k; ++i)
            { M = a_ik/a_kk
              subtract M times equation (k) from equation (i); }
```

To satisfy the second criterion, reorder equations as necessary. Sometimes, the first criterion is strengthened by requiring that the first nonzero coefficient in any equation be 1. That's easy to achieve by dividing the equation by the coefficient.

Matrix Computations   281

The process of eliminating unknowns above the diagonal as well as below is called *Gauss-Jordan elimination*. It could be carried out during the downward pass, simply by altering the loop in line 5 of the pseudocode, in Figure 8.2 for Gauss elimination. However, because it's used here only for nonsquare or singular systems, the Gauss-Jordan step is treated separately. Now, as an example, it's applied to the upper triangular system discussed earlier in Section 8.2:

$$\begin{cases} 7x_1 + 3x_2 + 8x_3 \quad\quad - 15x_5 - 12x_6 = -18 & (21) = (16) \\ \quad\quad\quad 3x_2 + 2x_3 - 2x_4 \quad\quad - 2x_6 = -5 & (22) = (17) \\ \quad\quad\quad\quad\quad\quad - 4x_4 \quad\quad - 6x_6 = -8 & (23) = (18) \\ \quad\quad\quad\quad\quad\quad - 2x_4 + 5x_5 + 3x_6 = 3 & (24) = (19) \\ \quad\quad\quad\quad\quad\quad\quad\quad\quad - 5x_5 - 6x_6 = -7 & (25) = (20) + (19) \end{cases}$$

$$\begin{cases} 7x_1 \quad\quad + 6x_3 + 2x_4 - 15x_5 + 10x_6 = -13 & (26) = (21) - (22) \\ \quad\quad 3x_2 + 2x_3 - 2x_4 \quad\quad - 2x_6 = -5 & (27) = (22) \\ \quad\quad\quad\quad\quad\quad - 4x_4 \quad\quad - 6x_6 = -8 & (28) = (23) \\ \quad\quad\quad\quad\quad\quad - 2x_4 + 5x_5 + 3x_6 = 3 & (29) = (24) \\ \quad\quad\quad\quad\quad\quad\quad\quad\quad - 5x_5 - 6x_6 = -7 & (30) = (25) \end{cases}$$

$$\begin{cases} 7x_1 \quad\quad + 6x_3 \quad\quad - 10x_5 - 7x_6 = -10 & (31) = (26) + (29) \\ \quad\quad 3x_2 + 2x_3 \quad\quad - 5x_5 - 5x_6 = -8 & (32) = (27) - (29) \\ \quad\quad\quad\quad\quad\quad\quad\quad - 10x_5 - 12x_6 = -14 & (33) = (28) - 2 \cdot (29) \\ \quad\quad\quad\quad\quad\quad - 2x_4 + 5x_5 + 3x_6 = 3 & (34) = (29) \\ \quad\quad\quad\quad\quad\quad\quad\quad\quad - 5x_5 - 6x_6 = -7 & (35) = (30) \end{cases}$$

$$\begin{cases} 7x_1 \quad\quad + 6x_3 \quad\quad\quad\quad + 5x_6 = 4 & (36) = (31) - 2 \cdot (35) \\ \quad\quad 3x_2 + 2x_3 \quad\quad\quad\quad + x_6 = -1 & (37) = (32) - (35) \\ \quad\quad\quad\quad\quad\quad\quad\quad\quad\quad 0 = 0 & (38) = (33) - 2 \cdot (35) \\ \quad\quad\quad\quad\quad\quad - 2x_4 \quad\quad - 3x_6 = -4 & (39) = (34) + (35) \\ \quad\quad\quad\quad\quad\quad\quad\quad\quad - 5x_5 - 6x_6 = -7 & (40) = (35). \end{cases}$$

To satisfy the second requirement for a reduced echelon system, just move the $0 = 0$ equation (38) to the end.

Equations such as (38) of the form $0 = 0$ can be called *nonrestrictive*, since they have no effect on the solutions. But you should realize that the occurrences of the two zeros are almost accidental.

For example, a slight change in the right-hand sides of the original equations would have yielded instead of (38) an equation

$$0 = b \quad (38*)$$

where $b \neq 0$. Since no values of the unknowns can satisfy a system containing (38*), neither the reduced echelon nor the original system would have any solution. They would be *inconsistent*.

However, when no equation of the form (38*) occurs in the reduced echelon system, you *can* find a solution. This process is best described in terms of the "new" unknowns that you see in each equation as you climb from bottom to top through the reduced echelon system. Here are the "new" unknowns in the equations of the reduced echelon example just considered:

$$\begin{aligned}
(36) &\quad x_1 \\
(37) &\quad x_2, x_3 \\
(39) &\quad x_4 \\
(40) &\quad x_5, x_6 .
\end{aligned}$$

Except for the equations $0 = 0$, each equation will have at least one new unknown. To construct a solution, proceed from bottom to top, skipping the equations $0 = 0$. If an equation has more than one "new" unknown, assign arbitrary values to all but one of them and solve the equation for the remaining one. This process is also called an *upward pass*. Here, then, is a solution for the earlier example:

$$\left\{\begin{aligned}
7x_1 \quad\quad + 6x_3 \quad\quad\quad\quad + 5x_6 &= 4 & (41) &= (36) \\
3x_2 + 2x_3 \quad\quad\quad\quad + x_6 &= -1 & (42) &= (37) \\
- 2x_4 \quad\quad - 3x_6 &= -4 & (43) &= (39) \\
- 5x_5 - 6x_6 &= -7 & (44) &= (40) \\
0 &= 0 & (45) &= (38)
\end{aligned}\right.$$

$$\begin{aligned}
(44) \quad & x_6 = \text{arbitrary value} \\
& x_5 = \frac{-7 + 6x_6}{-5} = \tfrac{7}{5} - \tfrac{6}{5}x_6 \\
(43) \quad & x_4 = \frac{-4 + 3x_6}{-2} = 2 - \tfrac{3}{2}x_6 \\
(42) \quad & x_3 = \text{arbitrary value} \\
& x_2 = \frac{-1 - 2x_3 - x_6}{3} = -\tfrac{1}{3} - \tfrac{2}{3}x_3 - \tfrac{1}{3}x_6 \\
(41) \quad & x_1 = \frac{4 - 6x_3 - 5x_6}{7} = \tfrac{4}{7} - \tfrac{6}{7}x_3 - \tfrac{5}{7}x_6 \quad .
\end{aligned}$$

Here's a summary of the preceding discussion of the solution of nonsquare or singular systems:

Suppose that the downward pass of Gauss elimination and then Gauss-Jordan elimination have been performed on a linear system, resulting in a reduced echelon system. If an equation of the form $0 = b$, where b is not zero, occurs in the reduced echelon system, then it and the original system are inconsistent: They have no solution. On the other hand, if no equation of that form occurs, then the system has at least one solution, and perhaps infinitely many, which can be computed by an upward pass.

Function GaussJordan

For completeness, MSP contains routine GaussJordan for computing reduced echelon forms:

```
                            // Return  rank A  and the reduced
    template<class Scalar>  // echelon form of  AX = B.  A  is
    int GaussJordan(        // mxn  and  X , B  are  nx1 .
        Matrix<Scalar>& A,  // Scalars  x  with  |x| < T  are
        Vector<Scalar>& B,  // regarded as zero.  Set status
               double  T,   // Code = 1 :  many solutions,
                 int& Code);//        0 :  one solution,
                            //       -1 :  no solution.
```

The *rank* of a matrix is the number of nonzero rows in its reduced echelon form. The tolerance parameter is necessary to recognize unrestrictive rows of the form $0 = 0$ and inconsistent rows of the form $0 = b$, where $b \neq 0$. The source code for GaussJordan appears in Figures 8.11 and 8.12. The code resembles that of the basic Gauss elimination routine GaussElim through the downward pass: There's a minor adjustment to implement Gauss-Jordan elimination. Since only the reduced echelon form is computed, not the solution itself, there's no upward pass. However, some messy code is required to inspect and process the equation to get the rank and the final form of the output. Here are the high points:

- After the elimination steps are complete, coefficients with norm $<$ T are replaced by zeros. (The appropriate tolerance for round-off error may change from one application to another, so the client is given control.)

- Equations corresponding to nonzero rows of A are normalized by dividing by their first nonzero coefficients. This makes it easier to construct the solutions from the reduced echelon form. At the same time, the number of these equations is tallied to get the rank of A.

- Equations are shuffled so that nonrestrictive $0 = 0$ equations come last, preceded by any inconsistent equations.

In Section 8.8, function GaussJordan is tested on an eigenspace analysis problem.

```
template<class Scalar>                  // Return  rank A  and the
  int GaussJordan(                      // reduced echelon form of
      Matrix<Scalar>& A,                // AX = B.  A  is  mxn
      Vector<Scalar>& B,                // and  X , B  are  nx1 .
            double   T,                 // Regard  Scalars  x
              int&   Code) {            // with  |x| < T  as  0 .
  try {                                 // Set  Code =
    int Low  = A.LowIndex(),            //  2 :  many solutions,
        m    = A.HighRowIndex(),        //  1 :  one solution,
        n    = A.HighColIndex();        //  0 :  no solution.
    if (Low != B.LowIndex() ||          // If the index ranges are
        m   != B.HighIndex() ||         //   inconsistent,
        m   <  Low || n < Low ) {       //   report and
      cerr << "\nException  IndexError";//   throw an MSP
      throw(Exception(IndexError)); }   //   exception.
    Code = (m < n ? 2 : 1);
    int i,k; for (k = Low; k <= min(m,n); ++k) {      // Downward
      for (i = k;                                     // pass.
           i <= m && A[i][k] ==
                     Scalar(0); ++i);
      if (i <= m) {
        if (i > k) {
          Vector<Scalar> T;  Scalar t;           // Swap
          T    = A[i];       t    = B[i];        // Equa-
          A[i] = A[k];       B[i] = B[k];        // tions
          A[k] = T;          B[k] = t; }         // i & k .
        for (i = Low; i <= m; ++i) if (i != k) { // Gauss-
          Scalar M = A[i][k]/A[k][k];            // Jordan
          A[i] -= A[k]*M;                        // elimina-
          B[i] -= B[k]*M; }}                     // tion.
        else Code = 2; }                // 0  on diagonal.
```

Figure 8.11. GaussEl function GaussJordan
Part 1 of 2; continued in Figure 8.12.

Homogeneous Linear Systems

One type of linear system is particularly important in theoretical considerations needed later. A *homogeneous* linear system has the form $A\xi = 0$ —that is,

$$\begin{cases} a_{11}x_1 + \cdots + a_{1n}x_n = 0 \\ \vdots \qquad \vdots \quad \vdots \\ a_{m1}x_1 + \cdots + a_{mn}x_n = 0. \end{cases}$$

```
      int Rank = 0,                          // Rank will be the
          Old_m = m;                         // number of nonzero
      for (k = Low; k <= m; ++k) {           // rows A[k] .
        int L = Low - 1;                     // L will indicate
        for (int j = k; j <= n; ++j) {       // the first nonzero
          if (abs(A[k][j]) < T)              // A[k]  entry.
            A[k][j] = 0;
          else if (L < Low) L = j; }         // If there is one,
        if (L >= Low) {                      // increment Rank
          ++Rank;                            // and divide Eq. k
          Scalar t = 1/A[k][L];              // by A[k][L] to
          A[k] *= t;                         // normalize it. If
          B[k] *= t; }                       // not, move Eq. k
        else {                               // to the end. If
          Scalar Bk = B[k];                  // it's inconsistent,
          if (abs(Bk) < T) {                 // report that with
            Bk = 0;  L = Old_m; }            // Code and use
          else {                             // Old_m and L to
            Code = 0; L = m; }               // make it precede
          for (int i = k; i < L; ++i) {      // any 0 = 0 equa-
            A[i] = A[i+1];                   // tions. Arrange to
            B[i] = B[i+1]; }                 // avoid reprocessing
          A[L].MakeZero();                   // any equation just
          B[L] = Bk;                         // moved, and to redo
          --k;  --m; }}                      // Eq. k if a new
      return Rank; }                         // one was moved up.
    catch(...) {
      cerr << "\nwhile solving  AX = B  by Gauss-Jordan elimina-"
             "\ntion, where for  A ,  Low,HighRow,HighCol = "
          << A.LowIndex() << ',' << A.HighRowIndex()
                          << ',' << A.HighColIndex()
          << "\nand   for  B ,  Low,High = "
          << B.LowIndex() << ',' << B.HighIndex() ;
      throw; }}
```

Figure 8.12. GaussEl function GaussJordan
Part 2 of 2; continued from Figure 8.11.

A homogeneous system always has the trivial solution $x_1 = x_2 = \cdots = x_n = 0$. To search for nontrivial solutions, perform the downward pass of Gauss elimination. Suppose $m < n$ or $m = n$ and a zero falls on the diagonal of the resulting equivalent upper triangular system. Now perform Gauss-Jordan elimination. In these two cases, the resulting reduced echelon system will contain at least one equation with more than one "new" unknown—that is, more than one unknown that does not occur in lower equations. Solution of that equation during the upward pass involves at least one arbitrary choice. Therefore, the system has infinitely many solution vectors. In summary:

Every homogeneous system of m linear equations in n unknowns $x_1, \ldots, x_n$ has at least one solution, the trivial one $x_1 = \cdots = x_n = 0$. If $m < n$ or if $m = n$ and a zero falls on the diagonal of the upper triangular system resulting from the downward pass of Gauss elimination, then the system has infinitely many solutions.

8.6 MATRIX INVERSES

> **Concepts**
>
> *Invertible matrices*
> *A matrix is invertible just when it's nonsingular*
> *Uniqueness of inverses*
> *Computing inverses by solving linear systems*
> *MSP function* `Inverse`
> *Computing matrix powers*

Many applications of linear systems involve the inverse matrix concept. An *inverse* of a square matrix A is a matrix J, such that $AJ = JA = I$, an identity matrix. If A has an inverse, it's said to be *invertible*. You'll see later that when A has an inverse, it has only one, so it's valid to speak of *the* inverse of A, which is denoted by A^{-1} when it exists.

Inverse matrices facilitate describing solutions of linear systems. For example, if A is invertible, then $X = A^{-1}B$ is the solution of the linear system or matrix equation $AX = B$, because $AX = A(A^{-1}B) = (AA^{-1})B = IB = B$. This idea is neat, but not always practical, because if A is $m \times m$ and B is $m \times 1$, then computing X directly via Gauss elimination generally takes only one mth of the time that it takes to compute A^{-1}. On the other hand, if you need to compute solutions X for many different B vectors, it may be very helpful to compute A^{-1} first, so that each solution requires merely the matrix multiplication $A^{-1}B$. This is particularly beneficial when A and A^{-1} represent mutually inverse coordinate transformations.

What matrices are invertible? The identity I is self-inverse since $II = I$. The zero matrix O is not invertible since $OJ = O \neq I$ for any matrix J. Many nonzero matrices fail to be invertible, too. For example,

$$\begin{bmatrix} 1 & 0 \\ 0 & 0 \end{bmatrix} \begin{bmatrix} a & b \\ c & d \end{bmatrix} = \begin{bmatrix} a & b \\ 0 & 0 \end{bmatrix} \neq \begin{bmatrix} 1 & 0 \\ 0 & 1 \end{bmatrix}$$

for *any* scalars a, b, c, and d; hence, the matrix on the left has no inverse.

If A is a nonsingular matrix, then you can solve the equation $AX = I$ to obtain a solution $X = R$, which satisfies half the criterion for an inverse: $AR = I$. What about the other equation,

$RA = I$? Arbitrarily chosen matrices A and R almost never satisfy the equation $AR = RA$. But these two aren't arbitrary: You know $AR = I$. In fact, that equation entails the other one, $RA = I$; but the argument, based on the last result in Section 8.5 about homogeneous linear systems, is delicate. First, remember that the transposition operator $\sim$ reverses the order of matrix multiplication, so $I = I^\sim = (AR)^\sim = R^\sim A^\sim$. This equation implies that no linear system of the form $A^\sim X = B$ can have more than one solution. *Proof:* If $A^\sim X_1 = B$ and $A^\sim X_2 = B$, then $X_1 = IX_1 = (R^\sim A^\sim)X_1 = R^\sim(A^\sim X_1) = R^\sim B = R^\sim(A^\sim X_2) = (R^\sim A^\sim)X_2 = IX_2 = X_2$. Therefore $A^\sim$ must be nonsingular, since otherwise the homogeneous system $A^\sim X = O$ would have more than one solution. Thus, you can solve the system $A^\sim X = I$ to obtain a solution $X = S$, so that $A^\sim S = I$. Transpose again: $I = I^\sim = (A^\sim S)^\sim = S^\sim A^{\sim\sim} = S^\sim A$. Finally, $R = IR = (S^\sim A)R = S^\sim(AR) = S^\sim I = S^\sim$; hence, $RA = I$. This paragraph has shown that every nonsingular matrix is invertible.

You can argue the converse as well: If A is invertible, then it must be nonsingular. *Proof:* If A were singular, then the homogeneous system $AX = O$ would have distinct solutions X_1 and X_2; but then $X_1 = IX_1 = (A^{-1}A)X_1 = A^{-1}(AX_1) = A^{-1}O = A^{-1}(AX_2) = (A^{-1}A)X_2 = IX_2 = X_2$.

Speaking of *the* inverse of a matrix A is justified by two propositions. First: If $AX = I$ then $X = A^{-1}$. *Proof:* $X = IX = (A^{-1}A)X = A^{-1}(AX) = A^{-1}I = A^{-1}$. Second: If $XA = I$, then $X = A^{-1}$. You can prove that for yourself.

Here's a summary of the previous paragraphs:

- A matrix is invertible just in case it's nonsingular; and then
- you can compute an inverse $X = A^{-1}$ by solving the system $AX = I$; and
- A^{-1} is the only matrix X satisfying either of the equations $AX = I$ or $XA = I$.

These results imply some others about invertible matrices A and B:

- Since $A^{-1}A = I$, A^{-1} is invertible and its inverse is A;
- Since $A^\sim(A^{-1})^\sim = (A^{-1}A)^\sim = I^\sim = I$, $A^\sim$ is invertible and its inverse is $(A^{-1})^\sim$;
- Since $(AB)(B^{-1}A^{-1}) = I$, AB is invertible and its inverse is $B^{-1}A^{-1}$.

MSP includes a function template for computing matrix inverses. Its straightforward source code is displayed in Figure 8.13. The next section describes a means of testing this routine.

Integral powers of matrices occur frequently in applications. MSP overloads the `^` operator to compute them, using for nonnegative powers the same algorithm it employed in Section 6.2 for polynomials, and inverting the result to get negative powers. *Danger:* Operator `^` has lower precedence than that accorded exponentiation in conventional mathematical notation—for example, to get `(A^m) + B` you *must* use the parentheses! The source code is shown in Figure 8.14. To avoid repeated error messages when it encounters an error situation during a recursive invocation, this code is structured as a pair of templates, similar to the analogous `Polynom` code in Figure 6.6. MSP also implements the corresponding replacement operator `^=` via a template in its header file `General.H`.

```
template<class Scalar>                      // Return  A
  Matrix<Scalar> Inverse(                   // inverse.
      const Matrix<Scalar>& A) {
    try {
      int m   = A.HighColIndex(),
          Low = A.LowIndex();
      Matrix<Scalar> I(m,m,Low);
      I.MakeIdentity();
      int Code = 0;                         // Use maximal column
      return Solve(A,I,Code); }             // pivoting.
    catch(...) {
      cerr << "\nwhile inverting a matrix";
      throw; }}
```

Figure 8.13. GaussEl template Inverse

```
template<class Scalar>                          // Return  A^n .
  Matrix<Scalar> _Power(                        // Recursion:
      Matrix<Scalar> A,                         // A^0 = I ;  if
              int  n) {                         // n  is odd,
    if (n == 0)      return A.MakeIdentity();   // then  A^n =
    if (n % 2 != 0) return A * _Power(A,n-1);   // A * A^(n-1);
    Matrix<Scalar> P = _Power(A,n/2);           // else  A^n =
    return P * P; }                             // (A^½n)^2 .


template<class Scalar>                          // Return  A^n .  This
  Matrix<Scalar> operator^(                     // shell  lets  try/catch
      const Matrix<Scalar>& A,                  // avoid reporting an ex-
                   int   n) {                   // ception at each return.
    try {
      if (n < 0)                                // If  n < 0 ,
        return Inverse(_Power(A,-n));           // then  A^n =
      return _Power(A,n); }                     // (A^(-n))^(-1) .
    catch(...) {
      cerr << "\nwhile raising a matrix to power  " << n;
      throw; }}
```

Figure 8.14. GaussEl matrix power operator

8.7 TESTING WITH HILBERT MATRICES

> **Concepts**
> Testing linear algebra software
> Hilbert matrices H
> Formulas for det H and $A = H^{-1}$
> Testing with the inverse Hilbert matrix A

Every software development project needs a set of problems with known solutions for testing. At least some test problems should not have been developed specifically for that purpose, and some should be really difficult.

A simple but not very challenging test would be to generate an $n \times n$ random matrix H with entries h in the range $0 \leq h < 1$ and compute $D = \det H$ and $A = H^{-1}$ using MSP software. Then you can check how close the approximations $HA \approx I$ and $AH \approx I$ are by evaluating the row norms $\|HA - I\|$ and $\|AH - I\|$. Next you can compute $\det A$ and the L, U factorization of A and check how close the approximations $\det A \approx 1/D$ and $LU \approx A$ are by evaluating $|\det A - 1/D|$ and $\|LU - A\|$. (The principal minors of a random matrix are usually nonsingular, so you should expect $LU \approx A$.) Finally, you can compute $X = A^{-1}$ and check how close the approximation $X \approx H$ is by evaluating $\|X - H\|$. Using MSP scalars of type double, all of these tests come out about the same, depending slightly on the size of n — for example,

n	Error norms $\approx$
6	10^{-15}
13	10^{-14}

However, randomly selected problems generally don't provide severe enough tests. Moreover, you don't know the *exact* solutions, so there's uncertainty in analyzing the results. On the other hand, the *Hilbert* matrices

$$H^{(n)} = \begin{bmatrix} 1 & \frac{1}{2} & \frac{1}{3} & \cdots & \frac{1}{n} \\ \frac{1}{2} & \frac{1}{3} & \frac{1}{4} & \cdots & \frac{1}{n+1} \\ \frac{1}{3} & \frac{1}{4} & \frac{1}{5} & \cdots & \frac{1}{n+2} \\ \vdots & \vdots & \vdots & & \vdots \\ \frac{1}{n} & \frac{1}{n+1} & \frac{1}{n+2} & \cdots & \frac{1}{2n-1} \end{bmatrix}$$

for $n = 1, 2, 3, \ldots$ provide classic linear algebra software test problems, whose solutions are known exactly. They appear naturally in many areas of mathematics, and even for moderately sized n are *ill-conditioned*: Round-off errors build up rapidly when you attempt to invert them, and the errors

```
Matrix<double> Hilbert(int n) {         // Return the
  try {                                  // nxn  Hilbert
    if (n <= 0) {                        // matrix  H .
      cerr << "\nException  IndexError";
      throw(Exception(IndexError)); }
    Matrix<double> H(n,n);
    int i,j;
    for (i = 1; i <= n; ++i)
      for (j = 1; j <= n; ++j)
        H[i][j] = 1.0/(i + j - 1);       // Convert to
    return H; }                          // double!
  catch(...) {
    cerr << "\nwhile computing Hilbert matrix  Hn  with  n = "
         << n;
    throw; }}
```

Figure 8.15. GaussEl function `Hilbert`

can dwarf the entries of the exact results. Linear algebra algorithms and software implementations are sometimes compared on the basis of how accurately they perform calculations on Hilbert matrices. MSP includes the simple function `Matrix<double> Hilbert(int n)` to generate Hilbert matrices; its source code is displayed in Figure 8.15.

Algebraic formulas are available [6] for the determinant of the $n \times n$ Hilbert matrix $H^{(n)}$ and for the k,lth entry $a_{kl}^{(n)}$ of the inverse $A^{(n)}$ of $H^{(n)}$:

$$\det H^{(n)} = \frac{\prod_{i<j}(i-j)^2}{\prod_{i,j}(i+j-1)}$$

$$a_{kl}^{(n)} = \frac{\prod_{i}(i+k-1)\prod_{j\neq k}(l+j-1)}{\prod_{i\neq l}(i-l)\prod_{j\neq k}(i-k)}$$

$$= (-1)^{k+1}(k+l-1)\binom{n+k-1}{n-l}\binom{n+l-1}{n-k}\binom{k+l-2}{k-1}^2.$$

As an example:

$$A^{(5)} = (H^{(5)})^{-1} = \begin{bmatrix} 25 & -300 & 1050 & -1400 & 630 \\ -300 & 4800 & -18900 & 26880 & -12600 \\ 1050 & -18900 & 79380 & -117600 & 56700 \\ -1400 & 26880 & -117600 & 179200 & -88200 \\ 630 & -12600 & 56700 & -88200 & 44100 \end{bmatrix}.$$

Most of the work in deriving these formulas lies in the determinant computation; the last two come from the first through straightforward algebra. The first formula shows that the determinant is not zero, so the matrix is invertible. The last one shows that all entries of the inverse matrix are integers, and its factor $(-1)^{k+l}$ produces the alternate ± pattern.

The second formula is the basis for MSP function

```
Matrix<double> InvHilbert(int n)
```

which returns the inverse of the n × n Hilbert matrix. This simply computes the entries $a_{kl}^{(n)}$ according to the formula; its source code is on the optional diskette. The inverse provides better test problems than the Hilbert matrix itself. Its entries are all integers; hence, for moderately sized n, you can enter them with no round-off error. Errors in the results of test computations are due to round-off in the computation, not in the input.

Proceeding in analogy with the earlier test with random matrices, the $n \times n$ Hilbert matrix H and its inverse A were generated for $n = 1, 2, 3, \ldots$ using routines Hilbert and InvHilbert. Then $D \approx \det A$ and the L,U factorization of A were computed with routine LU. (Since the principal minors of a Hilbert matrix are just smaller Hilbert matrices, they're nonsingular, so you should expect $LU \approx A$.) Finally, an approximation $X \approx H$ to A^{-1} was computed using routine Inverse. The resulting error norms are substantially different from those reported earlier for random test matrices H. To condense the resulting data, selected norms are tabulated in Figure 8.16. They show that determinants of the inverse matrices A are *huge* compared to the size of their entries: for $n = 13$, $\det A \approx 4 \times 10^{91}$, whereas $\|A\| \approx 4 \times 10^{17}$. This means that the determinant of a Hilbert matrix H is tiny compared to the size of its entries, so it's nearly singular, hence A is nearly singular, too.

For even the rather small n values tabulated, most of the results are very inaccurate. Accuracy of $\det A$ drops to about five significant digits at $n = 10$: $\det A \approx 5 \times 10^{52}$, but the error is about 2×10^{47}. For the computations $HA, AH, AX \approx I$ the results are about the same. XA loses accuracy much faster, falling to five significant digits even when $n < 8$. On the other hand, X is a better

n	6	8	10	12	13
$\|A\|$	1×10^7	1×10^{10}	1×10^{-13}	1×10^{-16}	4×10^{17}
$\det A$	2×10^{17}	4×10^{32}	5×10^{52}	4×10^{77}	4×10^{91}
$\|D - \det A\|$	1×10^5	1×10^{24}	2×10^{47}	3×10^{75}	3×10^{91}
$\|HA - I\|$	2×10^{-10}	2×10^{-7}	2×10^{-4}	2×10^{-1}	8×10^1
$\|AH - I\|$	3×10^{-10}	4×10^{-7}	3×10^{-4}	3×10^{-1}	2×10^2
$\|LU - A\|$	9×10^{-10}	6×10^{-7}	1×10^{-3}	8×10^{-1}	3×10^2
$\|AX - I\|$	2×10^{-10}	2×10^{-7}	1×10^{-4}	1×10^{-1}	2×10^2
$\|XA - I\|$	3×10^{-7}	2×10^{-3}	8×10^{-2}	7×10^6	4×10^9
$\|X - H\|$	4×10^{-12}	6×10^{-9}	2×10^{-6}	2×10^{-4}	4×10^0

Figure 8.16. Experimental results with Hilbert matrices

approximation to H, falling to five significant digits somewhat after $n = 10$. The approximation $LU \approx A$ is quite accurate: for $n = 13$, the norm of the error is only about 3×10^2, whereas $\|A\| \approx 4 \times 10^{17}$.

These tests show that even when you're using high-precision software such as the Borland C++ double arithmetic routines, *you must take extreme care with ill-conditioned linear algebra problems.* See Forsythe and Moler [14] for further information.

8.8 EIGENVALUES

> **Concepts**
>
> *Eigenvalues of a matrix A*
> *Example: compartmental analysis*
> *Characteristic polynomial $p_A(x)$*
> *Trace of A*
> *$p_{g(A)}(x)$, where g is a polynomial*
> *Newton's formula*
> *Computing characteristic polynomials with Leverrier's algorithm*
> *Computing eigenvalues and their multiplicities*
> *Eigenvectors and eigenspace analysis with function* `GaussJordan`

An *eigenvalue* of an $n \times n$ matrix A is a scalar t such that $tI - A$ is singular, where I is the $n \times n$ identity matrix. Eigenvalue problems arise from diverse applications, usually after considerable mathematical analysis. The entries of A are usually real scalars, but to perform the analysis, you must allow t and $tI - A$ to be complex. This section implements Leverrier's algorithm for computing eigenvalues. It's not the most common one, because it requires more arithmetic operations than some others. For that reason, it's also sometimes less accurate. But it's completely general, it's based on general mathematical principles that are easy to describe, and it's implemented with standard MSP software tools.

MSP routines involved with computing eigenvalues are contained in module Eigenval. As with other modules, this one consists of header and source code files Eigenval.H and Eigenval.CPP. For convenient reference, the header file is listed in Appendix A.2 and included on the diskette accompanying this book. The source code for all the routines is displayed and discussed later in this section. You'll also find it on the optional diskette. These routines work with double or complex scalars.

Example Application: Compartmental Analysis

As an example application, consider a system of compartments C_k for $k = 1, ..., n$, each filled with liquid containing a solute; the solute concentrations $y_k(t)$ vary with the time t, but are often written simply as y_k. Suppose some compartments are adjacent, and the rate of solute diffusion from C_j to C_k is proportional to the difference $y_j - y_k$ of their concentrations; call the proportionality constant c_{jk}. Set $c_{jk} = 0$ if the corresponding compartments aren't adjacent. Finally, suppose that if any compartment C_k were isolated from its neighbors, the concentration y_k would grow or decrease exponentially with rate $c_{kk} y_k$. (Thus, the total amount of solute in the system could grow or decrease.) The concentrations satisfy the system of n linear ordinary differential equations (ODE)

$$y_j' = c_{jj} y_j + \sum_{k \neq j} c_{jk}(y_j - y_k) = \sum_{k=1}^{n} a_{jk} y_k$$

for $j = 1, ..., n$. If you assemble the resulting coefficients a_{jk} into a matrix A and the functions $y_1, ..., y_n$ and $y_1', ..., y_n'$ into columns η and η', you can write the ODE system as $\eta' = A\eta$ or $\eta'(t) = A\eta(t)$. A solution of the system is determined once you specify the value $\eta(t_0)$ at any initial time t_0. Clearly, one solution is the column $\eta(t)$ of functions with constant value zero. The system is called *asymptotically stable* if every solution approaches that one as $t \to \infty$. That means

$$\lim_{t \to \infty} \|\eta\| = 0.$$

for some vector norm $\|\eta\|$. On the other hand, the system is called *unstable* if

$$\lim_{t \to \infty} \|\eta\| = \infty .$$

In an asymptotically stable system, the concentrations all decrease to zero; in an unstable system, some concentration increases without limit. It can be shown that the system is asymptotically stable if all eigenvalues of matrix A have negative real part, and unstable if any eigenvalue has positive real part. Oscillating solutions are possible if some eigenvalues are purely imaginary but none has positive real part. Thus, complex eigenvalue computations can play an important role even in application problems that, on the surface, involve neither eigenvalues nor complex numbers.

Characteristic Polynomial

In Section 8.3 you saw that a matrix is singular just in case its determinant is zero. Thus, the eigenvalues of an $n \times n$ matrix A are the scalars t such that $\det(tI - A) = 0$ — that is, the roots of the function $p_A(x) = \det(xI - A)$. This determinant is the sum of signed products of n entries, one from each row of $xI - A$, and exactly one of these products is the product of all the diagonal entries $x - a_{kk}$. Thus, $p_A(x)$ is an nth degree polynomial, called the *characteristic polynomial* of A:

$$p_A(x) = x^n + p_{n-1} x^{n-1} + \cdots + p_1 x + p_0 .$$

Clearly, $p_0 = p_A(0) = \det(-A) = (-1)^n \det A$. Moreover, each of the signed products that make up the determinant contains exactly one factor from each row of $xI - A$, so none contains exactly $n - 1$ diagonal terms $x - a_{kk}$. Thus, the x^{n-1} term in the characteristic polynomial is the same as the corresponding term in the product

$$\prod_{k=1}^{n} (x - a_{kk}) .$$

That is,

$$p_{n-1} = \sum_{k=1}^{n} a_{kk} = -\operatorname{tr} A .$$

The sum tr A of the diagonal entries of a matrix A is called its *trace*.

Factor the characteristic polynomial to display its roots $t_1, \ldots, t_n$—the eigenvalues of A. (Some t_k may coincide.) Then you can express the coefficients p_0 and p_{n-1} another way:

$$p_A(x) = \prod_{k=1}^{n}(x - t_k) \qquad p_0 = (-1)^n \prod_{k=1}^{n} t_k \qquad p_{n-1} = -\sum_{k=1}^{n} t_k .$$

More complicated formulas for the other coefficients will be derived later.

Leverrier's algorithm requires you to relate the eigenvalues $t_1, \ldots, t_n$ of A to those of powers A^m for $m = 0, 1, \ldots$. Most linear algebra texts show that $t_1^m, \ldots, t_n^m$ are eigenvalues of A^m, but that result isn't sufficient: It doesn't say anything about *all* the eigenvalues of A^m or about their multiplicity. Gantmacher [16, Section 4] includes the following stronger and more general result: For any polynomial g with degree m,

$$p_{g(A)}(x) = \prod_{k=1}^{n}(x - g(t_k)) ,$$

This equation says that when you expand the product on the right, you get a polynomial with the same coefficients as $p_{g(A)}(x)$. In this equation, and its proof, you must interpret g on the right as a polynomial over the complex numbers and on the left as a polynomial over complex matrices—for example, if $g(x) = x^m - 3x + 2$, then $g(A) = A^m - 3A + 2I$. The proof strategy is to show that

$$p_{g(A)}(t) = \prod_{k=1}^{n}(t - g(t_k))$$

for an arbitrary scalar t. Then, since the values of the two nth degree polynomials coincide for more than n arguments, their coefficients must all agree. First, Gantmacher factors the mth degree polynomial $g(x) - t$ completely:

$$g(x) - t = \prod_{j=1}^{m}(x - u_j).$$

Then the desired result follows if you regard both sides as matrix polynomials, evaluate them at $x = A$, and take determinants:

$$g(A) - tI = \prod_{j=1}^{m}(A - u_j I)$$

$$\det(g(A) - tI) = \prod_{j=1}^{m} \det(A - u_j I)$$

$$(-1)^n p_{g(A)}(t) = \prod_{j=1}^{m}(-1)^n p_A(u_j) = \prod_{j=1}^{m}\left((-1)^n \prod_{k=1}^{n}(u_j - t_k)\right)$$

$$= \prod_{k=1}^{n}\prod_{j=1}^{m}(t_k - u_j) = \prod_{k=1}^{n}(g(t_k) - t)$$

$$= (-1)^n \prod_{k=1}^{n}(t - g(t_k)).$$

Newton's Formula

Leverrier's algorithm also requires you to use Newton's formula, which relates the coefficients of a polynomial to the sums of powers of its roots. Detailed treatment of such a general result doesn't really belong in a discussion of eigenvalue computations, but it plays a crucial role and is used nowhere else in this book. Moreover, although it's mentioned in most algebra texts, it's rarely proved and in fact is sometimes misstated. The treatment here is adapted from Uspensky [54, Section 11.2].

For a positive integer n, let $t_1, \ldots, t_n$ and $p_0, \ldots, p_n$ be scalars such that

$$\sum_{j=0}^{n} p_j x^j = p_n \prod_{k=1}^{n}(x - t_k) .$$

Then *Newton's formula*

$$-m p_{n-m} = \sum_{j=0}^{m-1} p_{n-j} s_{m-j}$$

relates the coefficients p_j to the sums

$$s_m = \sum_{k=1}^{n} t_k^m .$$

To prove the formula, first differentiate $p(x)$:

$$p'(x) = p_n \sum_{k=1}^{n} \prod_{j \neq k} (x - t_j) \qquad \frac{p'(x)}{p(x)} = \sum_{k=1}^{n} \frac{1}{x - t_k}$$

$$p'(x) = \sum_{k=1}^{n} \frac{p(x)}{x - t_k} = \sum_{k=1}^{n} q_k(x).$$

You can find the coefficients of the $(n-1)$st-degree quotient polynomial $q_k(x)$ by using Horner's method to evaluate $p(t_k)$:

$$q_k(x) = q_{k0} + q_{k1}x + \cdots + q_{k,n-1}x^{n-1}$$
$$q_{k,n-1} = 1$$
$$q_{k,n-m-1} = p_{n-m} + q_{k,n-m} t_k \qquad \text{for } m = 1, \ldots, n-1.$$

Newton's formula results if you consider the coefficient of x^{n-m-1} in the previous equation for $p'(x)$:

$$(n-m)p_{n-m} = \sum_{k=1}^{n} q_{k,n-m-1} = \sum_{k=1}^{n} \sum_{j=0}^{m} p_{n-j} t_k^{m-j} = \sum_{j=0}^{m} p_{n-j} \sum_{k=1}^{n} t_k^{m-j}$$

$$= \sum_{j=0}^{m} s_{m-j} p_{n-j} = \sum_{j=0}^{m-1} s_{m-j} p_{n-j} + n p_{n-m}.$$

Computing Characteristic Polynomials

The functions declared by this Eigenval.H template return the characteristic polynomial of a square matrix A:

```
template<class Scalar>               // Return the
  Polynom<Scalar> CharPoly(          // characteristic
    const Matrix<Scalar>& A);        // polynomial of A .
```

If A isn't square, CharPoly throws an MSP Exception object. Its source code is displayed in Figure 8.17. CharPoly implements *Leverrier's algorithm*, which is based on the mathematics covered previously in this section. First, it allocates memory for the coefficients $p_0, \ldots, p_n$ of the characteristic polynomial, for a matrix Am that will contain A^m for $m = 1, \ldots, n$, and for the array

```
template<class Scalar>                      // Return the charac-
  Polynom<Scalar> CharPoly(                 // teristic poly-
      const Matrix<Scalar>& A) {            // nomial p of A.
    try {
      Polynom<Scalar> p;
      int m = A.LowIndex(),
          n = A.HighRowIndex();             // If A isn't
      if (n != A.HighColIndex()) {          // square,
        cerr << "\nException IndexError";   // report and
        throw(Exception(IndexError)); }     // throw an MSP
                                            // exception.
      n -= m - 1; p.SetUp(n);  p[n] = 1;    // n = degree of
      Matrix<Scalar> Am = A;                // p . Am will
      Vector<Scalar> s(n);                  // be A^m .
      for (m = 1; m <= n; ++m) {            // s[m] will contain
        s[m] = Trace(Am);                   // Trace(Am) = sum of
        Scalar Sum = 0;                     // the mth powers
        for (int j = 0; j <= m-1; ++j)      // of the roots
          Sum += s[m-j]*p[n-j];             // of p .
        p[n-m] = -Sum/m;                    // Newton's formula.
        if (m < n) Am *= A; }
      return p; }
    catch(...) {
      cerr << "\nwhile computing the characteristic polynomial"
              " of a matrix";
      throw; }}
```

Figure 8.17. Eigenval function CharPoly

$s_1, \ldots, s_n$ of sums of powers of the eigenvalues of A. Each entry s_m is the sum of the mth powers, hence the sum of the eigenvalues of Am, and it can be computed as the trace of Am. CharPoly initializes $p_n = 1$ and Am = A and then executes a loop for indices $m = 1, \ldots, n$. The mth iteration calculates s_m = tr Am and uses previously computed values s_k and p_{n-k} with Newton's formula to calculate the coefficient p_{n-m}.

Computing Eigenvalues

You can now compute all real eigenvalues of a real matrix A by using first CharPoly and then the Polynom function RealRoots. Better, you can find *all* eigenvalues, real or complex, by using CharPoly and then Roots. You can use Polynom function Polish to polish eigenvalues and

```
template<class Scalar>                 // Return a vector con-
  Vector<complex> Eigenvalues(         // sisting of the eigen-
          Vector<int>&    M,           // values of  A ;  enter
    const Matrix<Scalar>& A,           // into vector  M  their
                  double  T,           // multiplicities.  Use
                  int&    Code) {      // T , Code  like  Roots .
  try {
    Polynom<complex> p;  p = CharPoly(A);
    Vector<complex>  V = Roots(p,T,Code);
    if (Code == 0) Polish(V,p,T,Code);
    M = Multiplicities(V,p,T);
    return V; }
  catch(...) {
    cerr << "\nwhile computing the eigenvalues of a matrix";
    throw; }}
```

Figure 8.18. MSP template Eigenvalues

function Multiplicities to ascertain their multiplicities. The MSP Eigenvalues template incorporates all these features; its source code is displayed in Figure 8.18.

Figure 8.19 shows a test program for Eigenvalues, with sample output. The matrix shown has characteristic polynomial

$$x^4 + 2x^3 + 2x^2 + 2x + 1 = (x + 1)^2(x + i)(x - i),$$

a double eigenvalue -1, and two single eigenvalues $\pm i$.

Eigenvectors and Eigenspaces

If t is an eigenvalue of a matrix A, then $tI - A$ is singular, so $(tI - A)\xi = O$ for some nonzero vector ξ—that is, $A\xi = t\xi$. A nonzero vector ξ with this property is called an *eigenvector* of A *corresponding to t*. If s is any scalar and ξ is an eigenvector, then so is $s\xi$, because $A(s\xi) = s(A\xi) = s(t\xi) = t(s\xi)$. If ξ and η are eigenvectors corresponding to t, then so is $\xi + \eta$, because $A(\xi + \eta) = A\xi + A\eta = t\xi + t\eta D = t(\xi + \eta)$. Thus, the eigenvectors of A that correspond to t constitute a linear space called the *eigenspace* of t.

Some applications need information about the eigenspace corresponding to an eigenvalue t of a matrix A: perhaps a single eigenvector, the dimension m of the space, or even a basis consisting of m linearly independent eigenvectors corresponding to t. You can use function GaussJordan in

```
template<class Scalar>
  void Test3(Scalar t) {
    cout << "\nEnter an  nxn  " << TypeName(t)
         << " matrix A . n = ";
    int n;  cin >> n;
    Matrix<Scalar> A(n,n);
    A.KeyIn("A");  cout << endl;
    A.Show ("A");
    cout << "\nCoefficients of characteristic polynomial:\n";
    CharPoly(A).Show();
    cout << "\nEigenvalues, with multiplicities:\n";
    Vector<int> M;  int Code = 0;
    Eigenvalues(M,A,1e-10,Code).Show();
    M.Show("",True,"%19i ");
    Pause(); }
void main() {
  double x = 1;
  Test3(x); }
```

Output

```
A[1] :    0.0         0.0         0.0         -1.0
A[2] :    1.0         0.0         0.0         -2.0
A[3] :    0.0         1.0         0.0         -2.0
A[4] :    0.0         0.0         1.0         -2.0

Coefficients of characteristic polynomial:
  1.0         2.0         2.0         2.0           1.0

Eigenvalues, with multiplicities:
(-1.0    , 0.0    )   ( 5.7e-18, 1.0    )   ( 2.2e-17,-1.0    )
                2                     1                      1
```

Figure 8.19. Testing `Eigenvalues` template

module `GaussE1` to provide such data—for example, the test program displayed with sample output in Figures 8.20 and 8.21 analyzes matrix

$$A = \begin{bmatrix} 3 & 2 & 1 & 0 \\ 0 & 1 & 0 & 1 \\ 0 & 2 & 1 & 0 \\ 0 & 0 & 0 & 1 \end{bmatrix}.$$

It finds eigenvalues $t_1 = 1$ and $t_2 = 3$ with multiplicities 3 and 1. The reduced echelon form of $t_1 I - A$ is

$$\begin{bmatrix} 1 & 0 & 0.5 & 0 \\ 0 & 1 & 0 & 0 \\ 0 & 0 & 0 & 1 \\ 0 & 0 & 0 & 0 \end{bmatrix};$$

hence, the components $x_1, \ldots, x_4$ of the corresponding eigenvectors are determined by equations

$$\begin{cases} x_1 = -0.5 x_3 \\ x_2 = 0 \\ x_4 = 0. \end{cases}$$

The eigenspace has dimension 1 and consists of all multiples of the vector $[x_1, x_2, x_3, x_4] =$ [-0.5, 0, 1, 0]. You can interpret the second reduced echelon form similarly.

You'll find the source code for the test programs shown in Figure 8.19 through 8.21 in file EigenTst.CPP on the accompanying diskette.

```
void Test5() {                                      // Program to
  cout << "\nEnter an  nxn  double  matrix  A ."    // demonstrate
       << "\nn = ";                                 // GaussJordan
  int n;  cin >> n;                                 // eigenspace
  Matrix<double> A(n,n);                            // analysis.
  A.KeyIn("A");     cout << endl;                   //   T is
  A.Show ("A");                                     //   toler-
  Vector<int> M;   double T = 1e-10;   int Code = 0; //  ance in
  Vector<complex> L = Eigenvalues(M,A,T,Code);      //   Eigen-
  Matrix<complex> I(n,n);   I.MakeIdentity();       //   values &
  Matrix<complex> AA;  AA = A;                      //   0 level
  Vector<complex> O(n);    O.MakeZero();            //   in Gauss
```

Figure 8.20. Testing functions Eigenvalues, Multiplicity, and GaussJordan
Part 1 of 2; continued in Figure 8.21.

```
      for (int k = 1; k <= L.HighIndex(); ++k) {         // Jordan .
        cout << "\nEigenvalue  t = " << L[k]             // C =
             << "\nMultiplicity  = " << M[k];            // complex
        Matrix<complex> C; C = L[k]*I - AA;  Code = 0;   // tI - A .
        int Rank = GaussJordan(C,O,T,Code);
        cout << "\nEigenspace of  t has dimension  " << n - Rank
             << "\nReduced echelon form of  tI - A :\n\n";
        C.Show();
        Pause(); }}
```

Output (slightly reformatted)

```
A[1] =    3.0         2.0         1.0         0.0
A[2] =    0.0         1.0         0.0         1.0
A[3] =    0.0         2.0         1.0         0.0
A[4] =    0.0         0.0         0.0         1.0

Eigenvalue  t = (1, 0)
Multiplicity  = 3
Eigenspace of  t has dimension  1
Reduced echelon form of  tI - A :

( 1.0 , 0.0 )    ( 0.0 , 0.0 )    ( 0.50, 0.0 )    ( 0.0 , 0.0 )
( 0.0 , 0.0 )    ( 1.0 , 0.0 )    ( 0.0 , 0.0 )    ( 0.0 , 0.0 )
( 0.0 , 0.0 )    ( 0.0 , 0.0 )    ( 0.0 , 0.0 )    ( 1.0 , 0.0 )
( 0.0 , 0.0 )    ( 0.0 , 0.0 )    ( 0.0 , 0.0 )    ( 0.0 , 0.0 )

Eigenvalue  t = (3, 0)
Multiplicity  = 1
Eigenspace of  t has dimension  1
Reduced echelon form of  tI - A :

( 0.0 , 0.0 )    ( 1.0 , 0.0 )    ( 0.0 , 0.0 )    ( 0.0 , 0.0 )
( 0.0 , 0.0 )    ( 0.0 , 0.0 )    ( 1.0 , 0.0 )    ( 0.0 , 0.0 )
( 0.0 , 0.0 )    ( 0.0 , 0.0 )    ( 0.0 , 0.0 )    ( 1.0 , 0.0 )
( 0.0 , 0.0 )    ( 0.0 , 0.0 )    ( 0.0 , 0.0 )    ( 0.0 , 0.0 )
```

Figure 8.21. Testing functions `Eigenvalues`, `Multiplicity`, and `GaussJordan`
Part 2 of 2; continued from Figure 8.20.

9

Iterative Solution of Systems of Equations

Let D be a set of n-tuples of scalars. Often it's convenient to gather n scalar-valued functions $g_1, \ldots, g_n$ on D into a single vector-valued function

$$y(\xi) = \begin{bmatrix} g_1(\xi) \\ \vdots \\ g_n(\xi) \end{bmatrix}, \qquad \xi = \begin{bmatrix} x_1 \\ \vdots \\ x_n \end{bmatrix}$$

defined for all vectors ξ in D. A vector α in D is called a *fixpoint* of y if $y(\alpha) = \alpha$.

Fixpoints occur naturally in many applications, often through differential equations—for example, consider the second-order boundary value problem (BVP):

$$y'' = f(t, y, y') \qquad y(a) = y_0 \qquad y(b) = y^\star.$$

Given the function f, interval endpoints a and b, and boundary values y_0 and $y^\star$, the problem is to find a function y that's defined for all t in the interval $a \le t \le b$ and satisfies those three equations. A common solution technique, known as *Liebmann's method*, approximates the differential equation by a fixpoint problem. First, you divide the interval into n subdivisions:

$$a = t_0 < t_1 < \cdots < t_n = b$$

$$t_k = a + kh \quad \text{for} \quad k = 0, \ldots, n \qquad h = \frac{b-a}{n}.$$

For $j = 1, \ldots, n-1$ let y_k denote an approximation to $u(t_k)$, where $y = u(t)$ is the exact solution of the differential equation. The problem is to find good approximations $y_1, \ldots, y_{n-1}$. For notational convenience, let $y_n = y^*$. Now, consider two ways to approximate the derivative y' in the BVP by differential quotients:

$$y'(t_j) \approx \frac{y_j - y_{j-1}}{h} \qquad y'(t_j) \approx \frac{y_{j+1} - y_{j-1}}{2h}$$

for $j = 1, \ldots, n$. At the end of Section 9.1, you'll see that the left-hand formula isn't appropriate for Liebmann's method; the symmetric difference formula on the right is used instead. But the left-hand formula, applied three times, does yield a good approximation of the second derivative:

$$y''(t_j) \approx \frac{y'(t_{j+1}) - y'(t_j)}{h} \approx \frac{1}{h}\left[\frac{y_{j+1} - y_j}{h} - \frac{y_j - y_{j-1}}{h}\right] = \frac{y_{j+1} - 2y_j + y_{j-1}}{h^2}$$

for $j = 1, \ldots, n - 1$. The BVP is then approximated by this difference equation:

$$\frac{y_{j+1} - 2y_j + y_{j-1}}{h^2} = f\left(t_j, y_j, \frac{y_{j+1} - y_{j-1}}{2h}\right).$$

You can rewrite the difference equation as

$$y_j = \tfrac{1}{2} y_{j+1} + \tfrac{1}{2} y_{j-1} - \tfrac{1}{2} h^2 f\left(t_j, y_j, \frac{y_{j+1} - y_{j-1}}{2h}\right) = g_j(y_1, \ldots, y_{n-1})$$

for $j = 1, \ldots, n - 1$. Thus, the original boundary value problem is approximated first by the difference equation and then by the fixpoint problem

$$\eta = \gamma(\eta) = \begin{bmatrix} g_1(\eta) \\ \vdots \\ g_{n-1}(\eta) \end{bmatrix} \qquad \eta = \begin{bmatrix} y_1 \\ \vdots \\ y_{n-1} \end{bmatrix}.$$

Remember, the constants a, b, y_0, y^*, and n are all involved in this problem, as well as the original function f.

If the boundary value problem is linear—that is, if

$$f(t, y, y') = p(t) + q(t)y + r(t)y'$$

for some functions p, q, and r—then the fixpoint equations $y_k = g_k(y_1, \ldots, y_{n-1})$ can be rewritten as a linear system. Linear problems are simpler than the general case, and are considered in

Section 9.2. In Section 9.1, fixpoint iteration software is developed for the general case and demonstrated with a typical nonlinear boundary value problem. Performance is poor and motivates an analysis of the approximation error.

What's involved in the error analysis? Make two assumptions: There exist unique solutions $y = u(t)$ of the BVP and $y_1, \ldots, y_{n-1}$ of the difference equation. (You can find in the literature conditions that guarantee existence and uniqueness of solutions.) There are two separate error questions:

- How well do $y_1, \ldots, y_{n-1}$ approximate $u(t_1), \ldots, u(t_{n-1})$?
- How accurately can you compute $y_1, \ldots, y_{n-1}$?

These questions are related. The answer to the first generally involves the number n of subdivisions—the more, the better. But large n values yield large systems of fixpoint equations, which can require much time and space for accurate solution. If your application must limit the size of n, there will be a limit on the accuracy: No matter how accurately you compute $y_1, \ldots, y_{n-1}$, they'll remain a certain distance from the true solution values $u(t_1), \ldots, u(t_{n-1})$. For details of this analysis, see [5, Sections 11.1–11.4] and the literature cited there.

For this reason, extreme accuracy is often not important in solving multidimensional fixpoint problems. In Section 9.1, single dimensional fixpoint iteration is modified to fit the new situation and demonstrated both with a simple example and with a typical nonlinear second-order boundary value problem. You'll see that speed is even more of a problem than it was with single dimensional fixpoint iteration. That will motivate an analysis of the rate of convergence, which explains the situation.

In Section 9.3 the Newton-Raphson method is generalized to multidimensional problems. It can speed up computations considerably.

MSP module Equate2 implements algorithms for solving systems of equations. Like other MSP modules, it has header and source code files Equate2.H and Equate2.CPP. You'll find the header file on the diskette accompanying this book, and it is listed in Appendix A.2. Various figures in this chapter display all the source code file. You'll find both files, of course, on the optional diskette. As with the modules described earlier, Equate2 is implemented with templates, and works equally well with real or complex equations.

9.1 FIXPOINT ITERATION

Concepts
Limits and continuity in R^n
Cubes; open, closed, and convex subsets of R^n
Implementing the fixpoint algorithm

> *Examples: the need for error analysis*
> *Liebmann's method for second-order boundary value problems*
> *Jacobians*
> *Chain rule*
> *Mean-value theorem*
> *Lipschitz constants*
> *Convergence conditions and error analysis*

This section's topics involve some advanced calculus concepts that may not be familiar to you. A few are needed just to discuss the introductory material, which will motivate or facilitate the more detailed study later. These are the concepts of limit and continuity, and open, closed, and convex sets in n-dimensional space $\boldsymbol{R}^n$.

First, you can express the *convergence* of a sequence of n-dimensional vectors $\xi^{(0)}, \xi^{(1)}, \xi^{(2)}, \ldots$ in two equivalent ways:

$$\alpha = \lim_{k \to \infty} \xi^{(k)} \Leftrightarrow \lim_{k \to \infty} \left\| \xi^{(k)} - \alpha \right\| = 0 \Leftrightarrow a_j = \lim_{k \to \infty} x_j^{(k)} \quad \text{for} \quad j = 1, \ldots, n.$$

That is, $\xi^{(k)}$ approaches α if and only if the max norm $\| \xi^{(k)} - \alpha \|$ approaches 0, and that happens just in case each component $x_j^{(k)}$ of $\xi^{(k)}$ approaches the corresponding component a_j of α. A set D of vectors is called *closed* if it contains the limit of any convergent sequence of vectors in D.

Vector limits are compatible with vector and matrix algebra, just as in the single-dimensional case. For example, these equations hold for any convergent sequences of vectors $\xi^{(k)}$ and $\eta^{(k)}$, any scalar t, and any matrix A:

$$\lim_{k \to \infty} (\xi^{(k)} + \eta^{(k)}) = \lim_{k \to \infty} \xi^{(k)} + \lim_{k \to \infty} \eta^{(k)}$$

$$\lim_{k \to \infty} (t \xi^{(k)}) = t \lim_{k \to \infty} \xi^{(k)} \qquad \lim_{k \to \infty} (A \xi^{(k)}) = A \lim_{k \to \infty} \xi^{(k)}.$$

An *open cube* centered at an n-dimensional vector α is a subset of $\boldsymbol{R}^n$ of the form $\{\xi : \| \xi - \alpha \| < r\}$ for some constant $r > 0$. (This definition uses the max norm.) A set D is called *open* if each member of D is the center of an open cube entirely within D. Any open set containing a set S of vectors is called a *neighborhood* of S. It's easy to show that a set D is open just in case its complement $\boldsymbol{R}^n - D$ is closed. A subset of the form $\{\xi : \| \xi - \alpha \| \leq r\}$ is closed; it's called a *closed cube*.

Consider a function $y : D \to \boldsymbol{R}^m$, where D is a neighborhood of an n-dimensional vector α. y is said to be *continuous* at α if $y(\xi^{(k)})$ approaches $y(\alpha)$ whenever a sequence $\xi^{(0)}, \xi^{(1)}, \ldots$ in D

approaches α. You can show that y is continuous in this sense just when each component function g_j is continuous at α.

A vector η is said to lie *between* two vectors ξ and ζ if $\eta = t\xi + (1 - t)\zeta$ for some t in the unit interval. Finally, a subset D of $\mathbf{R}^n$ is called *convex* if it contains every vector η that lies between two members of D — for example, open and closed cubes are convex.

All the n-dimensional concepts just described can be extended, if necessary, to apply to complex vectors. This general terminology is used next to introduce the basic multidimensional fixpoint iteration and to show the need for detailed study of related convergence problems.

Implementing the Multidimensional Fixpoint Algorithm

In this chapter's introduction, you saw how a fixpoint problem—solving the equation

$$\xi = y(\xi) = \begin{bmatrix} g_1(\xi) \\ \vdots \\ g_n(\xi) \end{bmatrix}, \qquad \xi = \begin{bmatrix} x_1 \\ \vdots \\ x_n \end{bmatrix}$$

for a fixpoint $\xi = \alpha$—can arise naturally in applications, especially through differential equations. The equation's form suggests the following *multidimensional fixpoint iteration* algorithm: Start with an initial estimate $\xi^{(0)}$ of α, and then compute

$$\xi^{(k+1)} = y(\xi) = \begin{bmatrix} g_1(\xi^{(k)}) \\ \vdots \\ g_n(\xi^{(k)}) \end{bmatrix}, \qquad \xi^{(k)} = \begin{bmatrix} x_1^{(k)} \\ \vdots \\ x_n^{(k)} \end{bmatrix}$$

for $k = 0, 1, \ldots$. If the sequence $\xi^{(0)}, \xi^{(1)}, \xi^{(2)}, \ldots$ converges, then the desired solution should be

$$\alpha = \lim_{k \to \infty} \xi^{(k)}.$$

That is, the equation

$$y(\alpha) = y(\lim_{k \to \infty} \xi^{(k)}) = \lim_{k \to \infty} y(\xi^{(k)}) = \lim_{k \to \infty} \xi^{(k+1)} = \alpha$$

will hold if y is continuous on a closed set D containing your initial estimate $\xi^{(0)}$, and fixpoint iteration yields a convergent sequence of vectors $\xi^{(1)}$, $\xi^{(2)}$, ... in D.

This algorithm is easy to implement. But before rushing to do so, pause a moment to question the validity of what you're doing:

- Can you predict whether a solution exists?
- If the sequence converges to a fixpoint, is that necessarily the solution you're looking for? (There might be more than one fixpoint.)
- What conditions ensure convergence?
- How fast can you expect the sequence to converge?

You may want to be sure that a fixpoint exists before attempting to compute one. Section 4.3 presented a condition guaranteeing existence in the single-dimensional case, $n = 1$: If a continuous function g maps a nonempty closed interval $I = \{x : a \leq x \leq b\}$ into itself, then g has a fixpoint in I. That was easy to see by considering the graph of g. An analogous result—called *Brouwer's fixpoint theorem*—is true for any dimension n: If a continuous function y maps a closed cube I into itself, then y has a fixpoint in I. Unfortunately, there's no clear picture to illustrate that, and the theorem is hard to prove. (Usually, it's found in books on combinatorial topology.) In practical applications it's hard to find the appropriate cube; fortunately, however, the existence of a fixpoint is usually apparent from the problem setting. The main question is whether you can use fixpoint iteration to compute it.

Later in this section, conditions are presented that guarantee convergence of fixpoint iteration, provided your initial estimate is close enough to the desired fixpoint. Unfortunately, you usually don't know how close that must be until you execute the algorithm. Often, multidimensional functions have multiple fixpoints, and fixpoint iteration can lead you to an unanticipated solution—not the one you want. You must analyze situations like that individually.

As in the single-dimensional case of Section 4.3, the theory that leads to a convergence criterion can also predict the rate of convergence.

Figure 9.1 displays `Equate2` function template `Fixpoint`, an implementation of the multi-dimensional fixpoint algorithm. It's structured exactly the same as the single-dimensional `Equate1` template `Fixpoint`, shown in Figure 4.12. Compare their declarations:

```
———— from Equate1.H ————              ———— from Equate2.H ————
template<class Scalar>                template<class Scalar>
  Scalar Fixpoint(                      Vector<Scalar> Fixpoint(
    Scalar                                Vector<Scalar> G(
      g(Scalar),                            const Vector<Scalar>&),
    Scalar x,                             Vector<Scalar> X,
    double T,                             double T,
    int& Code);                           int& Code);
```

```
template<class Scalar>                      // Return an
  Vector<Scalar> Fixpoint(                  // approximation to a
      Vector<Scalar> G(                     // fixpoint of  G ,
        const Vector<Scalar>&),             // computed by fix-
      Vector<Scalar> X,                     // point iteration.
            double   T,                     // Use initial
              int&   Code) {                // estimate  X ,
    try {                                   // and stop when
      double M;  int n = 0;                 // ||X - G(X)|| < T .
      do {                                  // Set Code = 0 to
          Vector<Scalar> P = G(X);          // report success, or
          M = MaxNorm(X - P);               // = -1  if the de-
          X = P; }                          // sired accuracy is
        while (++n < 1000 && M >= T);       // not achieved after
      Code = (M < T ? 0 : -1);              // 1000 iterations.
      return X; }
    catch(...) {
      cerr << "\nwhile executing  Fixpoint";
      throw; }}
```

Figure 9.1. Equate2 template Fixpoint

The only significant source code differences you'll find between the two figures consist in changing Scalar parameters to Vector<Scalar> and using the max norm instead of absolute value. As with several Equate1 function definitions displayed in Chapter 4, the Fixpoint version shown in Figure 9.1 has been stripped of intermediate output features, because they obscure the structure of the code. Those are included in the version on the accompanying diskette. If Code = 0 on entering Fixpoint, you'll get no intermediate output. Set Code = 1 to see a report of the number of iterations. With Code > 1, you'll get a report of the successive values of M = MaxNorm(X - G(X)).

Examples: The Need for Error Analysis

As a first example of multidimensional fixpoint iteration, consider the nonlinear system

$$\begin{cases} x = \sin(x+y) \\ y = \cos(x-y) . \end{cases}$$

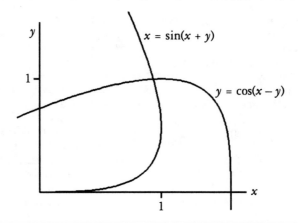

Figure 9.2. The curves $x = \sin(x + y)$ and $y = \cos(x - y)$

From its graph, shown in Figure 9.2 you can see that it has a solution near the point $x, y = 1,1$. The solution is a fixpoint of function

$$y(\xi) = \begin{bmatrix} \sin(x + y) \\ \cos(x - y) \end{bmatrix}, \qquad \xi = \begin{bmatrix} x \\ y \end{bmatrix}.$$

The fixpoint was computed by a test program, shown in Figure 9.3; its execution is described in Figure 9.4. With tolerance T = 10^{-6}, Fixpoint required nine iterations. That's comparable with the single-dimensional example in Figure 4.13. You'll find the test program source code in file Fig9_03.CPP on the accompanying diskette.

The Liebmann method for second-order boundary value problems (BVP), described in the introduction to this chapter, provides many example fixpoint problems. A BVP

$$y'' = f(t, y, y') \qquad y(a) = y_0 \qquad y(b) = y^*$$

leads to the equation $\eta = y(\eta)$, where

$$y(\eta) = \begin{bmatrix} g_1(\eta) \\ \vdots \\ g_{n-1}(\eta) \end{bmatrix} \qquad \eta = \begin{bmatrix} y_1 \\ \vdots \\ y_{n-1} \end{bmatrix} \qquad y_n = y^*$$

$$g_j(\eta) = \tfrac{1}{2} y_{j+1} + \tfrac{1}{2} y_{j-1} - \tfrac{1}{2} h^2 f\left(t_j, y_j, \frac{y_{j+1} - y_{j-1}}{2h}\right).$$

```
#include "Equate2.H"

Vector<double> G(const Vector<double>& X) {
  Vector<double> Y(2);
  Y[1] = sin(X[1] + X[2]);
  Y[2] = cos(X[1] - X[2]);
  return Y; }

void main() {
  double T = 1e-6;
  randomize(); Vector<double> X(2);  X.MakeRandom();
  cout << "\nSolving  x = sin(x + y)    by fixpoint iteration"
          "\n         y = cos(x - y)                          \n";
  cout << "\nTolerance            : "; Show(T);
  X.Show( "\nInitial estimate   x y",True);
  int Code = 2;  X = Fixpoint(G,X,T,Code);
  X.Show( "Solution   x y",True, "%.6f");
  cout << "Code = " << Code; }
```

Figure 9.3. Fixpoint test routine

```
Solving  x = sin(x + y)    by fixpoint iteration
         y = cos(x - y)

Tolerance            :   1.0e-06
Initial estimate  x y :   0.75      0.60

Itera-
tion     |X - G(X)|
   0     0.39
   1     0.053
   2     0.016
   3     0.0045
   4     0.0011
   5     0.00028
   6     6.7e-05
   7     1.5e-05
   8     3.4e-06
   9     7.3e-07

Solution  x y :  0.935082   0.998020
Code = 0
```

Figure 9.4. Testing two-dimensional fixpoint iteration

Liebmann's method is tested by the MSP client program shown in Figure 9.5. You'll find that code in file Fig9_05.CPP on the accompanying diskette. Function main first calls BVPSetUp to set up vectors T and U consisting of $t_1, \ldots, t_{n-1}$ and the corresponding values $u_1, \ldots, u_{n-1}$ of the exact solution $y = u(t)$. It also sets up the initial estimate $Y = \eta^{(0)}$ for the fixpoint iteration: This vector consists of values of the linear function that satisfies the boundary conditions. Then main calls Fixpoint to compute a solution vector Y and compares that with the vector U of exact solution values.

As an example, consider the nonlinear second-order boundary value problem

$$y'' = \frac{2}{y' + 1} \qquad y(1) = 1/3 \qquad y(4) = 20/3.$$

Its solution is the function $u(t) = 4/3 \, t^{3/2} - t$. To compute an approximate solution η via Liebmann's method, the program shown in Figure 9.5 defines

```
double f(double t, double y, double yp) {
  return 2/(yp + 1); },
```

double constants a = 1, y0 = $1/3$, b = 4, and yStar = $20/3$, and inputs the number n of subdivisions and the Fixpoint tolerance parameter T. For this example, the program was executed with T = 0.0001 and several n values, yielding the following results:

Number n of t steps	Iterations required	Error
5	36	.0017
10	122	.0022
20	375	.0083

The tabulated error value is the max norm of the error vector Y - U. As you can see, there was no point in using such a small tolerance T, because the solution of the fixpoint equation is evidently farther than that distance from the true solution of the BVP. Moreover, if you only need a few evenly spaced solution values, there's no point in making a finer subdivision, because in this case accuracy decreases as the number of subdivisions grows.

As a final example, consider the linear second-order BVP

$$y'' = 2y' - y - 3 \qquad y(0) = -3 \qquad y(2) = 1.$$

```
#include "Equate2.H"              // n  t-steps of size  h  from
                                  // a  to  b . Linear initial
double   a,b,y0,yStar,h,k,T;      // estimate  Y  with step size
int      n;                       // k . Tolerance  T  for
Vector<double> t,Y,U;             // approximating the exact BVP
                                  // solution  U . Return ODE's
double f(double t, double y,                // right
         double yp) { return 2/(yp + 1); }  // side &
double u(double t)   { return t*(4*sqrt(t)/3 - 1); }  // exact
                                                      // solu-
void BVPSetUp() {                                     // tion.
  h =     ( b - a )/n;            // Set up  t-
  k = (yStar - y0)/n;             // steps, linear
  t.SetUp(n-1);      Y.SetUp(n-1);   // initial esti-
                     U.SetUp(n-1);   // mate  Y ,
  for (int j = 1; j <= n-1; ++j) {   // and exact
    t[j] = a + j*h;   Y[j] = y0 + j*k;  // solution U .
                      U[j] = u(t[j]); }}

Vector<double> G(const Vector<double>& Y) {  // Fixpoint
  int    m = Y.HighIndex();                  // function for
  Vector<double> V(m);                       // Liebmann's
  for (int k = 1; k <= m; ++k) {             // method.
    double Ykm1 = (k == 1 ? y0    : Y[k-1]),  // Y[0] = y0 .
           Ykp1 = (k == m ? yStar : Y[k+1]),  // Y[n] = y* .
           Ypk  = (Ykp1 - Ykm1)/(2*h);        // Ypk ≈
    V[k] = (Ykp1 + Ykm1 - h*h*f(t[k],Y[k],Ypk))/2; }  // y' at
  return V; }                                        // t[k] .

void main() {
  a = 1;   b = 4;   y0 = 1./3.;   yStar = 20./3.;
  cout << "Solving by                y\" = 2/(y' + 1)        \n"
          "Fixpoint iteration     y(1) = 1/3  y(4) = 20/3\n\n"
          "Number of subdivisions: ";  cin >> n;  BVPSetUp();
  cout << "Tolerance:              ";  cin >> T;
  int Code = 1;   Y = Fixpoint(G,Y,T,Code);
  cout << "||Y - U|| = ";  Show(MaxNorm(Y - U)); }
```

Figure 9.5. Testing Liebmann's method

Its exact solution is $u(t) = -3 + 2te^{t-2}$. Although you might expect a linear problem to be more tractable, similar phenomena occur:

Number n of t steps	Iterations required	Error
5	42	.073
10	137	.016
17	331	.00081
20	431	.0037

To obtain these results, Figure 9.5 functions `main`, `f`, and `u` were adapted for this problem and then run with the same tolerance T = 0.0001.

These examples show the need for error analysis, in order to predict the behavior of problems for which true solutions are not known, and to suggest methods for arranging fixpoint problems so that the iterations converge more quickly. The analysis has two aspects, corresponding to these questions:

- How well does the true solution to the fixpoint equation approximate that of the BVP?
- How accurately can you compute the solution to the fixpoint equation?

The first of these is a deep problem in the numerical analysis of differential equations; beyond the scope of this book. The second question will be discussed later sections in this chapter.

Jacobians

Another advanced calculus concept that's needed for this section's error analysis is the Jacobian. It often plays the role of a multidimensional derivative. Consider a function $y : D \to \mathbf{R}^m$, where D is a set in $\mathbf{R}^n$:

$$y(\xi) = \begin{bmatrix} g_1(\xi) \\ \vdots \\ g_m(\xi) \end{bmatrix}, \qquad \xi = \begin{bmatrix} x_1 \\ \vdots \\ x_n \end{bmatrix}.$$

Suppose all the partial derivatives of the components g_j of y exist at a point ξ. Then the *Jacobian* of y at ξ is the matrix

$$Jy(\xi) = \begin{bmatrix} \dfrac{\partial g_1}{\partial x_1} & \cdots & \dfrac{\partial g_1}{\partial x_n} \\ \vdots & & \vdots \\ \dfrac{\partial g_m}{\partial x_1} & \cdots & \dfrac{\partial g_m}{\partial x_n} \end{bmatrix}.$$

For example, consider a linear function $y(\xi) = A\xi + \beta$, where A is an $m \times n$ matrix and β is an n-dimensional vector. The components of y and their partials are given by the equations

$$g_j(\xi) = \sum_{k=1}^{n} a_{jk} x_k + b_j, \qquad \frac{\partial g_j}{\partial x_k} = a_{jk}.$$

That is, the Jacobian of a linear function $y(\xi) = A\xi + \beta$ is the constant A.

Another familiar example occurs when $m = 1$. In that case, $y(\xi)$ has just the single component $g_1(\xi)$, and its Jacobian is just the row of partial derivatives:

$$Jy(\xi) = \left[\frac{\partial g_1}{\partial x_1}, \cdots, \frac{\partial g_1}{\partial g_n} \right].$$

This row is usually called the *gradient* $\nabla y(\xi)$.

Jacobians are particularly handy for stating the multidimensional *chain rule*, as follows. Suppose

- $y: D \to \mathbf{R}^m$, where D is an open set in $\mathbf{R}^n$, and all partial derivatives of y exist and are continuous there;
- $\varphi: y[D] \to \mathbf{R}^l$, and all partial derivatives of φ exist and are continuous in $y[D]$.

Let η denote the composition of φ and y—that is, $\varphi(y(\xi)) = \eta(\xi)$ for all vectors ξ in D. Then

- All the partial derivatives of η exist and are continuous at all vectors ξ in D, and $J\eta(\xi) = J\varphi(y(\xi))Jy(\xi)$.

The entries of $J\varphi(y(\xi))$ are the partial derivatives of the components of the function φ evaluated at points

$$y(\xi) = \begin{bmatrix} g_1(\xi) \\ \vdots \\ g_m(\xi) \end{bmatrix},$$

and are often written

$$\frac{\partial f_i}{\partial g_j}.$$

With that notation, the entries of $J\eta(X)$ take the familiar form

$$\frac{\partial h_i}{\partial x_k} = \sum_{j=1}^{m} \frac{\partial f_i}{\partial g_j} \frac{\partial g_j}{\partial x_k}.$$

A Multidimensional Mean Value Theorem

The mean-value theorem played a major role in the error analysis discussed in Section 4.3 for single-dimensional fixpoint iteration. For the multidimensional theory, you need an analogous result, which may be unfamiliar:

> Suppose a real-valued function g is defined on an open convex subset D of $\mathbf{R}^n$ and has continuous partial derivatives there, and suppose ξ and ζ are vectors in D. Then
>
> $$g(\xi) - g(\zeta) = \nabla g(\eta)(\xi - \zeta)$$
>
> for some vector η between, but different from, ξ and ζ.

(Here the gradient $\nabla g(\eta)$ is an n-dimensional row and $\xi - \zeta$ is an n-dimensional column.) The theorem is simple to prove. Define a function $\lambda : \mathbf{R} \to \mathbf{R}^n$ and a real-valued function f on the unit interval I by setting

$$\lambda(t) = \zeta + t(\xi - \zeta) \qquad f(t) = g(\lambda(t)).$$

Since λ is linear, $J\lambda(t) = \xi - \zeta$. By the chain rule, f has a continuous derivative on I; hence

$$g(\xi) - g(\zeta) = f(1) - f(0) = f'(t)(1 - 0) = f'(t)$$

for some t such that $0 < t < 1$, by the single-dimensional mean-value theorem. Again, by the chain rule,

$$f'(t) = Jg(\lambda(t))J\lambda(t) = \nabla g(\zeta)(\xi - \zeta)$$

where $\zeta = \lambda(t)$.

This mean-value theorem is often used to estimate the difference in values of a function $y : D \to \mathbf{R}^m$, where D is an open convex subset D of $\mathbf{R}^n$:

Suppose all components of y have continuous partials in D; then, for any vectors ξ and ζ in D, there's a vector η in D such that

$$\|y(\xi) - y(\zeta)\| \leq \|Jy(\eta)\| \|\xi - \zeta\|.$$

All the norms here are matrix row norms; since the vectors are columns, their row norms are the same as max norms.

The proof is simple: for some index j^* and some vector η between but different from ξ and ζ,

$$\|y(\xi) - y(\zeta)\| = \max_{j=1}^{m} |g_j(\xi) - g_j(\zeta)| = |g_{j^*}(\xi) - g_{j^*}(\zeta)|$$

$$= \|\nabla g_{j^*}(\eta)\| \|\xi - \zeta\| = \left[\sum_{k=1}^{n} \left|\frac{\partial g_{j^*}}{\partial x_k}\right|\right] \|\xi - \zeta\|$$

$$\leq \|Jy(\eta)\| \|\xi - \zeta\|.$$

At one step here, the row norm inequality, $\|AB\| \leq \|A\| \|B\|$ was used; it's valid for any matrices A and B that you can multiply.)

Convergence Condition, Speed of Convergence

As in the single-dimensional discussion in Section 4.3, the inequality just demonstrated is used to obtain a Lipschitz constant, which will indicate whether and how fast multidimensional fixpoint iteration converges. Consider a function $y : D \to \mathbf{R}^m$, where D is a subset of $\mathbf{R}^n$. A *Lipschitz constant* for y on D is a number L such that

$$\|y(\xi) - y(\zeta)\| \leq L \|\xi - \zeta\|$$

for all vectors ξ and ζ in D. Suppose D is open and convex and all components of y have continuous partials there. By the previous inequality, any number $L \geq \|Jy(\eta)\|$ for all η in D is a Lipschitz constant. (In this paragraph, the vector norms are max norms, and the matrix norm is a row norm.)

Now it's possible to state a condition guaranteeing convergence of fixpoint iteration:

Suppose $y : D \to D$, where D is an open convex subset D of $\mathbf{R}^n$ containing a fixpoint α of y. Suppose y has a Lipschitz constant $L < 1$ on D. Then D contains only one fixpoint, and fixpoint iteration $\xi^{(0)}, \xi^{(1)}, \xi^{(2)}, \ldots$ with $\xi^{(k+1)} = y(\xi^{(k)})$ converges to α for any initial estimate $\xi^{(0)}$ in D.

The proof is the same as that for the single-dimensional case: the *error vectors* $\varepsilon^{(k)} = \xi^{(k)} - \alpha$ satisfy the inequalities

$$\|\varepsilon^{(k+1)}\| \le L\|\varepsilon^{(k)}\| \qquad \|\varepsilon^{(k)}\| \le L^k\|\varepsilon^{(0)}\|$$

for all k. Convergence of this sort is called *linear*, as in the single-dimensional case.

You can use these error inequalities to study the convergence of the fixpoint iterations considered earlier in this section. In the first example,

$$y(\xi) = \begin{bmatrix} \sin(x+y) \\ \cos(x-y) \end{bmatrix}, \qquad \xi = \begin{bmatrix} x \\ y \end{bmatrix}$$

and the row norm of the Jacobian is $J = 2\max(|\sin(x+y)|,|\cos(x-y)|)$. At the fixpoint $x,y \approx 0.935, 0.998$ and $J \approx 0.708$, so you can use the Lipschitz constant $L = 0.71$. From Figure 9.2 you can see that $\|\varepsilon^{(0)}\| \le 0.1$ with the initial estimate $x,y = 1,1$. By the error inequality, $\|\varepsilon^{(k)}\| \le L^k\|\varepsilon^{(0)}\| \le 0.1 L^k$; hence, $\|\varepsilon^{(k)}\|$ will be less than the tolerance $T = 10^{-6}$ if $0.1 L^k \le T$. You can solve this inequality to get $k \ge 34$. The example iteration actually converged much faster (nine iterations in Figure 9.4) than this analysis predicts.

The boundary value problems (BVP)

$$y'' = f(t, y, y') \qquad y(a) = y_0 \qquad y(b) = y^*$$

considered earlier were approximated by fixpoint problems

$$\eta = y(\eta) = \begin{bmatrix} g_1(\eta) \\ \vdots \\ g_{n-1}(\eta) \end{bmatrix} \qquad \eta = \begin{bmatrix} y_1 \\ \vdots \\ y_{n-1} \end{bmatrix} \qquad y_n = y^*$$

$$h = \frac{b-a}{n} \qquad t_j = a + jh \quad \text{for} \quad j = 1, \ldots, n-1$$

$$g_j(\eta) = \tfrac{1}{2} y_{j+1} + \tfrac{1}{2} y_{j-1} - \tfrac{1}{2} h^2 f\left(t_j, y_j, \frac{y_{j+1} - y_{j-1}}{2h}\right) \quad \text{for} \quad j = 1, \ldots, n-1.$$

A solution η of the fixpoint problem should approximate the solution u of the BVP: $y_j \approx u(t_j)$ for $j = 1, \ldots, n-1$. The behavior of the fixpoint iteration $\eta^{(k+1)} = y(\eta^{(k)})$ is related to the value $\|Jy\|$ of the row norm of the Jacobian near the solution. That's not difficult to analyze. The j,kth entry of Jy is

$$\frac{\partial g_j}{\partial y_k} = 0 \quad \text{unless } k = j-1, j, \text{ or } j+1$$

$$\frac{\partial g_j}{\partial y_{j-1}} = \frac{1}{2} + \frac{h}{4}\frac{\partial f}{\partial y'} \qquad \frac{\partial g_j}{\partial y_j} = -\frac{h^2}{2}\frac{\partial f}{\partial y} \qquad \frac{\partial g_j}{\partial y_{j+1}} = \frac{1}{2} - \frac{h}{4}\frac{\partial f}{\partial y'}.$$

The row norm $\|Jy\|$ is the largest of the sums

$$\left|\frac{\partial g_1}{\partial y_1}\right| + \left|\frac{\partial g_1}{\partial y_1}\right|,$$

$$\left|\frac{\partial g_j}{\partial y_{j-1}}\right| + \left|\frac{\partial g_j}{\partial y_j}\right| + \left|\frac{\partial g_j}{\partial y_{j+1}}\right| \quad \text{for } j = 2, \ldots, n-2,$$

$$\left|\frac{\partial g_{n-1}}{\partial y_{n-2}}\right| + \left|\frac{\partial g_{n-1}}{\partial y_{n-1}}\right|.$$

The f partials in these equations are evaluated at

$$t_j, \; y_j, \; \frac{y_{j+1} - y_{j-1}}{2h}.$$

Assuming that the f partials are bounded in the region under study, the top and bottom sums approach ½ as n increases. The middle sums have the form

$$\left|\frac{1}{2} + x\right| + \left|\frac{1}{2} - x\right| + \frac{h^2}{2}\left|\frac{\partial f}{\partial y}\right|, \qquad x = \frac{h}{4}\frac{\partial f}{\partial y'}.$$

You can verify that

$$\left|\frac{1}{2} + x\right| + \left|\frac{1}{2} - x\right| \geq 1$$

for all x; hence, $\|Jy\| > 1$ for all h. In general, the Liebmann method applies fixpoint iteration in a borderline situation where convergence cannot be guaranteed; you should expect it to diverge sometimes, and converge sometimes, but never very fast.

Had the asymmetric formula

$$\frac{y_k - y_{k-1}}{h}$$

been used to approximate the derivative $y'(t_k)$ in deriving the Liebmann difference equation from the BVP, the component functions g_j and their partials would have formulas

$$g_j(\eta) = \tfrac{1}{2} y_{j+1} + \tfrac{1}{2} y_{j-1} - \tfrac{1}{2} h^2 f\left(t_j, y_j, \frac{y_j - y_{j-1}}{h}\right)$$

$$\frac{\partial g_j}{\partial y_{j-1}} = \frac{1}{2} + \frac{h}{2}\frac{\partial f}{\partial y'} \qquad \frac{\partial g_j}{\partial y_j} = -\frac{h^2}{2}\frac{\partial f}{\partial y} - \frac{h}{2}\frac{\partial f}{\partial y'} \qquad \frac{\partial g_j}{\partial y_{j+1}} = \frac{1}{2}.$$

The row norms of the first and last rows of the Jacobian Jy would still approach ½ as n increases. But the norms of the middle rows have the form

$$\left|\frac{1}{2} + 2x\right| + \left|2x\right| + \frac{1}{2} + \frac{h^2}{4}\left|\frac{\partial f}{\partial y}\right|, \qquad x = \frac{h}{2}\frac{\partial f}{\partial y'}.$$

You can verify that

$$\left|\frac{1}{2} + 2x\right| + \left|2x\right| + \frac{1}{2} \geq \left|\frac{1}{2} + x\right| + \left|\frac{1}{2} - x\right|$$

for all x. Thus, using the asymmetric differential quotient to approximate the first derivative in Liebmann's method would generally yield a Jacobian with a larger row norm. The symmetric quotient is more appropriate.

9.2 ITERATIVE SOLUTION OF LINEAR SYSTEMS

Concepts
Linear fixpoint iteration

Jacobi iteration

Strictly diagonally dominant matrices

Gauss-Seidel iteration

Example: a random system

Example: heat boundary value problem

Example: ODE boundary value problem

Tridiagonal matrices

This section is concerned with the solution of linear systems by linear fixpoint iteration. Given an $n \times n$ matrix A and an n-dimensional vector β, you'll convert equations $A\xi = \beta$ to an equivalent form $\xi = \gamma(\xi) = C\xi + \delta$, make an initial estimate $\xi^{(0)}$ and then construct a sequence of successive approximations using the scheme $\xi^{(k+1)} = \gamma(\xi^{(k)})$. According to Section 9.1, the approximations will converge to a unique solution $\xi = \alpha$ if $\|C\| = \|J\gamma(\xi)\| < 1$. In fact, $\|C\|$ is a Lipschitz constant for γ, and the error vectors, $\varepsilon^{(k)} = \xi^{(k)} - \alpha$ satisfy the inequalities

$$\|\varepsilon^{(k+1)}\| \leq \|C\|\|\varepsilon^{(k)}\| \qquad \|\varepsilon^{(k)}\| \leq \|C\|^k \|\varepsilon^{(0)}\|.$$

Thus, to solve a system $A\xi = \beta$ you'll want to find an equivalent fixpoint problem $\xi = C\xi + \delta$ with $\|C\| < 1$ and as small as possible.

Jacobi Iteration

Given an $n \times n$ linear system $A\xi = \beta$, *Jacobi's method* constructs an equivalent linear fixpoint problem $\xi = C\xi + \delta$ as in the following 3×3 example—convert the equations

$$\begin{cases} a_{11}x_1 + a_{12}x_2 + a_{13}x_3 = b_1 \\ a_{21}x_1 + a_{22}x_2 + a_{23}x_3 = b_2 \\ a_{31}x_1 + a_{32}x_2 + a_{33}x_3 = b_3 \end{cases}$$

to the fixpoint problem

$$\begin{cases} x_1 = \qquad\qquad -\dfrac{a_{12}}{a_{11}}x_2 - \dfrac{a_{13}}{a_{11}}x_3 + \dfrac{b_1}{a_{11}} \\ x_2 = -\dfrac{a_{21}}{a_{22}}x_1 \qquad\qquad - \dfrac{a_{23}}{a_{22}}x_3 + \dfrac{b_2}{a_{22}} \\ x_3 = -\dfrac{a_{31}}{a_{33}}x_1 - \dfrac{a_{32}}{a_{33}}x_2 \qquad\qquad + \dfrac{b_3}{a_{33}}, \end{cases}$$

that is,

$$\xi = C\xi + D^{-1}\beta \qquad C = -D^{-1}A + I,$$

where D is the diagonal matrix extracted from A. Of course, this process assumes that there's no zero on the diagonal of A. You can verify in general that this last equation is equivalent to $A\xi = \beta$: It holds just in case

$$\xi = -D^{-1}A\xi + \xi + D^{-1}\beta$$

$$0 = -D^{-1}A\xi + D^{-1}\beta$$

$$0 = -A\xi + \beta.$$

What criterion should A satisfy to ensure that $\|C\| < 1$? The row norm $\|C\| = \|-D^{-1}A + I\|$ is the largest sum

$$\sum_{j=1}^{n} |c_{ij}|;$$

hence, each of these sums must be < 1. But $c_{ii} = 0$ and $c_{ij} = -a_{ij}/a_{ii}$ when $j \neq i$, so A should satisfy the inequality

$$\sum_{\substack{j=1 \\ j \neq i}}^{n} \left|\frac{a_{ij}}{a_{ii}}\right| < 1, \qquad i.e., \sum_{\substack{j=1 \\ j \neq i}}^{n} |a_{ij}| < |a_{ii}|$$

for each i. Such a matrix A is called *strictly diagonally row dominant*. Note that $\|C\|$ approaches 1, hence, convergence deteriorates, as the two sides of this inequality get closer.

If you carry out the entire multidimensional fixpoint analysis with column norms in place of row norms, you'll get a corresponding result. Jacobi iteration also converges if A is strictly diagonally *column* dominant—for each j,

$$\sum_{\substack{i=1 \\ i \neq j}}^{n} \left|\frac{a_{ij}}{a_{ii}}\right| < 1, \qquad i.e., \sum_{\substack{i=1 \\ i \neq j}}^{n} |a_{ij}| < |a_{ii}|.$$

Strict diagonal dominance implies nonsingularity: If A is strictly diagonally row or column dominant, then $\|C\| < 1$ for the row or column norm; hence, Jacobi iteration converges to a unique fixpoint of the function $y(\xi) = C\xi + \delta$. This implies that the equations $A\xi = \beta$ have a unique solution for any β; hence, A is nonsingular.

The Equate2 template Jacobi implements this method. Shown in Figure 9.6, it's structured like the Section 9.1 template Fixpoint. Jacobi will be demonstrated later in this section.

```
template<class Scalar>                    // Return an approxi-
  Vector<Scalar> Jacobi(                  // mation to a solu-
    const Matrix<Scalar>& A,              // tion of  AX = B ,
    const Vector<Scalar>& B,              // computed by Jacobi
          Vector<Scalar>  X,              // iteration. Use
          double          T,              // initial estimate
          int&       Code) {              // X , and stop when
  try {                                   // || Xn - Xn-1 || <
    int    m = A.HighRowIndex(),          // T . Set  Code = 0
           n = 0;                         // to report success,
    double M;                             // or  = -1 if the
    do {                                  //             desired
      Vector<Scalar> P = B;               //             accuracy
      for (int i = 1; i <= m; ++i) {      //             is not
        for (int j = 1; j <= m; ++j)      //             achieved
          if (j != i) P[i] -= A[i][j]*X[j]; //           after
        P[i] /= A[i][i]; }                //             1000
      M = MaxNorm(P - X);                 //             itera-
      X = P; }                            //             tions.
    while (++n < 1000 && M >= T);
    Code = (M < T ? 0 : -1);
    return X; }
  catch(...) {
    cerr << "\nwhile doing Jacobi iteration";
    throw; }}
```

Figure 9.6. Equate2 template Jacobi

Gauss-Seidel Iteration

To compute $\xi^{(k+1)} = \gamma(\xi^{(k)})$ by Jacobi's algorithm, you use

- Entry $x_1^{(k)}$ in calculating $x_2^{(k+1)}, \ldots, x_n^{(k+1)}$,
- Entries $x_1^{(k)}, x_2^{(k)}$ in calculating $x_3^{(k+1)}, \ldots, x_n^{(k+1)}$,

$\vdots$

It might seem better to use

- $x_1^{(k+1)}$ in place of $x_1^{(k)}$ in calculating $x_2^{(k+1)}$,
- $x_1^{(k+1)}, x_2^{(k+1)}$ in place of $x_1^{(k)}, x_2^{(k)}$ in calculating $x_3^{(k+1)}$,

$\vdots$

```
template<class Scalar>                      // Return an approxi-
  Vector<Scalar> Seidel(                    // mation to a solu-
    const Matrix<Scalar>& A,                // tion of  AX = B ,
    const Vector<Scalar>& B,                // computed by Gauss-
          Vector<Scalar>  X,                // Seidel iteration.
          double   T,                       // Use initial esti-
          int&     Code) {                  // mate  X ,  and
  try {                                     // stop when
    int m = A.HighRowIndex(),               // || Xn - Xn-1 || <
        n = 0;                              // T .  Set  Code = 0
    double M;                               // to report success,
    do {                                    //   or  = -1  if
      Vector<Scalar> P = X;                 //   the desired
      for (int i = 1; i <= m; ++i) {        //   accuracy is
        X[i] = B[i];                        //   not achieved
        for (int j = 1; j <= m; ++j)        //   after  1000
          if (j != i) X[i] -= A[i][j]*X[j]; //   iterations.
        X[i] /= A[i][i]; }
      M = MaxNorm(P - X); }
    while (++n < 1000 && M >= T);
    Code = (M < T ? 0 : -1);
    return X; }
  catch(...) {
    cerr << "\nwhile doing Gauss-Seidel iteration";
    throw; }}
```

Figure 9.7. Equate2 template Seidel

because the approximations $x_1^{(k+1)}$, $x_2^{(k+1)}$, ... are supposedly more accurate than their predecessors. It's easy to make this modification. The resulting algorithm is called *Gauss-Seidel* iteration; it's implemented by the Equate2 template Seidel, shown in Figure 9.7.

In the Jacobi algorithm, each component $x_i^{(k+1)}$ is a linear combination of the components $x_1^{(k)}$, ..., $x_n^{(k)}$; hence, the same is true when $x_1^{(k+1)}$, ..., $x_i^{(k+1)}$ are used in place of $x_1^{(k)}$, ..., $x_i^{(k)}$ to compute $x_{i+1}^{(k+1)}$ in the Gauss-Seidel method. Thus, Gauss-Seidel iteration is also a special case of linear fixpoint iteration $\xi^{(k+1)} = C\xi^{(k)} + \delta$ with C and δ computed from the matrix and vector of the original system $A\xi = \beta$. You can analyze the convergence of Gauss-Seidel iteration in the same way as the Jacobi algorithm, and the same result applies: Gauss-Seidel iteration converges if the matrix A is strictly diagonally row or column dominant. Moreover, it *usually* converges faster than Jacobi iteration [5, Section 7.3]. However, it's possible for either method to converge but not the other, and even for either method to be faster when they both converge [55].

Example: A Random System

As a first example, the MSP client program shown in Figure 9.8 tests Jacobi and Gauss-Seidel iteration and shows how their rates of convergence depend on the degree of diagonal dominance. You'll find its source code on the accompanying diskette in file Fig9_08.CPP. For a given value d the program constructs an $n \times n$ matrix A with diagonal entries d. Its other entries, and those of an n-dimensional vector β, are selected at random from the unit interval. Using Gauss elimination, the program next computes an "exact" solution $\xi = \alpha$ of the system $A\xi = \beta$. Then it

```
#include "Equate2.H"

template<class Scalar>
  void J_or_GS(Scalar t) {
    cout << "Solving a  20x20  " << TypeName(t) <<
            "  system   AX = B\n"
            "by Jacobi  [1]  or Gauss-Seidel  [2]  iteration.\n"
            "Select method      [1/2] : ";  int    M;  cin >> M;
    cout << "Enter diagonal entry  d : ";  double d;  cin >> d;
    cout << "B  and the other entries of  A  are random.\n";
    int      n = 20;        double T = 1e-4;
    Vector<Scalar> X(n);    X.MakeZero();
    Vector<Scalar> B(n);    B.MakeRandom();
    Matrix<Scalar> A(n,n);  A.MakeRandom();
    for (int i = 1; i <= n; ++i) A[i][i] = d;
    int Code = 0;
    Vector<Scalar> Y = GaussElim( A,B,Code);
    Code = 1;
    if (M == 1)     X = Jacobi(A,B,X,T,Code);
    if (M == 2)     X = Seidel(A,B,X,T,Code);
    cout << "\nFor a Gauss elimination solution   Y ,"
            "\n|AY - B| = ";  Show(MaxNorm(A*Y-B));
    cout << "\n|AX - B| = ";  Show(MaxNorm(A*X-B);
    cout << "\n|  X - Y| = ";  Show(MaxNorm(  X-Y));  }

void main() {
  double x = 1;    J_or_GS(x);   cout << "\n\n";
  complex z = ii;
J_or_GS(z); }
```

Figure 9.8. Testing Jacobi and Gauss-Seidel iteration

	Jacobi's method			Gauss-Seidel method		
d	Iterations	$\|A\xi - \beta\|$	$\|\xi - \alpha\|$	Iterations	$\|A\xi - \beta\|$	$\|\xi - \alpha\|$
8	Diverges			7	9×10^{-5}	1×10^{-5}
10	152	9×10^{-4}	5×10^{-5}	6	1×10^{-4}	1×10^{-5}
15	15	7×10^{-4}	3×10^{-5}	5	5×10^{-5}	4×10^{-6}

Figure 9.9. Results for $n = 20$ using Jacobi and Gauss-Seidel iteration

computes approximate solutions ξ by Jacobi or Gauss-Seidel iteration, using the zero vector as initial estimate and a tolerance of 0.0001. It displays the number of iterations and the error measures $\|A\xi - \beta\|$ and $\|\xi - \alpha\|$. Figure 9.9 gives some results for $n = 20$.

The error measure $\|A\xi - \beta\|$ for the "exact" solution $\xi = \alpha$ is about 2×10^{-16}. Since the sum of the nondiagonal entries of any row of A is about $(n - 1) \times 0.5 = 9.5$, the input $d = 8$ should lead to a matrix that's not diagonally row dominant; hence, you can't *expect* either method to converge. Jacobi's method is the more sensitive to this criterion: It diverges. Jacobi's method converges when $d = 10$ and A is just barely row dominant. Both methods converge faster as A becomes more dominant.

As shown in Section 8.2, Gauss elimination requires about $2/3\, n^3 \approx 5000$ arithmetic operations to solve the system $A\xi = \beta$. Each step of an iterative method requires n additions and n multiplications for each row—hence, $2n^2$ operations in all. The most rapidly convergent example tabulated required five iterations—hence, about 4000 operations. Thus, if extreme accuracy is not important and an iterative method is known to converge very rapidly, you may save time by using that method rather than Gauss elimination.

You could also save time if matrix A were sparse and your algorithm for computing the successive approximations avoided operating with its zero entries, so that substantially fewer than $2n^2$ operations were required per iteration. That possibility isn't illustrated here, because it requires implementation of a new data structure for sparse matrices.

Example: PDE Boundary Value Problem

The heat conduction problem that introduced Section 8.1 provides a more practical test for Jacobi and Gauss-Seidel iteration. A square metal plate is divided by an evenly spaced grid into n^2 square cells, whose center temperatures $w_{11}, \ldots, w_{nn}$ are studied. (The relationship between this notation and that in Section 8.1 will become clear soon.) The top, bottom, left, and right boundary cell temperatures w_{1k}, w_{nk}, w_{j1}, and w_{jn} for $j, k = 1, \ldots, n$ are held at known constant values, and the other w_{jk} are given initial temperature zero. The plate is then allowed to reach an equilibrium temperature distribution, in which each interior temperature w_{jk} is the average of the four neighboring cell temperatures:

$$4 w_{j,k} - w_{j-1,k} - w_{j+1,k} - w_{j,k-1} - w_{j,k+1} = 0.$$

The equations for the interior cell temperatures, together with the boundary cell equations w_{jk} = constant, constitute an $n^2 \times n^2$ linear system $A\omega = \beta$.

This example is actually a *partial* differential equation BVP. Start with the equation

$$\frac{\partial^2 f}{\partial x^2} + \frac{\partial^2 f}{\partial y^2} = 0,$$

which governs the equilibrium heat distribution $w = f(x, y)$. Introduce coordinate axes parallel to the cell boundaries, so that the temperatures w_{jk} are measured at the n^2 intersections of evenly spaced grid lines $x = x_j$ and $y = y_k$ for $j,k = 1, \ldots, n$. Thus, $w_{jk} = f(x_j, y_k)$. Now approximate the second partials by difference quotients, as in the Liebmann method discussed earlier. You get the same system of linear equations for the unknown w_{jk} values:

$$\left.\frac{\partial^2 f}{\partial x^2}\right|_{x_j, y_k} \approx \frac{w_{j-1,k} - 2w_{j,k} + w_{j+1,k}}{h^2} \qquad \left.\frac{\partial^2 f}{\partial y^2}\right|_{x_j, y_k} \approx \frac{w_{j,k-1} - 2w_{j,k} + w_{j,k+1}}{h^2}$$

$$0 = \frac{\partial^2 f}{\partial x^2} + \frac{\partial^2 f}{\partial y^2} \approx \frac{w_{j-1,k} + w_{j+1,k} + w_{j,k-1} + w_{j,k+1} - 4w_{j,k}}{h^2}.$$

To describe this system, you'll need a way to correlate doubly indexed cell temperatures $w_{j,k}$ with singly indexed ω entries w_i. The scheme

$$i = j + (n - 1) k = \mathrm{sub}(j,k) \qquad w_i = w_{j,k}$$

is shown in Figure 8.1. The boundary cell equations have the simple form w_i = the constant boundary cell temperature: That is,

- If j or $k = 1$ or n, and $h = \mathrm{sub}(j,k)$, then $a_{hh} = 1$ and b_h = constant.

The interior cell equations have the form shown earlier: For other indices j and k,

- If $h = \mathrm{sub}(j,k)$, then $a_{hh} = 4$.
- If $i = \mathrm{sub}(j - 1, k), \mathrm{sub}(j + 1, k), \mathrm{sub}(j, k - 1)$, or $\mathrm{sub}(j, k + 1)$, then $a_{hi} = -1$.

All other entries a_{hi} and b_h are zero.

```
#include "Equate2.H"

int n = 7;
int sub(int j, int k) { return j + (k - 1)*n; }

void main() {
  cout << "Solving the heat BVP for an  nxn  plate          \n"
          "by Jacobi  [1]  or Gauss-Seidel iteration  [2]\n\n";
  cout << "Select the method   [1/2] : ";       int M;   cin >> M;
  cout << "Enter  n : ";                                 cin >> n;
  double T = .01;    int h,i,j,k,n2 = n*n;
  Vector<double> X(n2);       X.MakeZero();   {       // A , B
  Matrix<double> A(n2,n2);    A.MakeZero();           // are
  Vector<double> B(n2);       B.MakeZero();           // declared
  for (k = 1; k <= n; ++k) {                          // in this
    h = sub(1,k);  A[h][h] = 1;   B[h] = n - k;       // block,
    h = sub(n,k);  A[h][h] = 1;   B[h] = k - 1; }     // so they
  for (j = 2; j < n; ++j) {                           // die
    h = sub(j,1);  A[h][h] = 1;   B[h] = n - j;       // before
    h = sub(j,n);  A[h][h] = 1;   B[h] = j - 1;       // W is
    for (k = 2; k < n; ++k) {                         // built.
         h = sub(j,k);     A[h][h] =   4;
         i = sub(j-1,k);   A[h][i] = -1;
         i = sub(j+1,k);   A[h][i] = -1;
         i = sub(j,k-1);   A[h][i] = -1;
         i = sub(j,k+1);   A[h][i] = -1; }}
  int Code;   Code = 1;
  if (M == 1) X = Jacobi(A,B,X,T,Code);
     else     X = Seidel(A,B,X,T,Code);
  cout << "|AX - B| = ";
  Show(MaxNorm(A*X-B)); }
  Matrix<double> W(n,n);
  for (j = 1; j <= n; ++j)
    for (k = 1; k <= n; ++k) {
         i = sub(j,k);   W[j][k] = X[i]; }
  cout << "\n\n";  W.Show(); }
```

Figure 9.10. Program to solve the heat conduction problem

The MSP client program shown in Figure 9.10 sets up matrix A and vector β for $n = 7$ according to this scheme, using the Figure 8.1 boundary cell temperatures. Then it computes a solution vector ω by Jacobi or Gauss-Seidel iteration with tolerance $T = 0.01$ and outputs the error measure $\|A\omega - \beta\|$. Finally, it outputs the solution in array form compatible with Figure 8.1.

6.0	5.0	4.0	3.0	2.0	1.0	0.0
5.0	4.3	3.7	3.0	2.3	1.7	1.0
4.0	3.7	3.3	3.0	2.7	2.3	2.0
3.0	3.0	3.0	3.0	3.0	3.0	3.0
2.0	2.3	2.7	3.0	3.3	3.7	4.0
1.0	1.7	2.3	3.0	3.7	4.3	5.0
0.0	1.0	2.0	3.0	4.0	5.0	6.0

Figure 9.11. Solution to the heat conduction problem

You'll find the source code for this program in file `Fig9_10.CPP` on the accompanying diskette. Here are its results:

Method	Iterations	$\|A\omega - \beta\|$
Jacobi	31	0.036
Gauss-Seidel	19	0.019

Actually, matrix A isn't *strictly* diagonally row dominant: The absolute values of most of its diagonal entries are *equal to*, not greater than, the sums of the absolute values of the remaining entries in their rows. The iterative methods converge nevertheless, but slowly. Cutting the tolerance T to 0.001 increased the iteration counts to 47 and 27, respectively, and cut the error measures by a factor of 10. The computed solution is displayed in Figure 9.11.

The Gauss-Seidel method required $2(n^2)^2$ arithmetic operations per iteration times 19 iterations, or about 90,000 operations to compute this solution; Gauss elimination would have required about $2/3(n^2)^3 \approx 80,000$ operations. Thus, even if such a large tolerance is appropriate, the iterative methods are slower. You can only save time with the iterative methods if you can cut down the number of operations per iteration by avoiding arithmetic on the zero entries. In fact, that's quite simple with this example, but you lose the advantage of the ready-prepared iteration functions.

Example: Second-Order Linear ODEBVP

Earlier in this chapter you saw how to use the Liebmann method to convert a second-order ODE boundary value problem (BVP) into a difference equation suitable for solution by multidimensional fixpoint iteration. Starting with the problem

$$y'' = f(t, y, y') \qquad y(a) = y_0 \qquad y(b) = y^*,$$

you divide the interval $a \leq t \leq b$ into n subdivisions of length h:

$$h = \frac{b-a}{n} \qquad t_j = a + jh \text{ for } j = 0,\ldots,n.$$

Then you set $y_n = y^*$ and introduce approximations y_j to the solution $u(t)$ at $t = t_j$ for $j = 1, \ldots, n - 1$. These satisfy the difference equation

$$\frac{y_{j+1} - 2y_j + y_{j-1}}{h^2} = f\left(t_j, y_j, \frac{y_{j+1} - y_{j-1}}{2h}\right)$$

obtained when you approximate derivatives in the ODE by differential quotients. If f is linear with constant coefficients—that is,

$$f(t, y, y') = py' + qy + r$$

for some constants p, q and r—the difference equation is also linear:

$$\frac{y_{j+1} - 2y_j + y_{j-1}}{h^2} = p\frac{y_{j+1} - y_{j-1}}{2h} + qy_j + r.$$

Remember that y_0 and $y_n = y^*$ are constants specified in the problem. This equation can be rewritten as an $(n - 1) \times (n - 1)$ linear system $A\eta = \beta$, where A and β have entries

$$\begin{cases} a_{j,k} = 0 & \text{unless } k = j-1,\ j,\ \text{or } j+1 \\ a_{j,j-1} = \dfrac{1}{h^2} + \dfrac{p}{2h} \qquad a_{j,j} = -q - \dfrac{2}{h^2} \qquad a_{j,j+1} = \dfrac{1}{h^2} - \dfrac{p}{2h} \\ b_1 = q - a_{1,0}\, y_0 \\ b_j = q & \text{unless } j = 1 \text{ or } n-1 \\ b_{n-1} = q - a_{n-1,n}\, y_n. \end{cases}$$

A is called a *tridiagonal* matrix: All its nonzero entries lie on the diagonal, or immediately above or below it. You can solve a tridiagonal system $A\eta = \beta$ by Gauss elimination, by a special elimination algorithm for tridiagonal systems, or by the iterative methods introduced in this section. The special algorithm for tridiagonal systems is simply Gauss elimination, with the formulas adjusted to mention only the nonzero entries of A [5, Section 6.7]. With that method, you can use a variant of the Matrix data structure that stores the superdiagonal, diagonal, and subdiagonal entries of A in vectors. That technique avoids wasting space storing the zero entries and saves time by avoiding arithmetic operations with the zeros in the matrix. It's the most efficient way to handle problems of this type, but it is not covered further in this book because the tridiagonal matrix data type is not implemented here.

Success of the iterative methods depends on the diagonal row dominance of matrix A. In particular, when $1 < j < n - 1$, you want the condition

$$\left| a_{j,j} \right| > \left| a_{j,j-1} \right| + \left| a_{j,j+1} \right|$$

$$\left| q + \frac{2}{h^2} \right| > \left| \frac{1}{h^2} + \frac{p}{2h} \right| + \left| \frac{1}{h^2} - \frac{p}{2h} \right|$$

to hold for small positive h values. If $h < |2/p|$, the right-hand side of this inequality is just $2/h^2$. If $q < 0$, the inequality is false for sufficiently small h values. If $q > 0$, it's true for all h, but only barely, unless q is enormous. Thus, you should expect the iterative methods to converge very slowly, if at all.

Figures 9.12 and 9.13 show an MSP client program that compares the solutions of a linear ODEBVP obtained by the Gauss elimination, Jacobi, and Gauss-Seidel methods. Function main lets the user select the method, and inputs the number n of subdivisions and the tolerance T (if required). It calls BVPSetUp to set up the linear system corresponding to the linear BVP considered earlier in Section 8.1:

$$y'' = pu' + qu + r \qquad y(a) = y_0 \qquad y(b) = y^*$$
$$p = 2 \qquad q = -1 \qquad r = -3$$
$$a = 0 \qquad y_0 = -3$$
$$b = 2 \qquad y^* = 1$$

BVPSetUp builds the vector U, whose entries are the exact solution values $u(t) = -3 + 2te^{t-2}$ for $t = t_1, \ldots, t_{n-1}$. These correspond to the computed approximations $y_1, \ldots, y_{n-1}$. BVPSetUp constructs their initial estimates for the iterative methods. The initial entries of the solution vector Y are the values of the linear function that satisfies the boundary conditions; they're updated by each iteration. After computing the approximate solution Y the program reports the error $\|Y - U\|$. You'll find this source code in file Fig9_12.CPP on the accompanying diskette.

Figure 9.14 shows the results from the program in Figures 9.12 and 9.13. Clearly, the iterative methods are inappropriate. This is predictable because $q < 0$: The matrix is not diagonally row dominant. However, even if you change the problem to one with a positive q—for example, $y'' = y' + 2y + 3$—the results change little, because A becomes only slightly row dominant.

Because numerical solution of linear boundary value problems is so important in applications, several other versions of linear fixpoint iteration, comparable to the Gauss-Seidel method, have been investigated. They apply to particular types of differential equations, and are described in more specialized books. See [5, Section 7.3] and the literature cited there.

```
#include "Equate2.H"                    // Solving the ODEBVP
                                        //     u" = pu' + qu + r
                                        // y(a) = y0   y(b) = yStar
double p = 2, q      = -1, r = -3,
       a = 0, y0     = -3,              // h and k are the t
       b = 2, yStar  = 1, T,h,k;        // step and linear y step.
int    n;                               // T is the tolerance. n
Matrix<double> A;                       // is the number of  t
Vector<double> B,Y,U;                   // steps. AY = B is the
                                        // linear system. u and U
                                        // compute and store the
double u(double t) {                    // exact BVP solution, to be
  return -3 + 2*t*exp(t-2); }           // compared with the approx-
                                        // imate solution Y .

void BVPSetUp() {
  h = (b - a)/n;      k = (yStar - y0)/n;
  double Subdiagonal   = (1/h + p/2)/h,    // Set up the linear
         Diagonal      = -q - 2/(h*h),     // system.
         Superdiagonal = (1/h - p/2)/h;
  A.SetUp(n-1,n-1).MakeZero();
  for (int i = 1; i <= n-1; ++i) {
    if (i > 1) A[i][i-1]   = Subdiagonal;
    A[i][i]                = Diagonal;
    if (i < n-1) A[i][i+1] = Superdiagonal; }
  B.SetUp(n-1,1);   AllEntries(B,r);
  B[1]   -= Subdiagonal  *y0;              // Set up the linear
  B[n-1] -= Superdiagonal*yStar;           // initial estimate
  Y.SetUp(n-1,1);   U.SetUp(n-1,1);        // Y , and the exact
  for (int i = 1; i <= n-1; ++i) {         // solution U , to
    Y[i]     = y0 + i*k;                   // be compared with
    double t = a + i*h;                    // the approximate
    U[i]     = u(t); }}                    // solution Y .
```

Figure 9.12. Testing `GaussElim`, `Jacobi`, and `Seidel` with a linear second-order boundary value problem *Part 1 of 2; continued in Figure 9.13.*

```
void main() {
  cout << "Solving    y\" = 2y' - y - 3      \n"
          "           y(0) = -3  y(2) = 1  \n\n"
          "by Gauss Elimination       [1]  \n"
          "or Jacobi iteration        [2]  \n"
          "or Gauss-Seidel iteration [3]\n\n"
          "Enter method number [1/2/3] : ";
  int M;    cin >> M;
  cout << "n = number of subdivisions : ";  cin >> n;
  if (M > 1) { cout <<    "Tolerance  T : ";  cin >> T; }
  BVPSetUp();   int Code = 1;
  if (M == 1)   Y = GaussElim(A,B,    Code);      // Compute  Y ,
  if (M == 2)   Y = Jacobi    (A,B,Y,T,Code);     // compare with
  if (M == 3)   Y = Seidel    (A,B,Y,T,Code);     // exact
  cout << "\n||Y - U|| = ";                       // solution.
  Show(MaxNorm(Y - U)); }
```

Figure 9.13. Testing `GaussElim`, `Jacobi`, and `Seidel` with a linear second order boundary value problem
Part 2 of 2; continued from Figure 9.12.

Number of subdivisions	Gauss elimination	Jacobi's method		Gauss-Seidel method	
	Error	Iterations	Error	Iterations	Error
8	.028	92	.027	49	.028
16	.0069	298	.0021	165	.0046
32	.0017	905	.019	519	.0086

Figure 9.14. Test results

9.3 MULTIDIMENSIONAL NEWTON-RAPHSON ITERATION

> **Concepts**
> *Quadratic convergence*
> *Multidimensional Taylor theorem*
> *Multidimensional Newton-Raphson formula*
> *Complex Newton-Raphson formula*
> *Cauchy-Riemann equations*
> `Equate2` *template* `NR`
> *Examples*

In Section 9.1 you saw that fixpoint iteration $\xi^{(k+1)} = y(\xi^{(k)})$ converges if

- $y : D \to \mathbf{R}^n$ for some open set D of $\mathbf{R}^n$ containing a fixpoint $\alpha = y(\alpha)$ and your initial estimate $\xi^{(0)}$,
- all components $g_j(\xi)$ of $y(\xi)$ have continuous partial derivatives in D, and
- there's a Lipschitz constant $L < 1$, such that $\|Jy(\xi)\| \le L$ for all ξ in D.

Moreover, the error vectors $\varepsilon^{(k)} = \xi^{(k)} - \alpha$ satisfy the inequalities $\|\varepsilon^{(k+1)}\| \le L\|\varepsilon^{(k)}\|$ and $\|\varepsilon^{(k)}\| \le L^k\|\varepsilon^{(0)}\|$. This suggests that—as in the single-dimensional case—you should be able to expedite solution of an n-dimensional system of equations $\varphi(\xi) = 0$, where $\varphi : D \to \mathbf{R}^n$, by finding an equation $y(\xi) = \xi$ with the same solution α, such that $\|Jy(\xi)\|$ is as small as possible for ξ near α; that is, you want $Jy(\alpha) = 0$. Under rather general conditions this is in fact possible, and convergence will be *quadratic*: You can find M such that $\|\varepsilon^{(k+1)}\| \le M\|\varepsilon^{(k)}\|^2$. This result, and a precise statement of its conditions, is derived in this section.

Quadratic Convergence

Close analysis of the convergence of the iteration $\xi^{(k+1)} = y(\xi^{(k)})$ to approximate a fixpoint α where $Jy(\alpha) = 0$ requires use of a *multivariable Taylor theorem*:

Suppose g is a real-valued function defined on an open convex subset D of $\mathbf{R}^n$, with continuous second partials. Then for any vectors α and β in D,

$$g(\beta) = g(\alpha) + \nabla g(\beta)(\beta - \alpha) + \frac{1}{2}\sum_{i,j=1}^n \left.\frac{\partial^2 g}{\partial x_i \partial x_j}\right|_\xi (b_i - a_i)(b_j - a_j),$$

where

$$\begin{bmatrix} a_1 \\ \vdots \\ a_n \end{bmatrix} = \alpha \qquad \begin{bmatrix} b_1 \\ \vdots \\ b_n \end{bmatrix} = \beta,$$

and ξ is between, but distinct from, α and β.

To prove this, define functions η and f from the unit interval to $\mathbf{R}^n$ and $\mathbf{R}$ by setting

$$\eta(t) = \alpha + t(\beta - \alpha) = \begin{bmatrix} h_1(t) \\ \vdots \\ h_n(t) \end{bmatrix}, \qquad f(t) = g(\eta(t)),$$

so that

$$f'(t) = Jf(t) = Jg(\eta(t))\, J\eta(t) = \nabla g(\eta(t))(\beta - \alpha) = \sum_{j=1}^{n} \frac{\partial g}{\partial x_j}\bigg|_{\xi = \eta(t)} (b_j - a_j)$$

$$f''(t) = \sum_{j=1}^{n} \left[\frac{d}{dt} \frac{\partial g}{\partial x_j}\bigg|_{\xi=\eta(t)} \right] (b_j - a_j) = \sum_{j=1}^{n} \sum_{i=1}^{n} \left[\frac{\partial^2 f}{\partial x_i \partial x_j}\bigg|_{\xi=\eta(t)} h'_i(t) \right] (b_j - a_j)$$

$$= \sum_{i,j=1}^{n} \frac{\partial^2 g}{\partial x_i \partial x_j}\bigg|_{\xi=\eta(t)} (b_i - a_i)(b_j - a_j).$$

Now use the single variable Taylor theorem on function f:

$$g(\beta) = f(1) = f(0) + f'(0) + \tfrac{1}{2} f''(t) = g(\alpha) + \nabla g(\alpha)(\beta - \alpha) + \tfrac{1}{2} f''(t)$$

for some t between, but different from, 0 and 1.

You can apply the multidimensional Taylor theorem to consider fixpoint iteration $\xi^{(k+1)} = \gamma(\xi^{(k)})$ when $J\gamma(\xi) = 0$ at the fixpoint $\xi = \alpha$.

Suppose $\gamma: D \to D$ and its components have bounded continuous second partials in a convex neighborhood D of α:

$$g(\xi) = \begin{bmatrix} g_1(\xi) \\ \vdots \\ g_n(\xi) \end{bmatrix} \qquad \xi = \begin{bmatrix} x_1 \\ \vdots \\ x_n \end{bmatrix} \qquad \left| \frac{\partial^2 g}{\partial x_i \partial x_j} \right| \leq M$$

for a constant M and all ξ in D. Moreover, suppose $J\gamma(\alpha) = 0$. Then fixpoint iteration converges quadratically to α if its initial estimate lies in D. Moreover, the error vectors $\varepsilon^{(k)}$ satisfy the inequality

$$\|\varepsilon^{(k+1)}\| \leq \tfrac{1}{2} n^2 M \|\varepsilon^{(k)}\|^2.$$

To prove this, note that for each m, the mth row of $J\gamma(\alpha)$ is zero: $\nabla g_m(\alpha) = 0$. The desired inequality is then the result of a straightforward, if tedious, estimation of $\varepsilon^{(k+1)}$, using the multidimensional Taylor theorem:

$$\|\varepsilon^{(k+1)}\| = \|\xi^{(k+1)} - \alpha\| = \|\gamma(\xi^{(k)}) - \gamma(\alpha)\| = \max_{m=1}^{n} |g_m(\xi^{(k)}) - g_m(\alpha)|$$

$$= |g_m(\xi^{(k)}) - g_m(\alpha)| \quad \text{for some } m$$

$$= \left| \nabla g_m(\alpha)(\xi^{(k)} - \alpha) + \tfrac{1}{2} \sum_{i,j=1}^{n} \left. \frac{\partial^2 g_m}{\partial x_i \partial x_j} \right|_{\xi} (x_i^{(k)} - a_i)(x_j^{(k)} - a_j) \right| \quad \text{for some } m$$

$$= \tfrac{1}{2} \left| \sum_{i,j=1}^{n} \left. \frac{\partial^2 g_m}{\partial x_i \partial x_j} \right|_{\xi} (x_i^{(k)} - a_i)(x_j^{(k)} - a_j) \right|$$

$$\leq \tfrac{1}{2} \sum_{i,j=1}^{n} \left| \left. \frac{\partial^2 g_m}{\partial x_i \partial x_j} \right|_{\xi} \right| \left| (x_i^{(k)} - a_i) \right| \left| (x_j^{(k)} - a_j) \right|$$

$$\leq \tfrac{1}{2} \sum_{i,j=1}^{n} M \|\varepsilon^{(k)}\|^2 \leq \tfrac{1}{2} n^2 M \|\varepsilon^{(k)}\|^2.$$

Newton-Raphson Formula

Now the question is, how can you convert an n-dimensional system of equations

$$0 = \varphi(\xi) = \begin{bmatrix} f_1(\xi) \\ \vdots \\ f_n(\xi) \end{bmatrix} \qquad \xi = \begin{bmatrix} x_1 \\ \vdots \\ x_n \end{bmatrix},$$

with solution $\xi = \alpha$ into an equivalent fixpoint form $y(\alpha) = \xi$ with the same solution, so that $Jy(\alpha) = 0$ and fixpoint iteration with function y will converge quadratically? In the analogous single-dimensional case, the Newton-Raphson method converts the equation $f(x) = 0$ to fixpoint form $g(x) = x$, where

$$g(x) = x - \frac{f(x)}{f'(x)} = x - f'(x)^{-1} f(x).$$

That suggests setting

$$y(\xi) = \xi - J\varphi(\xi)^{-1} \varphi(\xi).$$

Of course, you must assume that the components of $\varphi(\xi)$ have first partials and the Jacobian $J\varphi(\xi)$ is nonsingular for ξ near the solution α. If $\varphi(\xi) = 0$, then clearly $y(\xi) = \xi$; and if $y(\xi) = \xi$, then $J\varphi(\xi)^{-1}\varphi(\xi) = 0$; and you get $\varphi(\xi) = 0$ after multiplying by $J\varphi(\xi)$. Thus, the equations $\varphi(\xi) = 0$ and $y(\xi) = \xi$ have the same solutions.

For the rest of this discussion, assume additionally that the components of φ have bounded continuous first, second, and third partials in a convex open set D containing the solution α and your initial estimate $\xi^{(0)}$. Then the first and second partials of the components of $y(\xi)$ are continuous and bounded on D, because they are rational functions of $x_1, \ldots, x_n, f_1(\xi), \ldots, f_n(\xi)$ and their first, second, and third partials. To verify that $Jy(\alpha) = 0$, note first that

$$J\varphi(\xi)y(\xi) = J\varphi(\xi)\xi - \varphi(\xi).$$

The kth entry of this vector is

$$\nabla f_k(\xi)y(\xi) = \nabla f_k(\xi)\xi - f_k(\xi)$$

$$\sum_{j=1}^{n} \frac{\partial f_k}{\partial x_j} g_j(\xi) = \sum_{j=1}^{n} \frac{\partial f_k}{\partial x_j} x_j - f_k(\xi).$$

Differentiating with respect to x_i and using the product rule, you get

$$\sum_{j=1}^{n} \frac{\partial^2 f_k}{\partial x_i \partial x_j} g_j(\xi) + \sum_{j=1}^{n} \frac{\partial f_k}{\partial x_j} \frac{\partial g_j}{\partial x_i} = \sum_{j=1}^{n} \frac{\partial^2 f_k}{\partial x_i \partial x_j} x_j + \frac{\partial f_k}{\partial x_i} - \frac{\partial f_k}{\partial x_i} = \sum_{j=1}^{n} \frac{\partial^2 f_k}{\partial x_i \partial x_j} x_j.$$

Evaluate this expression at $\xi = \alpha$ to get

$$\sum_{j=1}^{n} \left.\frac{\partial^2 f_k}{\partial x_i \partial x_j}\right|_{\xi=\alpha} \alpha_j + \sum_{j=1}^{n} \left.\frac{\partial f_k}{\partial x_j}\right|_{\xi=\alpha} \left.\frac{\partial g_j}{\partial x_i}\right|_{\xi=\alpha} = \sum_{j=1}^{n} \left.\frac{\partial^2 f_k}{\partial x_i \partial x_j}\right|_{\xi=\alpha}$$

$$\sum_{j=1}^{n} \left.\frac{\partial f_k}{\partial x_j}\right|_{\xi=\alpha} \left.\frac{\partial g_j}{\partial x_i}\right|_{\xi=\alpha} = 0.$$

But this is the k,ith entry of $J\varphi(\alpha)J\gamma(\alpha)$; hence, $J\varphi(\alpha)J\gamma(\alpha) = 0$. Finally, $J\gamma(\alpha) = 0$ because $J\varphi(\alpha)$ is nonsingular.

A more detailed analysis, beyond the scope of this book, shows that if the components of φ have bounded, continuous first, second, and third partials in a convex neighborhood D of α, and $J\varphi(\alpha)$ is nonsingular, then $\gamma : D' \to D'$ for some convex neighborhood D' of α contained in D [5, Section 10.2]. Therefore, Newton-Raphson iteration

$$\xi^{(k+1)} = \xi^{(k)} - J\varphi(\xi^{(k)})^{-1}\varphi(\xi^{(k)})$$

will converge quadratically to α if you select an initial estimate $\xi^{(0)}$ in D'. (Unfortunately, you usually don't know how small D' is—that is, how close you must start to the solution.)

For Newton-Raphson iteration, it's not necessary to calculate the inverse of $J\varphi(\xi^{(k)})$. You can save time by carrying out each iteration in several steps:

- Compute $A = J\varphi(\xi^{(k)})$.
- Compute $\beta = \varphi(\xi^{(k)})$.
- Solve the linear system $A\delta = \beta$ for δ.
- Compute $\xi^{(k+1)} = \xi^{(k)} - \delta$.

Complex Newton-Raphson Iteration

You've already seen *two*-dimensional Newton-Raphson iteration, perhaps without realizing it. Section 4.4 mentioned that the single-dimensional Newton-Raphson formula for solving equations $f(z) = 0$ for a root $z = a$ works with complex scalars as well as reals. No justification was given then, other than a remark that the calculus principles on which the method was based can be extended to apply to complex functions. The complex extension is justified here on the basis of the results just derived for two-dimensional *real* Newton-Raphson iteration. For these problems you may assume that f is a complex function defined on a convex neighborhood of a, $f'(a) \neq 0$, and $f'''(z)$ exists and is continuous and bounded in D. Given an initial estimate z_0 the single-dimensional Newton-Raphson formula yields a sequence $z_1, z_2, \ldots$ of successive approximations to a, where

$$z_{j+1} = z_j - \Delta z \qquad \Delta z = \frac{f(z_j)}{f'(z_j)}$$

If you write

$$z = x + iy \qquad f(z) = u(x, y) + iv(x, y),$$

then you can compute $f'(z)$ two ways using the partials of u and v:

$$f'(z) = \lim_{\Delta x \to 0} \left[\frac{u(x + \Delta x, y) - u(x, y)}{\Delta x} + i \frac{v(x + \Delta x, y) - v(x, y)}{\Delta x} \right] = \frac{\partial u}{\partial x} + i \frac{\partial v}{\partial x}$$

$$= \lim_{\Delta y \to 0} \left[\frac{u(x, y + \Delta y) - u(x, y)}{i\Delta y} + i \frac{v(x, y + \Delta y) - v(x, y)}{i\Delta y} \right] = -i \frac{\partial u}{\partial y} + \frac{\partial v}{\partial y}.$$

(In both calculations, $\Delta z = \Delta x + i\Delta y$. In the first, you keep $\Delta y = 0$; in the second, $\Delta x = 0$.) These two equations also yield the *Cauchy-Riemann equations*:

$$\frac{\partial u}{\partial x} = \frac{\partial v}{\partial y} \qquad \frac{\partial u}{\partial y} = -\frac{\partial v}{\partial x}.$$

Now reconsider the Newton-Raphson formula:

$$z_{j+1} = z_j - \Delta z \qquad z_j = x_j + iy_j$$

$$\Delta z = \Delta x + i\Delta y = \frac{f(z_j)}{f'(z_j)} = \frac{u + iv}{\frac{\partial u}{\partial x} + i\frac{\partial v}{\partial x}} = \frac{(u + iv)\left(\frac{\partial u}{\partial x} - i\frac{\partial v}{\partial x}\right)}{D}$$

$$D = \left(\frac{\partial u}{\partial x}\right)^2 + \left(\frac{\partial v}{\partial x}\right)^2$$

$$\Delta x = \frac{u\frac{\partial u}{\partial x} + v\frac{\partial v}{\partial x}}{D} \qquad \Delta y = \frac{v\frac{\partial u}{\partial x} - u\frac{\partial v}{\partial x}}{D}.$$

In these equations, u and v and their partials are evaluated at x_j, y_j. Using the Cauchy-Riemann equations, you get

$$D = \frac{\partial u}{\partial x}\frac{\partial v}{\partial y} - \frac{\partial v}{\partial x}\frac{\partial u}{\partial y} = \det \begin{bmatrix} \frac{\partial u}{\partial x} & \frac{\partial u}{\partial y} \\ \frac{\partial v}{\partial x} & \frac{\partial v}{\partial y} \end{bmatrix}$$

$$\Delta x = \frac{u\frac{\partial v}{\partial y} - v\frac{\partial u}{\partial y}}{D} = \frac{1}{D}\det \begin{bmatrix} u & \frac{\partial u}{\partial y} \\ v & \frac{\partial v}{\partial y} \end{bmatrix}$$

$$\Delta y = \frac{v\frac{\partial u}{\partial x} - u\frac{\partial v}{\partial x}}{D} = \frac{1}{D}\det \begin{bmatrix} \frac{\partial u}{\partial x} & u \\ \frac{\partial v}{\partial x} & v \end{bmatrix}.$$

These equations show that Δx and Δy are the solutions of the 2×2 linear system

$$\begin{bmatrix} \frac{\partial u}{\partial x} & \frac{\partial u}{\partial y} \\ \frac{\partial v}{\partial x} & \frac{\partial v}{\partial y} \end{bmatrix} \begin{bmatrix} \Delta u \\ \Delta y \end{bmatrix} = \begin{bmatrix} u \\ v \end{bmatrix}.$$

The left-hand matrix is exactly the Jacobian $Jf(z_j)$ that you get when you regard $f(z)$ as a function

$$f(z) = \begin{bmatrix} u(z) \\ v(z) \end{bmatrix} \qquad z = \begin{bmatrix} x \\ y \end{bmatrix}.$$

that is, the single-dimensional complex Newton-Raphson formula is the same as the corresponding two-dimensional real formula.

Equate2 Template NR

The multidimensional Newton-Raphson algorithm is implemented by `Equate2` template `NR`, shown in Figure 9.15. Its prototype is

```
template<class Scalar>                  // Return an approxi-
  Vector<Scalar> NR(                    // mation
    Vector<Scalar> FJF(const Vector<Scalar>& X,  // to a
                      Matrix<Scalar>& JF),       // root of
    Vector<Scalar> X,                            // F , com-
    double T,                                    // puted by Newton-
    int& Code);                                  // Raphson iteration.
```

To find a solution to an n-dimensional system of equations

$$0 = \varphi(\xi) = \begin{bmatrix} f_1(\xi) \\ \vdots \\ f_n(\xi) \end{bmatrix} \qquad \xi = \begin{bmatrix} x_1 \\ \vdots \\ x_n \end{bmatrix},$$

```
template<class Scalar>                  // Return an approximation
  Vector<Scalar> NR(                    // to a root of F , com-
    Vector<Scalar> FJF(                 // puted by Newton-Raphson
      const Vector<Scalar>& X,          // iteration. Use initial
            Matrix<Scalar>& JF),        // estimate X and update
    Vector<Scalar> X,                   // it for each successive
    double  T,                          // estimate. Stop when
    int&  Code) {                       // the distance
  try {                                 // M = ||dx||
    Vector<Scalar> F,dX;  Matrix<Scalar> JF;  // from the
    Scalar Det;           RowFinder R;         // previous X
    double M;             int n = 0;           // is  < T .
    do {
       F   = FJF(X,JF);                 // Solve the
       int C = 0;                       // linear system
       Det = Down(JF,R,C);              // JF(X) dX =
       dX  = Vector<Scalar>(Up(JF,-F,R)); // F(X) to get
       M   = MaxNorm(dX);               // dX . Set
       X -= dX; }                       // Code = 0 to
    while (++n < 1000 && M >= T);       // report suc-
    Code = (M < T ? 0 : -1);            // cess, or -1
    return X; }                         // if the de-
  catch(...) {                          // sired accuracy is
    cerr << "\nwhile executing  NR";    // not achieved after
    throw; }}                           // 1000 iterations.
```

Figure 9.15. Equate2 template NR

you must provide a function

```
Vector<Scalar> FJF(const Vector<Scalar>& X,
                         Matrix<Scalar>& JF),
```

where Scalar stands for double or complex. Given a vector X = ξ, FJF must calculate the entries of the Jacobian matrix JF = $J\varphi(\xi)$, and then return the vector $\varphi(\xi)$. You specify an initial estimate X and a tolerance T just as for the functions Fixpoint, Jacobi, and Seidel described in Sections 9.1 and 9.2. As with those functions, NR returns a status Code variable. The version of NR on the optional diskette also includes statements to provide intermediate output: just the number of iterations if you set Code = 1 on entry, but also a report of $\|\varphi(\xi^{(k)})\|$, det $J\varphi(\xi)$, and $\|\xi^{(k+1)} - \xi^{(k)}\|$ for each iteration if you set Code = 2. These statements were omitted from Figure 9.15 because they obscure the structure of the template, which exactly matches that of the implementations of fixpoint methods introduced earlier in this chapter and in Chapter 4.

NR performs Gauss elimination through the statements

```
F    = FJF(X,JF);
int C = 0;
Det  = Down(JF,R,C);
dX   = Vector<Scalar>(Up(JF,~F,R));
```

The first computes F = $\varphi(\xi)$ and JF = $J\varphi(\xi)$. The second performs the downward pass on the matrix JF = $J\varphi(\xi)$, returning the determinant Det and preparing JF and the RowFinder R for the upward pass. While the determinant isn't used by the NR version displayed in Figure 9.15, it's an optional intermediate output in the version on the optional diskette. (An extremely large or small determinant could reveal that a Jacobian is nearly singular, which could explain poor convergence properties in some examples.) The third statement performs the upward pass, solving the linear system $A\delta = \beta$ with $A = J\varphi(\xi)$ and $\beta = \varphi(\xi)$, returning the solution δ = dX. Because MSP function Up was written to solve systems of the form $AY = B$ with matrices Y and B, it's necessary to cast vector F to a Matrix type and the return value back to a Vector type. Because MSP Vector objects are regarded as rows, the first cast is performed by the transpose operator ~.

Besides the desirability of the determinant as an intermediate output, there's another reason to split the downward and upward passes in the Newton-Raphson algorithm. Often, the Jacobian doesn't vary much from one step to the next, so you can speed up execution by using the same matrix in the system $A\delta = \varphi(\xi)$ for several steps in succession. That saves the time required for several downward passes. You may want to experiment with that—for example, perform the downward pass only when the iteration count n is even.

Examples

The nonlinear system considered in Section 9.1 provides a good test for Newton-Raphson iteration—finding a root of the function

$$\varphi(\xi) = \begin{bmatrix} x - \sin(x+y) \\ y - \cos(x-y) \end{bmatrix} \qquad \xi = \begin{bmatrix} x \\ y \end{bmatrix}.$$

The components of φ and its Jacobian

$$J\varphi(\xi) = \begin{bmatrix} 1 - \cos(x+y) & -\cos(x+y) \\ \sin(x-y) & 1 - \sin(x-y) \end{bmatrix}$$

are computed by template FJF in Figure 9.16. The solution was computed by an MSP client program like the one in Figure 9.3, which tested fixpoint iteration. Their only essential difference is in the statements

```
X = Fixpoint(G ,X,T,Code)   // Fixpoint iteration
X = NR      (FJF,X,T,Code)  // Newton-Raphson iteration
```

You'll find the code in file Fig9_16.CPP on the accompanying diskette. It used the same initial estimate $x = y = 1$ and tolerance $T = 10^{-6}$, but required only three iterations to compute the same solution, compared with nine for fixpoint iteration.

```
template<class Scalar>                              // Compute func-
  Vector<Scalar> FJF(const Vector<Scalar>& X,       // tion value
                     Matrix<Scalar>& JF) { // F(X)  and
    Vector<Scalar> F(2);  JF.SetUp(2,2);            // Jacobian
    F[1] = X[1] - sin(X[1] + X[2]);                 // JF(X)  for
    F[2] = X[2] - cos(X[1] - X[2]);                 // the Newton-
    JF[1][2] = - cos(X[1] + X[2]);                  // Raphson test.
    JF[2][1] =   sin(X[1] - X[2]);
    JF[1][1] = 1 + JF[1][2];
    JF[2][2] = 1 - JF[2][1];
    return F; }
```

Figure 9.16. Template FJF for testing template NR

The Liebmann method for second-order boundary value problems (BVP), described earlier in this chapter, also provides a good test. The BVP

$$y'' = f(t, y, y') \qquad y(a) = y_0 \qquad y(b) = y^*$$

leads to the equations $\psi(\eta) = 0$, where

$$\psi(\eta) = \begin{bmatrix} p_1(\eta) \\ \vdots \\ p_{n-1}(\eta) \end{bmatrix} \qquad \eta = \begin{bmatrix} y_1 \\ \vdots \\ y_{n-1} \end{bmatrix} \qquad y_n = y^*$$

$$h = \frac{b-a}{n} \qquad t_j = a + jh \quad \text{for} \quad j = 0, \ldots, n$$

$$p_j(\eta) = y_{j+1} - 2y_j + y_{j-1} - h^2 f\left(t_j, y_j, \frac{y_{j+1} - y_{j-1}}{2h}\right) \quad \text{for} \quad j = 1, \ldots, n-1.$$

Liebmann's method is tested by an MSP client program like the one in Figure 9.5, which tested fixpoint iteration. They differ in the statements

```
X = Fixpoint(G ,X,T,Code)    // Fixpoint iteration
X = NR      (PJP,X,T,Code)    // Newton-Raphson iteration
```

You'll find its code in file `Fig9_17.CPP` on the accompanying diskette. NR requires a routine

```
Vector<double> PJP(const Vector<double>& Y,
                       Matrix<double>& JP)
```

to compute the components of $\psi(\eta)$ and the Jacobian $J\psi(\eta)$. The latter is the matrix whose i,jth entry is

$$\frac{\partial p_j}{\partial y_k} = 0 \quad \text{unless } k = j-1, j, \text{ or } j+1$$

$$\frac{\partial p_j}{\partial y_{j-1}} = 1 + \frac{h}{2}\frac{\partial f}{\partial y'} \qquad \frac{\partial p_j}{\partial y_j} = -2 - h^2 \frac{\partial f}{\partial y} \qquad \frac{\partial p_j}{\partial y_{j+1}} = 1 - \frac{h}{2}\frac{\partial f}{\partial y'}.$$

The f partials in these equations are all evaluated at

$$\left(t_j, y_j, \frac{y_{j+1} - y_{j-1}}{2h}\right).$$

```
Vector<double> PJP(const Vector<double>& Y,      // Newton-
                   Matrix<double>& JP) {         // Raphson
  int m = Y.HighIndex();                         // function for
  Vector<double> P(m);                           // Liebmann's
  JP.SetUp(m,m).MakeZero();                      // method.
  for (int k = 1; k <= m; ++k) {                 // Jacobian  JP
    double Ykm1 = (k == 1 ? y0    : Y[k-1]),     // is sparse.
           Ykp1 = (k == m ? yStar : Y[k+1]),     // Y[0] = y0 .
           Ypk  = (Ykp1 - Ykm1)/(2*h),           // Y[n] = y* .
           dfdy,dfdyp;                           // Ypk ≈
    P[k] = Ykp1 - 2*Y[k] + Ykm1                  // y' at  t[k] .
         - h*h*f(t[k],Y[k],Ypk,dfdy,dfdyp);      // Compute the
    if (k > 1) JP[k][k-1] = 1 + h*dfdyp/2;       // entries of  P
    JP[k][k] = -2 - h*h*dfdy;                    // and the non-
    if (k < m) JP[k][k+1] = 1 - h*dfdyp/2; }     // zero entries
  return P; }                                    // of the
                                                 // Jacobian.
```

Figure 9.17. Function PJP for Liebmann's method

PJP must in turn call a function

```
double f(double t, double   y, double   yp,
                   double& dfdy, double& dfdyp)
```

to return the right-hand side $f(t, y, y')$ of the ODE and compute its partials dfdy = $\partial f/\partial y$ and dfdyp = $\partial f/\partial y'$. Code for PJP is shown in Figure 9.17.

The example ODE in Section 9.1,

$$y'' = f(t,y,y') = \frac{2}{y'+1},$$

is implemented by the code

```
double f(double t ,                     // Return ODE's right
         double   y, double   yp,       // side and its par-
         double& dfdy, double& dfdyp) { // tials dfdy and
  double d = yp + 1;                    // dfdy' .
  dfdy = 0; dfdyp = -2/(d*d);
  return 2/d; }
```

Executed with the same boundary values, tolerance $T = 0.0001$ and linear initial estimate as the earlier fixpoint test, the Liebmann program with Newton-Raphson iteration yielded these results:

Number n of t steps	Iterations required	Error
5	3	.0012
10	3	.00032
20	3	.00008

The improvement is astounding!

Appendix: Selected Source Code

This appendix contains some excerpts from the MSP source code that seem inappropriate for the main part of the book, yet too important to relegate to the accompanying and optional diskettes. Each excerpt is discussed to some extent in the cited section of the book, and detailed commentary is provided in the source code itself.

The last section of the appendix describes the accompanying and optional diskettes, which contain the source code for all MSP functions, all demonstration programs discussed in the book, and a few more test routines used in developing the software.

A.1 VECTOR AND MATRIX DISPLAY OUTPUT FUNCTIONS

The Vector module display output function Show is described in Section 5.1. Like that of most display functions for providing neat row and/or column output, its source code is rather complicated, so it was not included in the main text. Because of its utility, however, it's shown here. It uses Borland C++ Library functions strlen, wherex, wherey, gotoxy, and the data type text_info, which are declared in header files String.H and ConIO.H. While strlen is a standard C Library function, the other features are specific to Borland C++.

```
template<class Scalar>                  // Standard output
  const Vector<Scalar>& Vector<Scalar>::  // using a  printf
    Show(char*   Name,                  // Format   string.
      Boolean RowFormat,                // Row or column format?
      char*   Format) const {           // Name to display.
```

```
try {
  if (Low > High) {                      // For an empty vector,
    if (!StrEmpty(Name))                 // print the name and the
      cout << Name << " : ";             // symbol  φ . NumL and
    (ostream&)cout << 'φ'; }             // L0  will be the max
  int NumL, L0 = strlen(Name);           // width of index  i  and
  if (L0 != 0) {                         // the left margin width.
    L0 += 4;                             // If the margin contains
    if (!RowFormat) {                    // Name ,  it also con-
      int DL = Digits(Low);              // tains  4  characters
      int DH = Digits(High);             // " : " .  In column
      NumL   = max(DL,DH);               // format, it contains
      L0    += NumL + 2; }}              // numeral  i  and  2
  text_info T;                           // characters  "[]" .
  int xx = wherex();                     // Find the screen width:
  gotoxy(10000,wherey());                // record where you are,
  T.screenwidth = wherex();              // move to the right edge,
  gotoxy(xx,wherey());                   // record where you stop,
  Boolean NL = False;                    // then move back.
  int     L  = L0;                       // L  is the number of

  for (int i = Low; i <= High;) {        // characters output.
    if (NL) {                            // Does this iteration
      cout << endl;                      // start a new line (with
      L = L0; }                          // left margin)?
    else                                 // If not, assume the next
      NL = True;                         // one does.
    if (L0 !=0 &&
       (!RowFormat || i == Low)) {       // If required, write
      cout << Name;                      // Name , and perhaps the
      if (!RowFormat)                    // index, in the left
        (ostream&)cout << '[' <<         // margin.
        setw(NumL)<< i << ']';
      cout << " : "; }
    else if (L == L0)                    // Otherwise, if at margin
      cout << setw(L0) << "";            // fill it with blanks.
    char* S;                             // Prepare output. This
    S = ForShow(This[i],Format);         // allocates  S  storage.
    if (!RowFormat) {                    // In column format, out-
      cout << S;  ++i; }                 // put a vector entry.
    else {
      int dL = strlen(S);                // In row format,
      if (L + dL < T.screenwidth) {      // if there's room,
        cout << S;  ++i;                 // output a vector
        if (i <= High) {                 // entry.  For each
          L += dL;                       // output except the
```

```
                   if (L+2 < T.screenwidth) {  // last, if there's
                     cout << " ";              // room, write two
                     L += 2;                   // blanks, and pre-
                     NL = False; }}}}          // pare to continue on the
       delete[] S; }                           // same line.  Release  S
       cout << endl;                           // storage.
       return This; }
     catch(...) {
       cerr << "\nwhile executing  Vector::Show to"
               "\ndisplay  " << Name << '[' << Low << "].."
            << Name << '[' << High  << "] with format  \""
            << Format << '"';
       throw; }}
```

The Matrix module display output function Show, described in Section 7.1, is based on the Vector function just displayed. It also uses standard C library functions strcpy, strcat, and sprintf, which are declared in header files String.H and StdIO.H. Here's its source code:

```
template<class Scalar>                       // Standard display
  const Matrix<Scalar>&                      // output using
    Matrix<Scalar>::Show(                    // printf  Format
      char* Name,                            // string.
      char* Format) const {
    char* S;   Boolean S_OK = False;         // For the
    char* T;   Boolean T_OK = False;         // catch .
    try {
      if (Low > HighRow) {                   // For an empty ma-
        if (!StrEmpty(Name))                 // trix, print the
          cout << Name << " : ";             // name and the
        (ostream&)cout << 'φ'; }             // empty set symbol.
      int L = Digits(Low);                   // m = maximum number
      int H = Digits(HighRow);               // of digits needed
      int m = max(L,H);                      // for the row index.
      S = new char[7];                       // Index numeral.
      T = new char[strlen(Name)+m+3];        // Row caption.
      S_OK = T_OK = True;
      for (int j = Low; j <= HighRow; ++j) {     // If  Name  is
        strcpy(T,Name);                          // specified,
        if (!StrEmpty(T)) {                      // construct for
          strcat(T,"[" );                        // each row a
          sprintf(S,"%d",j);                     // caption
          for (int k = 1; k <= m-strlen(S); ++k) // containing
            strcat(T," ");                       // Name  and
          strcat(T, S );                         // index.
          strcat(T,"]"); }                       // True  means
        This[j].Show(T,True,Format); }           // row format.
```

```
          delete[] S;                          // Release
          delete[] T;                          // scratchpad
          return This; }                       // memory.
    catch(...) {
       cerr << "\nwhile executing  Matrix::Show with  HighRow,"
               "HighCol,Low = "
            << HighRow << ','
            << HighCol << ',' << Low;          // Release
       if (S_OK) delete S;                     // allocated
       if (T_OK) delete T;                     // memory.
       throw; }}
```

A.2 MSP HEADER FILES

Some MSP header files are listed and described in detail in the book, but their listings sometimes comprise several figures, separated by several pages. Others were not listed because they contain only function prototypes. Those files provide a good index to the optional diskette, so they're listed here, in the order in which they're described in the text.

General.H   *See Section 3.1.*

```
    #ifndef   General_H
    #define   General_H

    #include  <IOManip.H>
    #include  <StdLib.H>

    //**********************************************************************
    // General programming aids
                                              // Shorthand for the
    #define This   (*this)                    // target of a class'
                                              // self pointer.

    typedef int      Boolean;                 // Simulation of the
    const   Boolean  True  = 1;               // two-valued Boolean
    const   Boolean  False = 0;               // data type.

    Boolean StrEmpty(const char* S);          // Is string S  empty?
    #define EmptyStr ""                       // Empty string.
    const   char     Blank = ' ';             // Blank character.
```

```
void Pause(                             // Display a prompt, then
    const char* Prompt                  // await and discard a
        = "\nPress any key...\n\n");    // keyboard character.

                                        // Number of characters
int Digits(int x);                      // in the shortest numeral
                                        // for  x .

int min(int t1,                         // Return the smaller of  t1
        int t2);                        // and  t2 .  The  StdLib.H
                                        // template doesn't suffice.

//******************************************************************
// Templates for derived functions and operators.  To facilitate
// use with any operand types, their bodies are in this header.

template<class Operand>                 // Generic absolute
    Operand abs(const Operand& x) {     // value function.
        return (x < 0 ? -x : x); }

template<class Operand>                 // Generic sign
    int Sign(const Operand& x) {        // function:  return
        return (x  < 0 ? -1 :           //      -1,0,1 for
                x == 0 ?  0 : 1); }     //      x <,=,> 0 .

template<class Operand>                 // Generic !=
    Boolean operator!=(const Operand& x,  // operator.
                       cons t Operand& y) {
        return !(x == y); }

template<class Operand>                 //     Generic singulary
    Operand operator+(const Operand& x) {  // + operator:  it
        return x; }                     // has no effect.

template<class LOperand,                // Generic +=
         class ROperand>                // operator.
    LOperand& operator+=(LOperand& x,
                         const ROperand& y) {
        return x = x + y; }

template<class LOperand,                // Generic -=
         class ROperand>                // operator.
    LOperand& operator-=(LOperand& x,
                         const ROperand& y) {
        return x = x - y; }
```

```cpp
template<class LOperand,                    // Generic  *=
         class ROperand>                    // operator.
  LOperand& operator*=(LOperand& x,
                const ROperand& y) {
    return x = x * y; }

template<class LOperand,                    // Generic  %=
         class ROperand>                    // operator.
  LOperand& operator%=(LOperand& x,
                const ROperand& y) {
    return x = x % y; }

template<class LOperand,                    // Generic  ^=
         class ROperand>                    // operator.
  LOperand& operator^=(LOperand& x,
                const ROperand& y) {
    return x = x ^ y; }

template<class LOperand,                    // Generic  /=
         class ROperand>                    // operator.
  LOperand& operator/=(LOperand& x,
                const ROperand& y) {
    return x = x / y; }

//*********************************************************************
// Exception handling

typedef enum {                              // MSP handles
    UnknownError, DivideError, EndOfFile,   // these
    FormatError,  IndexError,  IOError,     // exceptional
    MathError,    OtherError,  OutOfMemory }// situations.
  ExceptionType;

class Exception {                           // MSP exception class.
  public:
    ExceptionType Type;                     // What kind of situation?
    Exception(ExceptionType T               // Constructor.
              = UnknownError) {             // UnknownError ( 0 ) is
      Type = T; }};                         // used only as default.

#endif
```

Scalar.H   *See Section 3.9.*

```
#ifndef    Scalar_H
#define    Scalar_H

#include "General.H"
#include <Complex.H>

//*******************************************************************
// Logical features

#define Double  0                    // MSP supports these
#define Complex 1                    // Scalar  types,
#define Int     2                    // but  int  only partly.

template<class Scalar>               // Return  t  type.
  int Type(const Scalar& t);

template<class Scalar>               // Return  t  type name,
  char* TypeName(const Scalar& t);   // lower case, unpadded.

template<class Scalar>               // Return the shortest
  char* DefaultFormat(               // format.
    const Scalar& t);

template<class Scalar>               // Convert  t  to a string for
  char* ForShow(                     // output as if by  printf .
    const Scalar& t,                 // Call  DefaultFormat(t)  for
    char* Format = NULL);            // the default format.

template<class Scalar>               // Standard output;  return
  unsigned Show(                     // output length.  Call
    const Scalar& t,                 // DefaultFormat(t)  for the
    char* Format = NULL);            // default format.

//*******************************************************************
// Mathematical features

#define ii complex(0,1)              // i² = -1 .

double abs (const double& x);        // Return |x| .
double real(const double& x);        // Return the real and imagin-
double imag(const double& x);        // ary parts and conjugate
double conj(const double& x);        // ( x , 0 , and  x ) of  x .
```

```
void MakeRandom( double& x);     // Make  x , z  random:  0 <=
void MakeRandom( complex& z);    // x, real(z), imag(z) < 1 .
void MakeRandom(    int& k);     // Make  k  random.

#endif
```

ElemFunc.H   *See Section 3.10.*

```
#ifndef ElemFunc_H
#define ElemFunc_H

#include "Scalar.H"

long double Factorial(int n);    // n!
double Binomial(double x,        // Binomial coefficient  x
                int    m);       // over  m .
double Root     (double b,       // b^(1/r) .
                 double r);
double logbase (double b,        // Base  b  logarithm of  x .
                double x);

//******************************************************************
// Templates for functions defined similarly for real and complex
// arguments

template<class Scalar>                          // Cotangent.
   Scalar cot(Scalar t);
template<class Scalar>                          // Secant.
   Scalar sec(Scalar t);
template<class Scalar>                          // Cosecant.
   Scalar csc(Scalar t);

template<class Scalar>                          // Inverse
   Scalar acot(Scalar t);                       // cotangent,
template<class Scalar>                          //
   Scalar asec(Scalar t);                       // secant,
template<class Scalar>                          //
   Scalar acsc(Scalar t);                       // cosecant.

template<class Scalar>                          // Hyperbolic
   Scalar coth(Scalar t);                       // cotangent,
template<class Scalar>                          //
   Scalar sech(Scalar t);                       // secant,
template<class Scalar>                          //
   Scalar csch(Scalar t);                       // cosecant.
```

Appendix: Selected Source Code   355

```
template<class Scalar>                 // Inverse hyperbolic
  Scalar asinh (Scalar t);             // sine,
template<class Scalar>                 //
  Scalar acosh (Scalar t);             // cosine,
template<class Scalar>                 //
  Scalar atanh (Scalar t);             // tangent,
template<class Scalar>                 //
  Scalar acoth(Scalar t);              // cotangent,
template<class Scalar>                 //
  Scalar asech(Scalar t);              // secant,
template<class Scalar>                 //
  Scalar acsch(Scalar t);              // cosecant.

#endif
```

Equate1.H   *See Section 4.1.*

```
  #ifndef   Equate1_H
  #define   Equate1_H

  #include "Scalar.H"

  Boolean Bracket(double& xL,          // Find a bracketing in-
                  double& xR,          // terval [xL,xR] of
                  double  F(double),   // length <= T within
                  double  T);          // the specified one.
                                       // Return True to
                                       // report success.

  double Bisect(double f(double),      // Return an approximation
                double xL,             // x to a root of  f ,
                double xR,             // computed by the
                double T,              // bisection method.
                int&   Code);

  template<class Scalar>               // Return an approxima-
    Scalar Fixpoint(Scalar g(Scalar),  // tion to a fixpoint of
                    Scalar x,          // g , computed by
                    double T,          // fixpoint iteration.
                    int&   Code);

  template<class Scalar>               // Return an approximation
    Scalar NR(Scalar FdF(Scalar  x,    // to a root of  f ,
                         Scalar& dF),  // computed by Newton-
```

```
                Scalar x,                  // Raphson iteration.
                double T,
                int&   Code);

    double NRBisect(double f(double,double&),   // Return an approxi-
                double xL,                 // mation to a root
                double xR,                 // of  f  computed by the
                double T,                  // hybrid Newton-Raphson
                int&   Code);              // bisection method.

    template<class Scalar>                 // Return an approximation
      Scalar Secant(Scalar f(Scalar),      // to a root of  f ,
                Scalar x0,                 // computed by the secant
                Scalar x1,                 // method.
                double T,
                int&   Code);

#endif
```

Vector.H   *See Sections 5.1 and 5.2.*

```
    #ifndef   Vector_H
    #define   Vector_H

    #include "Scalar.H"

    template<class Scalar>
      class Vector {                       // The private parts of
                                           // class  Vector  are its
        private:                           // data components.
          int    Low;                      // Index bounds for the
          int    High;                     // vector entries.
          Scalar* Array;                   // Entry storage.

        public:

        //*****************************************************************
        // Selectors

          int LowIndex() const;            // Return the index lower
                                           // bound.

          int HighIndex() const;           // Return the index upper
                                           // bound.
```

```cpp
    Scalar& operator[](int k) const; // Entry selector:  adjust
                                     // for  Low != 0.

//******************************************************************
// Constructors and destructor

    Vector();                        // Default constructor:
                                     // set up an empty vector.

    Vector& SetUp(int Hi,            // Set index bounds and
                  int Lo=1);         // allocate storage.

    Vector(int Hi,                   // Construct a vector with
           int Lo = 1);              // specified index bounds.

   ~Vector();                        // Destructor.

//******************************************************************
// Copiers, assignment, and equality

    Vector(const Vector& Source);    // Copy constructor.

    Vector& Copy(int   Lo,           // Copy  Source[SourceLo]
        const Vector& Source,        // ..Source[SourceHi]  to
              int   SourceLo,        // target starting at
              int   SourceHigh);     // index  Lo .

    Vector& operator=(const          //
      Vector<double>& Source);       // Assignment, possibly
                                     // with type conversion.
    Vector& operator=(const          // But only
      Vector<complex>& Source);      // double-to-complex
                                     // conversion is
    Vector& operator=(const          // implemented.
      Vector<int>& Source);          //

    Boolean operator==(              // Equality.
      const Vector& W) const;

//******************************************************************
// Input/output
                                     // Prompted keyboard
    Vector& KeyIn(char* Name);       // input.
```

```
    const Vector& Show(                  // Standard output to
      char*   Name     = EmptyStr,       // display.
      Boolean RowFormat = True,
      char*   Format    = NULL) const;

    friend istream& operator>>(          // Stream input.
            istream& Source,
      Vector<Scalar>& Target);

//********************************************************************
// Member algebra functions

    Vector& MakeRandom();                // Make all entries
                                         // random.
    Vector& MakeZero();                  // Make all entries
                                         // zero.
    Vector& MakeUnit(int k);             // Set = the kth
                                         // unit vector.

    Vector operator-() const;            // Negative.

    Vector operator+(                    // Sum.
      const Vector& V) const;

    Vector operator-(                    // Difference.
      const Vector& V) const;

    Vector operator*(                    // Right scalar
      const Scalar& t) const;            // multiple.

    Scalar operator*(                    // Scalar product.
      const Vector& V) const;

    double operator|(                    // Inner product:
      const Vector& V) const;            // V * conj(W) .

    Vector operator%(                    // 3D cross product.
      const Vector& V) const;
                                         // End of the definition
    };                                   // of class  Vector .

//********************************************************************
// Unfriendly non-member logical function
```

```
template<class Stream, class Scalar>            // Stream
  Stream& operator<<(Stream& Target,            // output.
      const Vector<Scalar>& Source);

//*****************************************************************
// Unfriendly non-member vector algebra functions

template<class LScalar, class RScalar>          // Set all V
  Vector<LScalar>& AllEntries(                  // entries
    Vector<LScalar>& V,                         // = t .
    const RScalar& t);

template<class Scalar>                          // L-infinity
  double MaxNorm (const Vector<Scalar>& V);     // norm.

template<class Scalar>                          // L2  norm.
  double Euclid  (const Vector<Scalar>& V);

template<class Scalar>                          // L1  norm.
  double CityNorm(const Vector<Scalar>& V);

template<class Scalar>                          // Conjugate.
  Vector<Scalar> conj(const Vector<Scalar>& V);

template<class Scalar>              // Left scalar multiple.  Were
  Vector<Scalar> operator*(         // it a friend, this would be
    const       Scalar&  t,         // ambiguous with the analogous
    const Vector<Scalar>& V);       // Polynom  operator.

template<class Scalar>                          // 3D  triple
  Scalar Triple(const Vector<Scalar>& U,        // product:
                const Vector<Scalar>& V,        // U * (V % W) .
                const Vector<Scalar>& W);

#endif
```

Polynom.H   *See Sections 6.1 and 6.2.*

```
#ifndef  Polynom_H
#define  Polynom_H

#include "Vector.H"
```

```
template<class Scalar>
  class Polynom: public Vector<Scalar> {
    public:

      //*******************************************************************
      //  Logical functions

        friend int Degree(                // Return the degree: same
          const Polynom<Scalar>& P);      // as  P.HighIndex() .
                                          // Construct a zero poly-
        Polynom();                        // nomial. (Use  Vector
                                          // constructor.)
        Polynom& SetUp(int n);            // Set degree, allocate
                                          // memory.
        Polynom(const double&  t);        // Construct a constant
        Polynom(const complex& t);        // polynomial.

        Polynom(                          // Copy a vector to a
          const Vector<Scalar>& V);       // polynomial.

        Polynom& operator=(const          // Assignment, with
          Polynom<double>& Source);       // complex = double  con-
                                          // version. The double =
        Polynom& operator=(const          // complex   instance
          Polynom<complex>& Source);      // isn't implemented.

      //*******************************************************************
      //  Polynomial algebra member functions

        Polynom operator-() const;        // Polynomial negation.

        Polynom operator+(                // Polynomial addition.
          const Polynom& Q) const;        // The general case
                                          // doesn't cover left sca-
        friend Polynom<Scalar>            // lar addition. That's a
          operator+(                      // friend, to maintain
            const        Scalar&  t,      // consistency with left
            const Polynom<Scalar>& P);    // scalar multiplication.

        Polynom operator-(                // Polynomial subtraction.
          const Polynom& Q) const;        // The general case
                                          // doesn't cover left sca-
        friend Polynom<Scalar>            // lar subtraction. That
          operator-(                      // is a friend, to main-
            const        Scalar&  t,      // tain consistency with
            const Polynom<Scalar>& P);    // left scalar addition.
```

```
    Polynom operator*(              // Polynomial multiplica-
      const Polynom& P)             // tion. The general case
        const;                      // doesn't cover left
                                    // scalar multiplication.
    friend Polynom<Scalar>          // That must be a friend,
      operator*(                    // lest the compiler con-
        const           Scalar&  t, // fuse it with the analo-
        const Polynom<Scalar>& P);  // gous  Vector  operator.

    Polynom operator^(int n) const; // Return  This^n .

    void Divide(                    // Divide a copy of  This
      const Polynom& G,             // by  G , getting
            Polynom& Q,             // quotient  Q  and
            Polynom& R) const;      // remainder  R .

    Polynom operator/(              // Divide copies of  This
      const Polynom& G) const;      // by  G ;  return the
    Polynom operator%(              // quotient and the
      const Polynom& G) const;      // remainder.

    Scalar  Horner(Polynom& Q,      // Divide  This  by  x − t
            const Scalar& t)        // to get quotient  Q(x) ;
      const;                        // return the remainder.

    Scalar operator()(              // Evaluate  This
      const Scalar& t) const;       // polynomial at  t .

  };                // End of the definition of class  Polynom .

//*****************************************************************
//  Related unfriendly non-member functions

Polynom<double> xTo(int n);         // Return polynomial x^n .

template<class Scalar>              // Return a greatest
  Polynom<Scalar> GCD(              // common divisor of  P,
    Polynom<Scalar> P,              // Q .  Treat as  0
    Polynom<Scalar> Q,              // polynomials with
           double  T = 0);          // norm  <= T .

template<class Scalar>              // Return a least common
  Polynom<Scalar> LCM(              // multiple of  P, Q .
    const Polynom<Scalar>& P,       // Treat as  0  poly-
    const Polynom<Scalar>& Q,       // nomials with norm
              double  T = 0);       // <= T .
```

```
template<class Scalar>                  // Return the derivative
  Polynom<Scalar> Deriv(                 // P'(x) .
    const Polynom<Scalar>& P );
                                         // Return the anti-
template<class Scalar>                   // derivative of P(x)
  Polynom<Scalar> Integ(                 // with constant
    const Polynom<Scalar>& P );          // coefficient 0 .

template<class Scalar>                      // Return an approxi-
  Scalar NR(const Polynom<Scalar>& F,       // mation to a root
                    Scalar   x,             // of f , computed
                    double   T,             // by Newton-Raphson
                    int&     Code);         // iteration.

template<class Scalar>                   // Root contains
  void Polish(Vector<Scalar>& Root,      // approximations to
        const Polynom<Scalar>& F,        // roots of F .
                    double   T,          // Improve them,
                    int&     Code);      // using NR .

template<class Scalar>                   // Return Cauchy's
  double CauchyBound(                    // bound for the
    const Polynom<Scalar>& F );          // roots of P .

template<class Scalar>                   // Return a vector consis-
  Vector<int> Multiplicities(            // ting of the multiplici-
    const Vector<Scalar>& x,             // ties of the x[i] as
    const Polynom<Scalar>& P,            // roots of P . Regard
              double   E);               // z = 0 if |z| < E .

Boolean Bracket(double&  xL,             // Find a bracketing interval
                double&  xR,             // [xL,xR] of length <= T
    const Polynom<double>& F,            // within the specified one.
                double   T);             // Return True if successful.

double NRBisect(                         // Return an approximation
    const Polynom<double>& F,            // to a root of F ,
                double   xL,             // computed by the hybrid
                double   xR,             // Newton-Raphson
                double   T,              // bisection method.
                int&     Code);

Vector<double> RealRoots(                // Return a vector of all
  Polynom<double> F,                     // real roots of F , ap-
              double   S,                // proximated by
              double   T,                // NRBisect . Ignore any
              int&     Code);            // two separated by < S .
```

```
template<class Scalar>                    // Return an approximation
  Scalar NRStep(                          // to a root of  F , com-
    const Polynom<Scalar>& F,             // puted by Newton-Raphson
                  Scalar    x,            // iteration, controlling
                  double    T,            // overstep.
                  int&      Code);

template<class Scalar>                    // Return a vector of all
  Vector<complex> Roots(                  // the roots of  F , as
    Polynom<Scalar> F,                    // approximated by
              double  T,                  // NRStep .
              int&    Code);

Polynom<double> Legendre(int n);          // Return the  nth  degree
                                          // Legendre polynomial.

#endif
```

Matrix.H
See Sections 7.1 and 7.2.

```
#ifndef    Matrix_H
#define    Matrix_H

#include "Vector.H"

template<c lass Scalar>
  class Matrix {                          // The private parts of
                                          // class  Matrix  are its
    private:                              // data components.
      int             Low;                // Index bounds for the
      int             HighRow;            // matrix entries.
      int             HighCol;
      Vector<Scalar>* Row;                // Row vector storage.

    public:

      //*****************************************************************
      // Selectors

      int LowIndex()          const;      // Selectors for
      int HighRowIndex()      const;      // the index
      int HighColIndex()      const;      // bounds.

      Vector<Scalar>& operator [](        // Row selector.
                      int j) const;
      Vector<Scalar> Col(int k) const;    // Column selector.
```

```
//*********************************************************************
//   Constructors and destructor

Matrix();                           // Default constructor.

Matrix& SetUp(int HiRow,            // Set index bounds and
              int HiCol,            // allocate storage.
              int Lo = 1);

Matrix(int HiRow,                   // Construct a matrix with
       int HiCol,                   // specified index bounds.
       int Lo = 1);

~Matrix();                          // Destructor.

//*********************************************************************
//   Converters to and from type  Vector

operator Vector<Scalar>();          // Row, column to vector.

Matrix(const Vector<Scalar>& V);    // Vector to row matrix.

//*********************************************************************
//   Copiers, assignment, and equality

Matrix(const Matrix& Source);       // Copy constructor.

Matrix& Copy(int    LoRow,          // Copy the source matrix
             int    LoCol,          // rectangle with indicat-
       const Matrix& Source,        // ed corner indices to
             int    SourceLoRow,    // the target matrix so
             int    SourceLoCol,    // that its  Lo  corner
             int    SourceHiRow,    // falls at target posi-
             int    SourceHiCol);   // tion  (LoRow,LoCol) .

Matrix& operator=(const             //
  Matrix<double>& Source);          // Assignment, possibly
                                    // with type conversion,
Matrix& operator=(const             // but only
  Matrix<complex>& Source);         // double-to-complex
                                    // conversion is
Matrix& operator=(const             // implemented.
  Matrix<int>& Source);             //

Boolean operator==(                 // Equality.
  const Matrix& B) const;
```

```
//*******************************************************************
//  Input/output
                                    // Prompted keyboard
Matrix& KeyIn(char* Name);          // input.

const Matrix& Show(                 // Standard display
  char* Name   = EmptyStr,          // output using  printf
  char* Format = NULL) const;       // Format  string.

friend istream& operator>>(         // Stream input.
         istream& Source,
  Matrix<Scalar>& Target);

//*******************************************************************
// Member algebra functions

Matrix& MakeRandom();                   // Make all entries
                                        // random.
Matrix& MakeZero();                     // Make all entries
                                        // zero.
Matrix& MakeIdentity();                 // Set  This = iden-
                                        // tity matrix.

Matrix operator~() const;               // Transpose
                                        // and conjugate
Matrix operator!() const;               // transpose.

Matrix operator-() const;               // Negative.

Matrix operator+(                       // Sum.
  const Matrix& W) const;

Matrix operator-(                       // Difference.
  const Matrix& W) const;

Matrix operator*(                       // Right scalar
  const Scalar& t) const;               // multiple.

Matrix operator*(                       // Matrix * matrix.
  const Matrix& A) const;

Vector<Scalar> operator*(               // Matrix *
  const Vector<Scalar>& W) const;       // column vector.

};              // End of the definition of class  Matrix .
```

```cpp
//*******************************************************************
//   Unfriendly non-member logical functions

template<class Scalar>                          // Vector to
  Matrix<Scalar> operator~(                     // column.
    const Vector<Scalar>& V);

template<class Stream, class Scalar>            // Stream
  Stream& operator<<(Stream& Target,            // output.
      const Matrix<Scalar>& Source);

//*******************************************************************
//   Unfriendly non-member matrix algebra functions

template<class LScalar, class RScalar>          // Set all  A
  Matrix<LScalar>& AllEntries(                  // entries
    Matrix<LScalar>& A,                         // = t .
      const RScalar& t);

template<class Scalar>                          // Row norm.
  double RowNorm(const Matrix<Scalar>& A);

template<class Scalar>                          // Column norm.
  double ColNorm(const Matrix<Scalar>& A);

template<class Scalar>                          // Trace.
  Scalar Trace(const Matrix<Scalar>& A);

template<class Scalar>                          // Conjugate.
  Matrix<Scalar> conj(
    const Matrix<Scalar>& A);

template<class Scalar>                          // Left scalar
  Matrix<Scalar> operator*(                     // multiple.
    const       Scalar&  T,
    const Matrix<Scalar>& A);

template<class Scalar>                          // Row vector
  Vector<Scalar> operator*(                     //  * matrix.
    const Vector<Scalar>& V,
    const Matrix<Scalar>& A);

#endif
```

GaussEl.H   *See Section 8.4.*

```cpp
#ifndef GaussEl_H
#define GaussEl_H

#include "Matrix.H"

class RowFinder: public Vector<int> {
  public:
    RowFinder();                        // Default constructor.
    RowFinder& SetUp(int Hi,            // Set the index bounds
                int Lo = 1);            // and allocate storage.
    void Swap(int i,                    // Interchange entries
              int j); };                // i and j .

//*******************************************************************
//  Gauss elimination routines

template<class Scalar>                  // Return the solution  X
  Vector<Scalar> GaussElim(             // of the linear system
    Matrix<Scalar> A,                   // AX = B .  Set status
    Vector<Scalar> B,                   // report  Code .
          int& Code);
                                        // Downward pass on  A .
template<class Scalar>                  // Max. col. pivoting if
  Scalar Down(Matrix<Scalar>& A,        // Code = 0 .  Return
              RowFinder& Row,           // det A ,  A with stored
                 int& Code);            // multipliers, row find-
                                        // er.  Set status  Code .
template<class Scalar>
  Scalar det(Matrix<Scalar> A);         // Determinant of  A .

template<class Scalar>                  // Upward pass on  AX =
  Matrix<Scalar> Up(                    // B :  return  X .  A
    const Matrix<Scalar>& A,            // is  mxm ,  X , B are
          Matrix<Scalar>  B,            // mxn .  A , Row were
    const RowFinder& Row);              // prepared by  Down .

                                        // Return  X  such that
template<class Scalar>                  // AX = B ,  where  A  is
  Matrix<Scalar> Solve(                 // mxm  and  X , B  are
          Matrix<Scalar>  A,            // mxn .  Use  Code  like
    const Matrix<Scalar>& B,            // Down .  Throw an MSP
                int&  Code);            // Exception object if  A
                                        // is singular.
```

```cpp
template<class Scalar>                  // Return det A and  L,U
  Scalar LU(Matrix<Scalar>& L,          // decomposition of  A
            Matrix<Scalar>& U,          // if all principal minors
            Matrix<Scalar>  A,          // are nonsingular. Set
                       int& Code);      // Code like GaussElim .

                                        // Return rank A and the reduced
template<class Scalar>                  // echelon form of AX = B. A is
  int GaussJordan(                      // mxn and X , B are nx1 .
    Matrix<Scalar>& A,                  // Scalars x with |x| < T are
    Vector<Scalar>& B,                  // regarded as zero. Set status
           double   T,                  // Code = 1 :  many solutions,
              int&  Code);              //        0 :  one solution,
                                        //       -1 :  no solution.

template<class Scalar>
  Matrix<Scalar> Inverse(               // Return  A  inverse.
    const Matrix<Scalar>& A);

template<class Scalar>
  Matrix<Scalar> operator^(             // Return  A^n .
    const Matrix<Scalar>& A,
                     int  n);

Matrix<double> Hilbert   (int n);       // Return the  nxn
                                        // Hilbert matrix and its
Matrix<double> InvHilbert(int n);       // inverse.

#endif
```

Eigenval.H   *See Section 8.8.*

```cpp
#ifndef Eigenval_H
#define Eigenval_H

#include "Matrix.H"
#include "Polynom.H"

template<class Scalar>                  // Return the
  Polynom<Scalar> CharPoly(             // characteristic
    const Matrix<Scalar>& A);           // polynomial of  A .

template<class Scalar>                  // Return a vector con-
  Vector<complex> Eigenvalues(          // sisting of the eigen-
         Vector<int>&   M,              // values of  A ; enter
```

```
          const Matrix<Scalar>& A,         // into vector M  their
                  double     T,            // multiplicities. Use
                  int&    Code);           // T , Code  like  Roots .

     #endif
```

Equate2.H   *See the Chapter 9 introduction.*

```
     #ifndef    Equate2_H
     #define    Equate2_H

     #include "Gaussel.H"

     template<class Scalar>                   // Return an
       Vector<Scalar> Fixpoint(               // approximation to
          Vector<Scalar> G(const Vector<Scalar>&),// a fixpoint of  G ,
          Vector<Scalar> X,                   // computed by
                 double  T,                   // fixpoint
                    int& Code);               // iteration.

     template<class Scalar>                   // Return an
       Vector<Scalar> Jacobi(                 // approximation to
          const Matrix<Scalar>& A,            // a solution of
          const Vector<Scalar>& B,            // AX = B ,
                Vector<Scalar>  X,            // computed by
                 double  T,                   // Jacobi
                    int& Code);               // iteration.

     template<class Scalar>                   // Return an
       Vector<Scalar> Seidel(                 // approximation to
          const Matrix<Scalar>& A,            // a solution of
          const Vector<Scalar>& B,            // AX = B ,
                Vector<Scalar>  X,            // computed by
                 double  T,                   // Gauss-Seidel
                    int& Code);               // iteration.

     template<class Scalar>                      // Return an approxi-
       Vector<Scalar> NR(                        // mation
          Vector<Scalar> FJF(const Vector<Scalar>& X,   // to a
                             Matrix<Scalar>& JF),      // root of
          Vector<Scalar> X,                            // F , com-
                 double  T,                     // puted by Newton-
                    int& Code);                 // Raphson iteration.

     #endif
```

A.3 EXAMPLE MATHEMATICAL FUNCTION WITH INTERMEDIATE OUTPUT

The book often displays intermediate output from MSP functions to demonstrate convergence of iterative processes, failure of various methods, and so forth. The code required to produce intermediate output is elementary, but often so long that it overshadows more important aspects of the MSP functions. Therefore, the intermediate output features have been purged from source code shown in the book. The versions on the optional diskette, however, contain all those features. To show how they're constructed from those shown in the book, the diskette version of function template NR is displayed here. You can compare it with the version displayed in Section 4.4. No other example is more complicated.

```
char* Head[2][4] =                          // Column heads
   {{"x", "f(x)",  "f'(x)", "dx\n"},        // and widths for
    {"z","|f(z)|","|f'(z)|","dz\n"  }};     // intermediate
int w[2][4] =                               // NR  output.
   {{ 4,11, 9, 9},{10,16,10,13}};

template<class Scalar>                      // Return an approxi-
   Scalar NR(Scalar FdF(Scalar   x,         // mation to a root
                        Scalar& dF),        // of  f ,  computed
             Scalar x,                      // by Newton-Raphson
             double T,                      // iteration. Use
             int&   Code) {                 // initial estimate
   Boolean B = (Type(x) == Complex);        // x ,  update it for
   try {                                    // each successive
     if (Code > 1) {                        // estimate, and stop
       cout << "\nItera-\ntion";            // when the distance
       for (int j=0; j<4; ++j)              // to the next one
         cout << setw(w[B][j])              // is  < T .  If
              << Head[B][j]; }              // Code >1 ,  display
     int    n = 0;                          // intermediate
     Scalar Fx,dF,dx;                       // results, with a
     do {                                   // heading.  If
       Fx = FdF(x,dF);                      // Code =1 ,  report
       if (Fx == Scalar(0)) dx = 0;         // the number  n  of
         else dx = Fx/dF;                   // iterations.  Set
       if (Code > 1) {                      // Code = 0  to
         cout << setw(3) << n << " ";       // report success,
         Show(x);                           // or  = -1  if the
         if (!B) {                          // desired accuracy
           Show(Fx);                        // is not achieved
           Show(dF); }                      // after  1000
```

```
              else {                        // steps.
                Show(abs(Fx));
                Show(abs(dF)); }
            Show(dx);  cout << endl; }
          x -= dx; }
        while (++n < 1000 && abs(dx) >= T);
        if (Code > 1) cout << endl;
          else if (Code == 1) cout <<
            setw(4) << n << "  iterations\n";
        if (abs(dx) < T) Code = 0;
          else Code = -1;
        return x; }
      catch(...) {
        cerr << "\nwhile executing  NR ;   now  x,T = "
          << x << ',' << T;
        throw; }}
```

A.4 OPTIONAL AND ACCOMPANYING DISKETTES

You may use the coupon at the back of this book to obtain an optional diskette containing the source code for *all* MSP toolkit routines, all the demonstration programs in this book, and several more test programs. The root directory of the diskette, labeled MSP_SpV1g_A, contains

	Files
ReadMe	More information about what's on the diskette
MSP.Lib	MSP object code (large model, for linking with client programs)
	Subdirectories
Include	MSP header files (for inclusion with client programs)
Source	MSP source code (source for MSP.Lib)
Demos	Source code for all demonstration programs in this book
Tests	Source code for additional test programs used in developing MSP

All source code was tested, and MSP.Lib was compiled, by Borland C++ Version 5.02, *using the large memory model*. The code is self-documenting, in the style used in the book.

Before linking a client program, notice which MSP *.H header file(s) it includes. That tells you which MSP object code files you must link with. Of course, you can link with the whole object code library file MSP.Lib on the diskette. But if you want to compile and link with separate MSP *.Obj object code files for each module—for example, if you're revising some of them—you'll need to use them as shown in Figure A.1.

Including this *.H header	Requires linking with these *.Obj object code files					
General	General					
Scalar	Scalar	General				
ElemFunc	ElemFunc	Scalar	General			
Equate1	Equate1	Scalar	General			
Vector	VectorL	VectorM	Scalar	General		
Polynom	Polynom	VectorL	VectorM	Scalar	General	
Matrix	MatrixL	MatrixM	VectorL	VectorM	Scalar	General
GaussEl	GaussEl	MatrixL	MatrixM	VectorL	VectorM	Scalar	General
EigenVal	EigenVal	MatrixL	MatrixM	Polynom	VectorL	VectorM
	Scalar	General				
Equate2	Equate2	GaussEl	MatrixL	MatrixM	VectorL	VectorM
	Scalar	General				

Figure A.1. Linking with MSP *.Obj files

Figure A.2 lists all the files in the optional diskette's subdirectories. Demos and tests are grouped according to the header files they include. After compiling them *using the large memory model* you can link them with the object code Library file MSP.Lib, or with individually compiled object code files according to their header files and Figure A.1.

The diskette labeled CPPTkit4E&S, which accompanies this book, consists of the same files as the optional diskette, *minus* the Source directory, *plus* the Windows executable file Equ2Test.Exe. This file was compiled from the source code file Equ2Test.CPP in the Tests directory under the large memory model and linked with the library file MSP.Lib in the root directory. It tests all the routines in the MSP Equate2 module.

As they become suitable for distribution, I will make available MSP revisions and extensions. Revisions will be provided gratis via my Internet home page

    http:math.sfsu.edu/smith

If you have no access to these media, send e-mail to smith@math.sfsu.edu or write me at

    Department of Mathematics
    San Francisco State University
    1600 Holloway Avenue
    San Francisco, CA 94132
    U.S.A

MSP extensions will be announced the same way and will be provided at nominal cost.

Appendix: Selected Source Code

Include *file*	Source *files*	Demos	Tests	Text location	Comment
		Fig2_03.CPP		§2.3	
		Template.H			
		Template.CPP		§2.8	
		Fig2_04.CPP		§3.2	
		Fig3_06.CPP		§3.7	
		Fig3_12.CPP			
		Fig3_14.CPP			
		Fig3_15.CPP			
		Fig3_16.CPP		§3.8	
General.H	General.CPP		GenTest.CPP	§3.1	
		Fig3_10.CPP		§3.6	
		Fig3_11.CPP			
		Fig3_19.CPP		§3.8	
Scalar.H	Scalar.CPP	Fig3_25.CPP	ScaTest.CPP	§3.9	
			ErrTest.CPP		
			Riemann.CPP	§5.3	
ElemFunc.H	ElemFunc.CPP		ElemTest.CPP	§3.10	*Also* Polynom.H
		Legendre.CPP		§6.2	
Equate1.H	Equate1.CPP		Equ1Test.CPP	§4.1–4.4	
Vector.H	VectorL.CPP	Stokes.CPP	VecTestL.CPP	§5.1	
	VectorM.CPP		VecTestM.CPP	§5.2	
Polynom.H	Polynom.CPP		PolyTest.CPP	§6.1–6.3	*Also* ElemFunc.H
		Legendre.CPP		§6.2	
Matrix.H	MatrixL.CPP		MatTestL.CPP	§7.1	
	MatrixM.CPP		MatTestM.CPP	§7.2	
GaussEl.H	GaussEl.CPP		GausTest.CPP	§8.4–8.7	
			GausTest.In		*Input file*
			GausTest.In2		*Input file*
			GausTest.In3		*Input file*

Figure A.2. Contents of accompanying diskette
Continued on the following page.

Include *file*	Source *files*	Demos	Tests	*Text location*	Comment
EigenVal.H	EigenVal.CPP		EigenTst.CPP	§8.8	
Equate2.H	Equate2.CPP	Fig9_03.CPP Fig9_05.CPP Fig9_09.CPP Fig9_10.CPP Fig9_12.CPP Fig9_16.CPP Fig9_17.CPP	Equ2Test.CPP	§9.1–9.3	

Figure A.2. Contents of accompanying diskette
Continued from the previous page.

Bibliography

[1] Milton Abramowitz and Irene A. Stegin, eds., *Handbook of Mathematical Functions with Formulas, Graphs, and Mathematical Tables*, National Bureau of Standards Applied Mathematics Series, No. 55. Washington, DC: United States Government Printing Office, 1964.

[2] Joel Adams et al., *Turbo C++: An Introduction to Computing*. Englewood Cliffs, NJ: Prentice Hall, 1996.

[3] Borland International, Inc., *Turbo Assembler, Version 3.0: Quick Reference Guide, User's Guide*. Scotts Valley, CA, 1991.

[4] Borland International, Inc., *Borland C++ Development Suite, Version 5: Targets Windows 3.1, 95, NT, and DOS*. Scotts Valley, CA, 1998. (CD-ROM)

[5] Richard L. Burden and J. Douglas Faires, *Numerical Analysis*, 5th ed. Boston: PWS Kent, 1993.

[6] Choi Man-Duen, "Tricks or Treats with the Hilbert Matrix," *American Mathematical Monthly*, 90 (1983): 301–312.

[7] S. D. Conte and Carl de Boor, *Elementary Numerical Analysis: An Algorithmic Approach*, 3rd ed. New York: McGraw-Hill, 1980.

[8] R. Courant, *Differential and Integral Calculus*, 2d ed., 2 vols., trans. E. J. McShane. New York: Interscience, 1936–1937.

[9] Ray Duncan, *Advanced MS-DOS*. Redmond, WA: Microsoft Press, 1986.

[10] Dyad Software Corporation, *M++: The Complete Math Library for C++, Version 5.0*. New Castle, WA, 1995.

[11] Margaret A. Ellis and Bjarne Stroustrup, *The Annotated C++ Reference Manual*. Reading, MA: Addison-Wesley, 1990.

[12] Daniel T. Finkbeiner II, *Introduction to Matrices and Linear Transformations*, 3d ed. San Francisco: Freeman, 1978.

[13] B. H. Flowers, *An Introduction to Numerical Methods in C++*. Oxford: Clarendon Press, 1995.

[14] George E. Forsythe and Cleve B. Moler, *Computer Solution of Linear Algebraic Systems*. Englewood Cliffs, NJ: Prentice-Hall, 1967.

[15] George E. Forsythe, "Pitfalls in Computation, or Why a Math Book Isn't Enough," *American Mathematical Monthly*, 77 (1970): 931–955.

[16] F. R. Gantmacher, *The Theory of Matrices*, vol. 1. New York: Chelsea, 1977.

[17] Samuel P. Harbison and Guy L. Steele, Jr., *C: A Reference Manual*, 4th ed. Englewood Cliffs, NJ: Prentice-Hall, 1994.

[18] Einar Hille, *Analysis*, 2 vols. Waltham, MA: Blaisdell, 1964–1966.

[19] Charles D. Hodgman, et al., eds., *C.R.C. Standard Mathematical Tables*, 11th ed. Cleveland, OH: Chemical Rubber Publishing Co., 1957.

[20] A. S. Householder, *The Numerical Treatment of a Single Nonlinear Equation*. New York: McGraw-Hill, 1970.

[21] Institute of Electrical and Electronics Engineers, *An American National Standard: IEEE Standard for Binary Floating-Point Arithmetic*. New York, 1985.

[22] Intel Corporation, *iAPX 286 Programmer's Reference Manual (including the iAPX 286 Numeric Supplement)*. Santa Clara, CA, 1984.

[23] International Business Machines Corporation, *IBM Personal Computer Hardware Personal Reference Library Technical Reference*. Boca Raton, 1982.

[24] Eugene Isaacson and Herbert B. Keller, *Analysis of Numerical Methods*. New York: John Wiley & Sons, 1966.

[25] Richard Karpinski, "Paranoia: A Floating-Point Benchmark," *Byte*, February 1985: 223–235.

[26] Brian W. Kernighan and Dennis M. Ritchie, *The C Programming Language*, 2d ed. Englewood Cliffs, NJ: Prentice-Hall, 1988.

[27] R. Klatte et al., *C-XSC: A C++ Class Library for Extended Scientific Computing*, trans. G. F. Corliss et al. Berlin: Springer-Verlag, 1993.

[28] Donald E. Knuth, *Fundamental Algorithms*, 2d ed., vol. 1 of *The Art of Computer Programming*; *Seminumerical Algorithms*, vol. 2 of *The Art of Computer Programming*. Reading, MA: Addison-Wesley, 1973.

[29] Stanley B. Lippman, *C++ Primer*, 2d ed. Reading, MA: Addison-Wesley, 1991.

[30] Melvin J. Maron and Robert J. Lopez, *Numerical Analysis: A Practical Approach*, 3d ed. Pacific Grove, CA: Wadsworth, 1991.

[31] David B. Meredith, *X(Plore), DOS Version*. Englewood Cliffs, NJ: Prentice-Hall, 1993.

[32] Microsoft Corporation, *Microsoft MS-DOS Programmer's Reference, Version 6.0*. Redmond, WA, 1993.

[33] Microsoft Corporation, *Microsoft MS-DOS 6: User's Guide*. Redmond, WA, 1993.

[34] Microsoft Corporation, *Microsoft Windows 95 Resource Kit*. Redmond, WA, 1995.

[35] Microsoft Corporation, *Microsoft Windows User's Guide, Version 3.1*. Redmond, WA, 1990–1992.

[36] Microsoft Corporation, *The MS-DOS Encyclopedia*, ed. R. Duncan. Redmond, WA, 1988.

[37] Peter Norton, *Inside the PC*, 6th ed. Indianapolis, IN: Sams Publishing, 1995.

[38] Peter Norton, *Programmer's Guide to the IBM PC*. Redmond, WA: Microsoft Press, 1985.

[39] Paul J. Perry, et al., *Using Borland C++ 4*, special ed. Carmel, IN: Que Corporation, 1994.

[40] Charles Petzold, *Programming Windows 3.1*, 3d ed. Redmond, WA: Microsoft Press, 1992.

[41] P. J. Plauger, *The Standard C Library*. Englewood Cliffs, NJ: Prentice Hall, 1992.

[42] P. J. Plauger, *The Draft Standard C++ Library*. Upper Saddle River, NJ: Prentice Hall, 1995.

[43] William H. Press et al., *Numerical Recipes in C: The Art of Scientific Computing*. Cambridge: Cambridge University Press, 1989.

[44] Alain Reverchon and Marc Duchamp, *Mathematical Software Tools in C++*, trans. Véronique Fally and Sara Richter. New York: John Wiley & Sons, 1993.

[45] Rogue Wave Software, Inc., *Math.h++: Powerful C++ Types for Numerical Programming*. Corvallis, OR, 1994.

[46] Rogue Wave Software, Inc., *LAPACK.h++: A Toolbox of C++ Classes for Numerical Linear Algebra*. Corvallis, OR, 1994.

[47] Namir Clement Shammas, *Generic Programming for Borland C++*. New York: McGraw-Hill, 1991.

[48] James T. Smith, *Advanced Turbo C*. New York: Intertext/McGraw-Hill, 1989.

[49] James T. Smith, *C++ Applications Guide*. New York: Intertext/McGraw-Hill, 1992.

[50] James T. Smith, *C++ for Scientists and Engineers*. New York: Intertext/McGraw-Hill, 1991.

[51] James T. Smith, *Programmer's Guide to the IBM PC AT*. Englewood Cliffs, NJ: Prentice-Hall, 1986.

[52] Bjarne Stroustrup, *The C++ Programming Language*, 2d ed. Reading, MA: Addison-Wesley, 1991.

[53] Bjarne Stroustrup, *The Design and Evolution of C++*. Reading, MA: Addison-Wesley, 1994.

[54] J. V. Uspensky, *Theory of Equations*. New York: McGraw-Hill, 1948.

[55] Stewart Venit, "The Convergence of Jacobi and Gauss-Seidel Iteration," *Mathematics Magazine*, 48 (1975): 163–167.

[56] J. H. Wilkinson, *Rounding Errors in Algebraic Processes*. Englewood Cliffs, NJ: Prentice-Hall, 1963; New York: Dover Publications, 1994.

[57] Wolfram Research, Inc., *Mathematica: A System for Doing Mathematics by Computer: User's Guide for Microsoft Windows, Version 2.2.4*. Champaign, IL, 1995.

[58] Stephen Wolfram, *The Mathematica Book*, 3d ed. Cambridge: Cambridge University Press, 1996.

[59] C. Ray Wylie, *Advanced Engineering Mathematics*, 4th ed. New York: McGraw-Hill, 1975.

Index

A

Absolute value, integer arithmetic, 68
Abstract data types (ADTs)
   C++ adaptation of, 38
   explanation of, 37
   parallel with higher mathematics, 38
Addition
   matrix algebra, 252
   performing operation, 66–67
   polynomial algebra, 197–198
   vectors, 175–176
Addition-multiplication rule, 84
American National Standards Institute (ANSI), floating-point arithmetic standard, 72
Arbitrarity long integers, 70
Arithmetic
   complex library functions, 81–89
   floating-point arithmetic, 70–76
   integer arithmetic, 65–70
   math library functions, 76–80
   modular arithmetic, 66, 67

Assignment
   Matrix module, 240–241
   Polynom module, 196
   vectors, 160–163
Assignment operators, 45–46
   overloading of, 45–46
Asymptotically stable system, 293–294
AT&T Bell Laboratories, 3, 27

B

Basic input/output system (BIOS), 22–23
   functions of, 23
Binomial coefficients, 118
Bisect function, 129–132
Bisection method, scalar equation solution, 126–132
Blank characters, in MSP, 61–62
Boolean type, in MSP, 60–61
Borland C++, 13–14, 26–27
Borland C++ compiler,
   reasons for use, 13

Borland C++ Library
  exceptions issued to MSP handlers, 51, 53
  information source about, 27
  math function, 69
  programming topics, 5–6
  services of, 26–27
Bracketing interval, in solving scalar equations, 125
Brouwer's fixpoint theorem, 308

C

C++
  class, 38–40
  constructors, 43–46
  destructors, 46
  exceptions, 48–53
  as high level language, 10–12
  information sources on, 16–19
  limitations of, 11
  member functions, 41–42
  new features, 29–33
  overloading, 33–35
  reasons for use of, 9–10
  selectors, 41–42
  and use of MSP, 56–57
  version 2 stream input/output, 31–33
Cam function, 126–127
Capitalization, identifier capitalization in MSP, 59–60
Cauchy's bound, polynomial roots, 214–215
C compilers, 12
Characteristic polynomial, 294–296
  computation of, 297–298
  Leverrier's algorithm, 297–298
City norms, 173, 174
Class, 38–40
  class inheritance, 47–48
  concept of, 10
  declaration of class, 38–39
  derived classes, 46–48
  Polynom module, 191–192
Class templates, 10, 39–40
  purposes of, 39–40
Client programs, as software layer, 28
Comments, function of, 29
Compartmental analysis, eigenvalues, 293–294
Complex arithmetic, 82–83
Complex class, 81–82
Complex.H functions, error handling, 98
Complex hyperbolic cosine/sine, 85–86
Complex library functions, 81–89
  complex arithmetic, 82–83
  complex casts, need for, 88–89
  complex class, 81–82
  complex exponential function, 84–85
  complex hyperbolic functions, 85–86
  elementary functions, 83–84
  inverse trigonometric functions, 88
  logarithms, 87–88
  square roots, 86–87
Conj function, for noncomplex scalars, 116
Conjugation operation, matrix algebra, 250–251
Constant, 31
Constant polynomial, 193
Constructors, 43–46
  copy constructors, 44
  default constructor, 234–235
  default constructors, 43–44
  Matrix module, 234–236
  Polynom module, 193–194
  vectors, 155–158, 172–173
Contours
  countour integrals, 182–185
  single closed contour, 181
  vector module construction, 180–182

Control word, 75, 102
Converters
 Matrix module, 237–238
 use of, 45
Copiers
 Matrix module, 238–240
 Polynom module, 195–196
 vectors, 159–160
Coprocessor exception masking, 102–104
Copy constructors, 44
Cosine
 complex hyperbolic cosine, 85–86
 principle inverse cosine, 88
Cramer's rule, determinants, 266–267
Cross product, in vector algebra, 179–180

D

Default constructors, 43–44
Deflation, all real roots, finding, 219–220
Degree selector, Polynom module, 192
Delete operator, function of, 29
Destructors
 function of, 46
 Matrix module, 236–237
 vectors, 158–159
Determinants, 266–267
 applications for, 266
 Cramer's rule, 266–267
 definition of, 266
Differentiation, polynomial algebra, 206–207
Digits function, in MSP, 63
Division
 integer arithmetic, 68–69
 polynomial algebra, 202–206
DOS/Windows
 levels of, 24
 services of, 25
 and use of assembly language, 25–26

Double variables, 73–74
Downward pass
 Gauss elimination, 261–264
 Gauss elimination software, 274–275

E

EasyWin, 14
Eigenvalues, 292–302
 characteristic polynomials, 294–296, 297–298
 compartmental analysis, 293–294
 computation of, 298–299
 eigenspaces, 299–300
 eigenvectors, 299–300
 Newton's formula, 296–297
 origins of eigenvalue problems, 292–293
Elementary functions, 83–84
ElemFunc module, 117–121
 factorials/binomial coefficients, 118
 hyperbolic functions, 119–120
 root and logbase functions, 118–119
 trigonometric functions, 120–121
Empty strings, in MSP, 61–62
Equality, vectors, 163
Equality operators, Matrix module, 241
Error handling, math error handling, 93–98
Euclidean norm, 173, 174
Exception-handling, 49, 50–53
 Borland C++ hardware exception handling, 104–105
 coprocessor exception masking, 102–104
 floating-point exceptions, 99–110
 hardware exceptions, 101
 and MSP library, 51, 53
 purpose of, 93
 SIGFPE service function FPError, 107–110
 signaling, 105–107

Exceptions, 48–53
  in MSP, 64
Exponential function, complex, 84–85

F

Factorial function, 69
Factorials, ElemFunc module, 118
Fixpoint iteration, 305–320
  Brouwer's fixpoint theorem, 308
  convergence condition/speed of
    convergence, 317–320
  error analysis, 309–314
  Jacobians, 314–316
  Liebmann method, 310–314
  limits/continuity, 306–307
  linear, 321
  Lipschitz constant, 317–319
  multidimensional fixpoint algorithm,
    307–309
  multidimensional mean value theorem,
    316–317
  scalar equation solutions, 133–138
Floating-point arithmetic, 70–76
  binary floating-point standard, 72–74
  floating-point numbers, 71–72
  IEEE floating-point standard, 72–74
  input/output, 76
  real numbers, 71
  rounding methods, 75
  standard arithmetic operations, 74–75
Floating-point numbers, 71–72
Friend functions, 42–43, 81–82
Functions, 26
  friend functions, 42–43, 81–82
  invocation of, 49–50
  mandatory function prototypes, 30
  member functions, 40, 41–42
  operator functions, 42

Function templates, 11
  and overloading, 34–35

G

Gauss elimination, 260–266
  downward pass, 261–264
  equivalent systems in, 260
  upward pass, 264–266
Gauss elimination software, 268–279
  basic routine, 268–271
  downward pass, 274–275
  LU factorization, 278–279
  pivoting, 271–272
  row finder, 273–274
  upward pass, 275–278
Gauss-Jordan elimination
  Gauss-Jordan routine, 283–284
  rectangular systems, 281–283
Gauss-Seidel iteration, 323–324
  convergence of, 324
  testing of, 325–329
Greatest common divisor, polynomials,
  203–206

H

Hardware exceptions, 101
  in Borland C++, 104–105
  triggering of, 101
Hilbert matrices, 289–292
  testing with, 289–292
Homogeneous linear systems, 284–286
Horner's algorithm
  polynomial algebra, 207–209
  synthetic division, 208
Hyperbolic functions
  complex, 85–86
  ElemFunc module, 119–120
  math library functions, 78–79

I

Identifier capitalization, in MSP, 59–60
#idndef and #endif, MSP, 58
Imag function, for noncomplex scalars, 116
Infimum, 249
Initialization
   Borland C++ initialization routine, 104
   placement of, 30
Institute of Electrical and Electronics
   Engineers (IEEE), floating-point
   standard, 72–74
Integer arithmetic, 65–70
   absolute value, 68
   addition/subtraction/multiplication, 66–67
   arbitrarily long integers, 70
   C++ integer types, 65–66
   division, 68–69
   factorial function, 69
   formatting output, 70
   invoking Math library function, 69
   order relations, 67–68
Integers, in C++, 65–66
Integration, polynomial algebra, 206–207
Interrupt, 99–101
   exception handling, 100–101
   purpose of, 99
Interrupt controller, 101
Interrupt number, 100
Interrupt Service Routine (ISR), 100
Interrupt vector, 100
Iterative solutions, Newton-Raphson iteration, 334–346
Iterative solutions of equations, fixpoint iteration, 305–320
Iterative solutions of linear systems
   Gauss-Seidel iteration, 323–324
   Jabobi iteration, 321–323
   linear fixpoint iteration, 321
   PDE boundary value problem, example of, 326–329
   random system, example of, 325–326
   second-order linear ODEBVP, example of, 329–333

J

Jacobi iteration, 321–323
   strict diagonal dominance, 322
   testing of, 325–329
Jacobians, fixpoint iteration, 314–316

L

Least common multiple, polynomials, 206
Legendre polynomials
   orthogonality, 211
   polynomial algebra, 210–212
   polynomial roots, 222
Lehmer's method, polynomial roots, 223
Leverrier's algorithm, characteristic polynomial computation, 297–298
Library code, templates for, 35–36
Library functions
   complex library functions, 81–89
   external validation, 89
   internal validation experiments, 90–91
   math library functions, 76–80
   testing of, 89–92
Liebmann method, 303–304
   fixpoint iteration, 310–314
   for second-order boundary value problems (BVP), 344
Linear equations, 257–260
   solving system of, 258–260
Linear fixpoint iteration, 321
Lipschitz constant, 134–136
   fixpoint iteration, 317–319

Local variable declarations, function of, 29
Logarithms
  complex, 87–88
  math library functions, 78–79
  principal logarithm, 87
  principle common logarithm, definition of, 87–88
Logbase function, ElemFunc module, 118–119
Logical functions, and vectors, 151
LU factorization, Gauss elimination software, 278–279

M

Manipulator, 32
Masking, coprocessor exception masking, 102–104
Mathematical Software Package (MSP), 2, 4, 27–28
  application program in, 22
  blank characters, 61–62
  Boolean type, 60–61
  design goals/guidelines, 55
  digits function, 63
  ElemFunc module, 117–121
  empty strings, 61–62
  and exception handling, 51, 53
  exceptions, 64
  floating-point arithmetic, 70–76
  general module, 57–58
  identifier capitalization, 59–60
  #idndef and #endif, 58
  integer arithmetic, 65–70
  matherr function, 95–97
  Matrix module, 229–256
  max and min function, 63–64
  modules of, 15–16
  ongoing development, 4
  output, inspection of, 62–63
  Polynom module, 189–227
  revisions/extension for, 9
  scalar module, 110–117
  source code, 347–373
  source code on diskettes, 8
  structure of, 27–28
  templates for derived functions/operators, 64
  this, definition of, 58–59
  and use of C++, 56–57
  vector classes, 151–188
Matherr function, MSP function, error handling, 95–97
Math error handling, 93–98
  complex.H functions, 98
  matherr function, 95–97
Math library functions, 76–80
  hyperbolic function, 78–79
  logarithms, 78–79
  real arithmetic functions, 77–78
  real powers, 78–79
  real trigonometric functions, 79–80
Matrix algebra
  addition, 252
  conjugate transpose operator, 251
  conjugation operation, 250–251
  mathematical components, 246–247
  matrix multiplication, 253–255
  matrix norms, 248–250
  plus and minus operators, 250
  replacement operators, 255
  scalar multiplication, 253
  special matrix construction, 248
  subtraction, 252
  trace, 252–253
Matrix computations
  determinants, 266–267
  Eigenvalues, 292–302
  Gauss elimination, 260–266

Gauss elimination software, 268–279
Hilbert matrices, 289–292
linear equations, 257–260
matrix inverses, 286–288
rectangular systems, 280–286
Matrix inverses, 286–288
invertible matrix, 286–287
Matrix module, 229–256
assignment, 240–241
constructors, 234–236
converters, 237–238
copiers, 238–240
data structure, 230–232
destructors, 236–237
equality operators, 241
keyboard input/display output, 241–243
selectors, 232–234
stream input/output, 243–245
Matrix multiplication, matrix algebra, 253–255
Matrix norms, 248–250
Max function, in MSP, 63–64
Maximal column pivoting, 271–272
Max norm, 173, 174
Member functions, 40–43
membership criteria, 42–43
Min function, in MSP, 63–64
Modular arithmetic, 66, 67
Multidimensional fixpoint algorithm, 307–309
Multiplication
matrix algebra, 253–255
performing operation, 66–67
polynomial algebra, 199–200
vectors, 177–179

N

Name spaces, 11
Negation, polynomial algebra, 197

New operator, function of, 29
Newton-Raphson iteration, 123, 139–149, 334–346
bouncing, avoiding, 146–147
complex Newton-Raphson iteration, 338–346
Equate2 template NR, 340–342
failures of, 144–145
graphical representation, 140–141
MSP implementation of, 141–143
multidimensional, 334–346
Newton-Raphson formula, 336–338
Newton's method, square root computation, 140
overshooting of, 224
polynomial roots, 213–214, 224, 226
and polynomials, 189
quadratic convergence, 140, 334–336
Newton's formula, eigenvalues, 296–297
Noncomplex scalars, 116
Norms
city norms, 173, 174
Euclidean norm, 173, 174
matrix norms, 248–250
max norm, 173
vectors, 173–174
Null character, to terminate strings, 61

O

Object-oriented programming (OOP), 15
and software development, 3
Operating system, 24–26
DOS/Windows, 24–26
Operator functions, purpose of, 42
Order relations, integer arithmetic, 67–68
Ordinary differential equation initial value problems (ODEIVP), 8
Ordinary differential equation (ODE), 293
Orthogonality, Legendre polynomials, 211

Overflow, 99
Overloading, 33–35
   of assignment operators, 45–46
   and function templates, 34–35
   purpose of, 33–34
   templates for library code, 35–36

P

Pananoia, 75
Parameter initialization, 11
Parameterization function, 180–181
Pivoting
   Gauss elimination software, 271–272
   maximal column pivoting, 271–272
Pivots, 261
Polar decomposition, of complex number, 83
Polynomial algebra, 190
   addition, 197–198
   differentiation, 206–207
   division, 202–206
   Horner's algorithm, 207–209
   integration, 206–207
   Legendre polynomials, 210–212
   multiplication, 199–200
   negation, 197
   powers, 200–201
   subtraction, 199
Polynomial roots
   all real roots, finding, 219–222
   Cauchy's bound, 214–215
   complex roots, 223–227
   Legendre polynomials, 222
   Lehmer's method, 223
   multiplicity of root, 215–217
   Newton-Raphson iteration, 213–214, 224, 226
   real roots, finding, 218–219
   Taylor's theorem, 215–217

Polynomials
   constant polynomial, 193
   construction in stages, 194
   Newton-Raphson approximation, 189
   and vectors, 190
   zero polynomial, 193
   See also Polynom module
Polynom module, 189–227
   assignment, 196
   classes, 191–192
   constructors, 193–194
   copiers, 195–196
   Degree selector, 192
   type conversion, 196
Powers, polynomial algebra, 200–201
Principal argument, 83
   syntax for, 41

Q

Quadratic convergence, 136, 140
   Newton-Raphson iteration, 334–336
Quadratic Formula, 88

R

Raising the flag, 106
Random scalars, 117
Real arithmetic functions, math library functions, 77–78
Real function, for noncomplex scalars, 116
Real numbers, 71
Real powers, math library functions, 78–79
Real trigonometric functions, math library function, 79–80
Rectangular systems, 280–286
   Gauss-Jordan elimination, 281–283
   Gauss-Jordan routine, 283–284
   homogeneous linear systems, 284–286

nonsquare/singular systems, 280–283
reduced echelon system, 280
Reduced echelon system, 280
GaussJordan computation of, 283–284
Reference parameters, 11
uses of, 30–31
Replacement operators, vectors, 180
Riemann sums, in vector modules, 181–185
Root function, ElemFunc module, 118–119
Rounding methods, floating-point arithmetic, 75
Row finder, Gauss elimination software, 273–274

S

Scalar equation solutions
bisection method, 126–132
bracketing intervals, 125
checking solutions, 125–126
fixpoint iteration, 133–138
module/header file, 124
Newton-Raphson iteration, 139–149
trace output, 124
Scalar module (MSP), 110–117
code for, 111–112
constant ii, 116
displaying scalars, 114–116
function template instances, codas for, 112–113
noncomplex scalars, 116
random scalars, 117
type identification functions, 113
Scalar multiplication, matrix algebra, 253
Secant method, scalar equation solution, 147–149
Second-order boundary value problems (BVP), Liebmann method for, 344

Selectors, 41–42
Matrix module, 232–234
vectors, 154–155
SIGFPE service function FPError, 107–110
Signal dispatcher, 105
Signaling, exception handling, 105–107
Signal service function, 106
Sine, complex hyperbolic sine, 85–86
Software layers
basic input/output system (BIOS), 22–23
Borland C++, 26–27
client programs, 28
Mathematical Software Package (MSP), 27–28
operating system, 24–26
Source code
mathematical function with intermediate output, 370–371
for MSP header files, 350–369
for vector/matrix display output functions, 347–350
Square roots
complex, 86–87
principle square root, 87
Stack
and exception handling, 50–53
invocation of, 49–50
Stokes' theorem, 186–188
Stream input/output, 11
of C++ version 2, 31–33
vectors, 166–168
Streams, types of, 31–32
Subtraction
matrix algebra, 252
performing operation, 66–67
polynomial algebra, 199
vectors, 175–176

Surfaces
   surface integrals, 185–186
   vector model construction, 181–182
   wedding band surface, 181
Syntactic devices, class templates as, 40
Synthetic division, Horner's algorithm, 208
System service function, 100

T

Taylor's theorem, polynomial roots, 215–217
Templates
   function templates, 34–35
   for library code, 35–36
   in MSP, for derived functions and operators, 64
This, in MSP, 58–59
Toolkits, 12
   basics in development of, 6
   use of, 5
Trace, matrix algebra, 252–253, 294
Tridiagonal matrix, 330–331
Trigonometric functions
   ElemFunc module, 120–121
   inverse, 88
   real, 79–80
Triple product, in vector algebra, 179–180
Try block, 50
Type conversion, Polynom module, 196

U

Upward pass
   Gauss elimination, 264–266
   Gauss elimination software, 275–278

V

Variable argument list, 107
Vector algebra functions
   addition/subtraction, 175–176
   cross product/triple product, 179
   multiplication, 177–179
   plus/minus/conjugate, 174–175
   replacement operators, 180
   special vector construction, 171–173
   vector norms, 173–174
Vector field, 182
Vector module demonstration, 180–188
   contours, 180–182
   parameterization function, 180–181
   Stokes' theorem, 186–188
   surface integrals, 185–186
   surfaces, 181–182
   three dimensional contour integrals, 182–184
Vectors
   assignment, 160–163
   coda for source code, 169
   constructors, 155–158
   copiers, 159–160
   data structure, 152–154
   destructors, 158–159
   equality, 163
   input/output, 163–166
   and logical functions, 151
   and mathematical functions, 151–152
   and polynomials, 190
   selectors, 154–155
   stream input/output, 166–168
Vector spaces, 38, 190

W

Wedding band surface, 181
Windows. *See* DOS/Windows

Z

Zero polynomial, 193

A diskette containing software for the book *C++ Toolkit for Engineers and Scientists* is available directly from the author. It includes all source code for the Mathematical Software Package (MSP) described in the book, as well as many MSP test programs. To order your copy, fill out and mail this card.

Yes, please send me a copy of the optional diskette for *C++ Toolkit for Engineers and Scientists*, by James T. Smith. I enclose a check or money order for $30 plus 7% sales tax if I live in California.

Sorry, credit card orders are not acceptable.

Name: _____

Firm: _____

Address: _____

City: _____   State: _____   ZIP: _____

James T. Smith
Mathematics Department
San Francisco State University
1600 Holloway Avenue
San Francisco, CA 94132

© 1999 Springer-Verlag New York, Inc.

This electronic component package is protected by federal copyright law and international treaty. If you wish to return this book and the CD-ROM to Springer-Verlag, do not open the CD-ROM envelope, or remove it from the book. Springer-Verlag will not accept any returns if the package has been opened and/or separated from the book. The copyright holder retains title to and ownership of the package. U.S. copyright law prohibits you from making any copy of the CD-ROM for any reason, without the written permission of Springer-Verlag, except that you may download and copy the files from the CD-ROM for your own research, teaching, and personal communications use. Commercial use without the written consent of Springer-Verlag is strictly prohibited. Springer-Verlag or its designee has the right to audit your computer and electronic components usage to determine whether any unauthorized copies of this package have been made.

Springer-Verlag or the author(s) makes no warranty or representation, either expressed or implied, with respect to this CD-ROM or book, including their quality, merchantability, or fitness for a particular purpose. In no event will Springer-Verlag or the author(s) be liable for direct, indirect, special, incidental, or consequential damages arising out of the use or inability to use the electronic CD-ROM or book, even if Springer-Verlag or the author(s) has been advised of the possibility of such damages.